TUNNELS AND UNDERGROUND CITIES: ENGINEERING AND
INNOVATION MEET ARCHAEOLOGY, ARCHITECTURE AND ART

PROCEEDINGS OF THE WTC2019 ITA-AITES WORLD TUNNEL CONGRESS, NAPLES, ITALY, 3-9 MAY, 2019

Tunnels and Underground Cities: Engineering and Innovation meet Archaeology, Architecture and Art

Volume 8: Public Communication and Awareness/Risk Management, Contracts and Financial Aspects

Editors

Daniele Peila
Politecnico di Torino, Italy

Giulia Viggiani
University of Cambridge, UK
Università di Roma "Tor Vergata", Italy

Tarcisio Celestino
University of Sao Paulo, Brasil

CRC Press
Taylor & Francis Group
Boca Raton London New York

CRC Press is an imprint of the
Taylor & Francis Group, an **informa** business

A BALKEMA BOOK

Cover illustration:

View of Naples gulf

First published in paperback 2024

First published 2020
by CRC Press/Balkema
4 Park Square, Milton Park, Abingdon, Oxon, OX14 4RN

and by CRC Press/Balkema
2385 NW Executive Center Drive, Suite 320, Boca Raton FL 33431

CRC Press/Balkema is an imprint of the Taylor & Francis Group, an informa business

Publisher's Note
The publisher has gone to great lengths to ensure the quality of this reprint but points out that some imperfections in the original copies may be apparent.

ISBN: 978-0-367-46873-6 (hbk)
ISBN: 978-1-03-283943-1 (pbk)
ISBN: 978-1-003-03165-9 (ebk)

DOI: 10.1201/9781003031659

Typeset by Integra Software Services Pvt. Ltd., Pondicherry, India

Visit the Taylor & Francis Web site at
http://www.taylorandfrancis.com

and the CRC Press Web site at
http://www.crcpress.com

Tunnels and Underground Cities: Engineering and Innovation meet Archaeology,
Architecture and Art, Volume 8: Public Communication and Awareness/Risk Management,
Contracts and Financial Aspects – Peila, Viggiani & Celestino (Eds)
© 2020 Taylor & Francis Group, London, ISBN 978-0-367-46873-6

Table of contents

*Tunnels and Underground Cities: Engineering and Innovation meet Archaeology,
Architecture and Art, Volume 8: Public Communication and Awareness/Risk Management,
Contracts and Financial Aspects – Peila, Viggiani & Celestino (Eds)
© 2020 Taylor & Francis Group, London, ISBN 978-0-367-46873-6*

Preface

The World Tunnel Congress 2019 and the 45th General Assembly of the International Tunnelling and Underground Space Association (ITA), will be held in Naples, Italy next May.

The Italian Tunnelling Society is honored and proud to host this outstanding event of the international tunnelling community.

Hopefully hundreds of experts, engineers, architects, geologists, consultants, contractors, designers, clients, suppliers, manufacturers will come and meet together in Naples to share knowledge, experience and business, enjoying the atmosphere of culture, technology and good living of this historic city, full of marvelous natural, artistic and historical treasures together with new innovative and high standard underground infrastructures.

The city of Naples was the inspirational venue of this conference, starting from the title Tunnels and Underground cities: engineering and innovation meet Archaeology, Architecture and Art.

Naples is a cradle of underground works with an extended network of Greek and Roman tunnels and underground cavities dated to the fourth century BC, but also a vibrant and innovative city boasting a modern and efficient underground transit system, whose stations represent one of the most interesting Italian experiments on the permanent insertion of contemporary artwork in the urban context.

All this has inspired and deeply enriched the scientific contributions received from authors coming from over 50 different countries.

We have entrusted the WTC2019 proceedings to an editorial board of 3 professors skilled in the field of tunneling, engineering, geotechnics and geomechanics of soil and rocks, well known at international level. They have relied on a Scientific Committee made up of 11 Topic Coordinators and more than 100 national and international experts: they have reviewed more than 1.000 abstracts and 750 papers, to end up with the publication of about 670 papers, inserted in this WTC2019 proceedings.

According to the Scientific Board statement we believe these proceedings can be a valuable text in the development of the art and science of engineering and construction of underground works even with reference to the subject matters "Archaeology, Architecture and Art" proposed by the innovative title of the congress, which have "contaminated" and enriched many proceedings' papers.

Andrea Pigorini   Renato Casale
SIG President   *Chairman of the Organizing Committee WTC2019*

Acknowledgements

REVIEWERS

The Editors wish to express their gratitude to the eleven Topic Coordinators: Lorenzo Brino, Giovanna Cassani, Alessandra De Cesaris, Pietro Jarre, Donato Ludovici, Vittorio Manassero, Matthias Neuenschwander, Moreno Pescara, Enrico Maria Pizzarotti, Tatiana Rotonda, Alessandra Sciotti and all the Scientific Committee members for their effort and valuable time.

SPONSORS

The WTC2019 Organizing Committee and the Editors wish to express their gratitude to the congress sponsors for their help and support.

Tunnels and Underground Cities: Engineering and Innovation meet Archaeology,
Architecture and Art, Volume 8: Public Communication and Awareness/Risk Management,
Contracts and Financial Aspects – Peila, Viggiani & Celestino (Eds)
© 2020 Taylor & Francis Group, London, ISBN 978-0-367-46873-6

WTC 2019 Congress Organization

HONORARY ADVISORY PANEL

Pietro Lunardi, President WTC2001 Milan
Sebastiano Pelizza, ITA Past President 1996-1998
Bruno Pigorini, President WTC1986 Florence

INTERNATIONAL STEERING COMMITTEE

Giuseppe Lunardi, Italy (Coordinator)
Tarcisio Celestino, Brazil (ITA President)
Soren Eskesen, Denmark (ITA Past President)
Alexandre Gomes, Chile (ITA Vice President)
Ruth Haug, Norway (ITA Vice President)
Eric Leca, France (ITA Vice President)
Jenny Yan, China (ITA Vice President)
Felix Amberg, Switzerland
Lars Barbendererder, Germany
Arnold Dix, Australia
Randall Essex, USA
Pekka Nieminen, Finland
Dr Ooi Teik Aun, Malaysia
Chung-Sik Yoo, Korea
Davorin Kolic, Croatia
Olivier Vion, France
Miguel Fernandez-Bollo, Spain (AETOS)
Yann Leblais, France (AFTES)
Johan Mignon, Belgium (ABTUS)
Xavier Roulet, Switzerland (STS)
Joao Bilé Serra, Portugal (CPT)
Martin Bosshard, Switzerland
Luzi R. Gruber, Switzerland

EXECUTIVE COMMITTEE

Renato Casale (Organizing Committee President)
Andrea Pigorini, (SIG President)
Olivier Vion (ITA Executive Director)
Francesco Bellone
Anna Bortolussi
Massimiliano Bringiotti
Ignazio Carbone
Antonello De Risi
Anna Forciniti
Giuseppe M. Gaspari

Giuseppe Lunardi
Daniele Martinelli
Giuseppe Molisso
Daniele Peila
Enrico Maria Pizzarotti
Marco Ranieri

ORGANIZING COMMITTEE

Enrico Luigi Arini
Joseph Attias
Margherita Bellone
Claude Berenguier
Filippo Bonasso
Massimo Concilia
Matteo d'Aloja
Enrico Dal Negro
Gianluca Dati
Giovanni Giacomin
Aniello A. Giamundo
Mario Giovanni Lampiano
Pompeo Levanto
Mario Lodigiani
Maurizio Marchionni
Davide Mardegan
Paolo Mazzalai
Gian Luca Menchini
Alessandro Micheli
Cesare Salvadori
Stelvio Santarelli
Andrea Sciotti
Alberto Selleri
Patrizio Torta
Daniele Vanni

SCIENTIFIC COMMITTEE

Daniele Peila, Italy (Chair)
Giulia Viggiani, Italy (Chair)
Tarcisio Celestino, Brazil (Chair)
Lorenzo Brino, Italy
Giovanna Cassani, Italy
Alessandra De Cesaris, Italy
Pietro Jarre, Italy
Donato Ludovici, Italy
Vittorio Manassero, Italy
Matthias Neuenschwander, Switzerland
Moreno Pescara, Italy
Enrico Maria Pizzarotti, Italy
Tatiana Rotonda, Italy
Alessandra Sciotti, Italy
Han Admiraal, The Netherlands
Luisa Alfieri, Italy
Georgios Anagnostou, Switzerland

Andre Assis, Brazil
Stefano Aversa, Italy
Jonathan Baber, USA
Monica Barbero, Italy
Carlo Bardani, Italy
Mikhail Belenkiy, Russia
Paolo Berry, Italy
Adam Bezuijen, Belgium
Nhu Bilgin, Turkey
Emilio Bilotta, Italy
Nikolai Bobylev, United Kingdom
Romano Borchiellini, Italy
Martin Bosshard, Switzerland
Francesca Bozzano, Italy
Wout Broere, The Netherlands
Domenico Calcaterra, Italy
Carlo Callari, Italy

Luigi Callisto, Italy
Elena Chiriotti, France
Massimo Coli, Italy
Franco Cucchi, Italy
Paolo Cucino, Italy
Stefano De Caro, Italy
Bart De Pauw, Belgium
Michel Deffayet, France
Nicola Della Valle, Spain
Riccardo Dell'Osso, Italy
Claudio Di Prisco, Italy
Arnold Dix, Australia
Amanda Elioff, USA
Carolina Ercolani, Italy
Adriano Fava, Italy
Sebastiano Foti, Italy
Piergiuseppe Froldi, Italy
Brian Fulcher, USA
Stefano Fuoco, Italy
Robert Galler, Austria
Piergiorgio Grasso, Italy
Alessandro Graziani, Italy
Lamberto Griffini, Italy
Eivind Grov, Norway
Zhu Hehua, China
Georgios Kalamaras, Italy
Jurij Karlovsek, Australia
Donald Lamont, United Kingdom
Albino Lembo Fazio, Italy
Roland Leucker, Germany
Stefano Lo Russo, Italy
Sindre Log, USA
Robert Mair, United Kingdom
Alessandro Mandolini, Italy
Francesco Marchese, Italy
Paul Marinos, Greece
Daniele Martinelli, Italy
Antonello Martino, Italy
Alberto Meda, Italy

Davide Merlini, Switzerland
Alessandro Micheli, Italy
Salvatore Miliziano, Italy
Mike Mooney, USA
Alberto Morino, Italy
Martin Muncke, Austria
Nasri Munfah, USA
Bjørn Nilsen, Norway
Fabio Oliva, Italy
Anna Osello, Italy
Alessandro Pagliaroli, Italy
Mario Patrucco, Italy
Francesco Peduto, Italy
Giorgio Piaggio, Chile
Giovanni Plizzari, Italy
Sebastiano Rampello, Italy
Jan Rohed, Norway
Jamal Rostami, USA
Henry Russell, USA
Giampiero Russo, Italy
Gabriele Scarascia Mugnozza, Italy
Claudio Scavia, Italy
Ken Schotte, Belgium
Gerard Seingre, Switzerland
Alberto Selleri, Italy
Anna Siemińska Lewandowska, Poland
Achille Sorlini, Italy
Ray Sterling, USA
Markus Thewes, Germany
Jean-François Thimus, Belgium
Paolo Tommasi, Italy
Daniele Vanni, Italy
Francesco Venza, Italy
Luca Verrucci, Italy
Mario Virano, Italy
Harald Wagner, Thailand
Bai Yun, China
Jian Zhao, Australia
Raffaele Zurlo, Italy

Public communication and awareness

Tunnels and Underground Cities: Engineering and Innovation meet Archaeology,
Architecture and Art, Volume 8: Public Communication and Awareness/Risk Management,
Contracts and Financial Aspects – Peila, Viggiani & Celestino (Eds)
© 2020 Taylor & Francis Group, London, ISBN 978-0-367-46873-6

Designing a sustainable railway infrastructure: Envision protocol and the carbon footprint

N. Antonias
Italferr S.p.A., Roma, Italy

ABSTRACT: Italferr develops design solutions based on principles of sustainability and adopts innovative methodologies that guarantee a holistic approach to the design and implementation stages of the infrastructure. Italferr is the first Engineering Firm in the world to have acquired the Certificate of Conformity to Regulations ISO 14064-1 for the Calculation methodology of greenhouse gas emissions caused by the construction of transport infrastructures as part of an eco-sustainable strategy. The skills and experience gained in the design of sustainable construction have led to the identification of an objective and useful tool for designing and building sustainable infrastructures: the Envision Protocol rating system. Italferr has developed the first Guidelines for applying the Envision protocol to railway infrastructures and promoting a reduction in the total amount of materials to be supplied during the construction phase, through a wider reuse of excavated materials in underground construction, which also allows an overall reduction of traffic flows for off-site transportation.

1 INTRODUCTION

A new strategic vision for the infrastructure development cannot be separated from an overall assessment of the effective environmental, social and economic sustainability of the works.

New sustainability methodologies and protocols represent operational tools to promote an innovative engineering concept that sees each project as an opportunity to dialogue with the communities involved, focusing on the local needs of each territory and environmental context, enhancing the reference territory and communicating to the community, in a clear and transparent way, the benefits deriving from the realization of the works.

The use of a sustainability protocol in infrastructure design and construction, even through an evaluation of the work lifecycle that aims to integrate the paradigms of the circular economy into the feasibility analysis of the specific project, represents a real mindset change that involves all parties: from the design team to the developer, from construction companies to their suppliers. The construction sector needs a strong leadership to promote a sustainability approach, which could boost design competences and investments and connect with the Italian tradition of sustainability. In this respect the railway sector is investing many resources to promote the use of sustainable practices in railway design.

In this work, after an introduction of the Envision Protocol, it will be showed the Methodology for calculation of greenhouse gas emissions and the Guidelines for applying the Envision protocol to railway infrastructures, both of which have been developed by Italferr.

2 ENVISION PROTOCOL

2.1 *Description*

Envision is the first rating system create to help designing and building sustainable infrastructures.

It is the result of the collaboration between ISI, Institute for Sustainable Infrastructure, a non-profit organization based in Washington created specifically to develop sustainability rating systems for civil infrastructures, and the Zofnass Program for Sustainable Infrastructure at the Harvard Graduate School of Design.

The system has a double nature: firstly, it can be considered a work tool to design sustainable works; secondly, a certification system which allows the assessment of the project by a third-party entity certifying its results in sustainability terms.

The protocol is organized in 60 credits divided into the following 5 categories (Fig. 1):

- Quality of life: assessment of the extent to which the project has a positive impact on the communities concerned;
- Leadership: collaboration and engagement in the project, exploitation of the possibilities to improve performances;
- Use of resources: reduction and efficiency in the use of resources, energy and water;
- Natural world: reduction of the environmental footprint and impact on the natural world;
- Climate and Risk: mitigation of global warming and reduction of air pollution, as well as increase in durability, flexibility and adaptation to different conditions of use.

The system establishes different levels of achievement for the specific requirements of each credit, with the granting of different scores based on the actual features of projects and actions put in place. The system ensures also an assessment of improvement actions and features rather than a mere conformity to obligations and regulatory standards, which is assumed to be a benchmark.

Lastly, the Envision system establishes four certification levels, based on the percentage of achievement compared to the maximum score applicable to the project.

To obtain the certification, the project is initially self-assessed by a qualified professional (Envision Sustainability Professional, ENV SP) who is a member of the project team.

The ENV SP plays a key role within the Envision certification process since, having acquired the professional, technical and operational skills related to the Protocol, he can apply it within the certification process of sustainable infrastructures and acting as an interface between the project team, the client and the third-party verifier issuing the final certification.

The project is then verified by an independent third-party Organization which, based on the information and objective evidence provided by the project team, may issue the corresponding certification once the actual attainment of the threshold is proven.

The Envision Protocol may be applied to all types of infrastructural projects, regardless of their size. Through an integrated and participated design process, it promotes the assessment of synergies, opportunities and efficiencies of the project as well as the infrastructural, environmental, economic and social context.

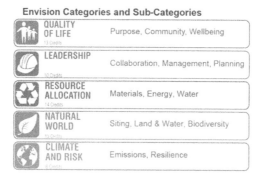

Figure 1.   Envision Categories and Sub-Categories.

It also provides methodological foundations for an innovative and sustainable design approach, guiding and justifying the decision-making process and reducing costs and operating risks.

2.2 *Focus on Protocol categories*

The different Protocol categories, through the related credits, drive the development of the project in the search for more effective and sustainable solutions.

You find below the addresses outlined by these 5 categories.

2.2.1 *Quality of life*
When developing a project, a fundamental aspect is to pay attention to the issues related to the communities involved and to the analysis of their needs, targets and features. An infrastructure project often has the potential to improve the quality of life of the neighboring areas, contributing to the development of local skills, the availability and variety of services and the general livability of places.

This is why it is necessary to carry out evaluations and define design strategies in terms of public health and safety, reduction of pollution and mitigation of the impact of the building and work site. A commitment is required from the project team, in collaboration with the community and the stakeholders involved, aimed at reducing risks and improving the availability and usability of the site as well as its wider territorial context. Through the analysis of the features of the place, such as natural and cultural resources as well as sites of historical interest, we can identify opportunities to enhance and restore local identity and wealth.

2.2.2 *Leadership*
A sustainable project requires also a fruitful communication and collaboration between the client, his project team and all the stakeholders involved (local authorities, associations, public administrations, users, etc.), according to a holistic vision extended to the entire useful lifecycle.

The "Leadership" section rewards projects in which an effective and collaborative leadership manifests itself in order to contribute positively to the redevelopment of the surrounding environment, together with an attention dedicated to the design choices immediately related to classical sustainability issues (energy, water, resources, etc.).

A sustainable project must include the involvement of stakeholders in order to create synergies and substantial opportunities for improvement and innovation. This requires the Client and the project team to pay particular attention to gathering, developing and integrating the various inputs from the various stakeholders in order to create an infrastructure that best meets the needs of the community.

A broader vision of the project can translate into a restraint of management costs, an increase in the useful lifecycle of a project and the prevention of potential future problems.

2.2.3 *Use of resources*
Resources are necessary assets both for the construction and operation phase of the infrastructures. For this reason, policies must be clearly identified and pursued right from the design phase for a correct identification and allocation of resources useful for the work aimed at the concrete reduction of the natural resources employed. In this regard we consider both the physical materials (materials/products), the energy required for construction, commissioning (operation and maintenance) and the use of water.

The minimization of the total amount of materials used for construction becomes a primary goal that can be translated into the use of materials available in the reference context as well as recycled materials and materials that can be reused at the end of the lifecycle.

A sustainable project is also a project that encourages, if possible, the use of energy from renewable sources, even through innovative technologies, as well as a rational water management, through, for example, collection and recycling systems.

2.2.4 *Natural world*

The Infrastructure projects have an impact on the natural environment, habitats, species and non-living natural systems. The way in which a project fits into these systems and the new elements that may be introduced must be carefully assessed. The Protocol drives the understanding and the minimization of negative impacts, identifying specific ways in which the infrastructure can interact with the natural systems in a positive and synergic way, reducing the impacts on animals and plants and their habitats, avoiding the introduction of invasive species and minimizing habitat fragmentation. The location of the infrastructure should take into account the presence of areas with high ecological value, avoiding areas of high eco-systemic value such as waterways and wetlands. Projects should also aim at preserving areas of geological and hydrogeological value, avoiding interruption of natural cycles. The choice of already developed and populated areas is optimal to prevent damages to the environment, increase the value of the areas and rehabilitate contaminated areas.

2.2.5 *Climate and Risk*

A sustainable project must include, among its goals, the reduction of emissions of pollutants and greenhouse gases as well as the ability to react by adapting to any natural phenomena and nearby changes.

An evaluation of the conditions and the short and long-term climate changes is useful to determine, in collaboration with the competent bodies, vulnerabilities and risks of the system. These estimates make it possible to foresee the necessary adaptability features of the project and to draw up plans for "reconfiguration" in response to the assumed climate changes.

2.3 *Guidelines for the application of the Envision Protocol to the Italian railway infrastructure*

Italferr, in collaboration with Rete Ferroviaria Italiana and ICMQ, has drawn up specific Guidelines to adapt the Protocol to the Italian regulatory context and to the specific type of infrastructure works.

The Guidelines have been developed to provide design firms, consultants, professionals and anyone else interested therein, with a simple and easy tool to use the Envision Protocol in the design of railway infrastructures. This is meant to give value to the priorities of the communities concerned by the project, to facilitate investments by banks and funds and, finally, to identify sustainable and innovative solutions.

The first part of the document introduces the general principles of the Envision Protocol, the description of the design process and the relevant regulatory framework, indicating both the general and specific standards that apply to the railway sector.

This is followed by a list of the most important terms and definitions that are typical of the railway sector, as well as the abbreviations used in the document.

The Guidelines then discuss each credit, summarizing what the Protocol requires: intent, assessment metrics, benchmark, levels of achievement and documents to be provided to support one's request. The same process is used to describe, for each credit, the benchmark and documents requested by RFI-Italferr for railway design.

It should be underlined that, concerning the design of infrastructures, the process for project approval in Italy is radically different from the American one. Indeed, in Italy there are many Entities in charge of overseeing specific aspects of the project and each one has a veto power over the latter. This implies the preparation by the designers of many explanatory reports, including the Environmental Impact Assessment, Geological reports and so on. Therefore, some of these benchmarks have been included as a minimum benchmark to be achieved.

The features of a railway infrastructure, in terms of territorial development, number of stakeholders involved, times, costs, difficulty to modify the routes or the conditions of execution, are one of the main issues that are dealt with in the Guidelines, with the intent to combine the spirit and strictness of the Envision protocol with the features of local infrastructures and regulations. To this end, industrial practices in Italian railway design have been adapted or enhanced according to the requirements of the protocol. An example can be given in relation to the QL 2.5 credit "Encourage alternative modes of transportation", where, in addition to emphasizing the role of railway infrastructures as well as providing design instructions to promote alternative transportation systems to reach railway junction stations, the Guidelines mention the potential enhancement and redevelopment of abandoned railway stretches, turning them into bike-pedestrian paths along itineraries that have not yet been urbanized and are thus of particular value. It is worth noting that this represents an heritage of more than 1,500 km along the Italian territory.

In relation to the NW 1.1 credit "Preserve prime habitat", the study of the impact that the infrastructure may have on the habitat determines the need to pay particular attention to issues such as the location of the railroad or the preservation of areas of geological and hydrogeological value. In addition, the Guidelines highlight the presence of high value agricultural areas, considering the agricultural vocation of Italian territory, with particular attention to DOC and DCG, PGIs etc., in the area interested by the project.

The Guidelines for the Sustainable Design of Railway Infrastructures promotes a universal approach to sustainability. As a sort of handbook, it describes, for each protocol criterion, a precise metric of evaluation, a benchmark to reach and a list of necessary support documentation to reach a certain level of achievement.

The Guidelines become a practical tool for promoting the design of transport infrastructures geared to environmental sustainability with a view to better incorporating the work into the reference territory.

The Guidelines are a useful reference for Stakeholders to gain a clear and complete picture of the benefits associated with the realization of the infrastructure work. It also provides an objective self-assessment of the work that allows the designer to improve the features of the intervention through a continuous feedback process aimed at refining the project solution with an eye to sustainable development as well as a tool for assessment of the sustainability of infrastructure projects by the competent authorities.

The Guidelines for application of the Envision Protocol to the railway infrastructures sector, developed by Italferr and validated by ISI - Institute for Sustainable Infrastructure, represent a recognized reference method to be adopted for sustainable transport infrastructures, both at national and international level.

3 THE VALUE OF SUSTAINABILITY IN UNDERGROUND CONSTRUCTIONS

3.1 *Examples of Envision's credits*

The realization of underground works requires a careful management of significant environmental aspects such as water and materials resulting from the excavation.

The use of the Protocol in the development of infrastructure work projects allows the designers to be guided in identifying effective choices and to communicate to the Stakeholders the assessments carried out from the early planning stages and oriented to the principles of sustainability.

As an example, provide for a rational management of the water resources interfered by the excavation of the tunnels allows to reduce the risk of impoverishing the acquifer through interventions of recirculation and reuse of the intercepted water within construction activities. The Protocol requires specific assessments through the credits of the category of "Resource Allocation".

Similarly, planning the reuse of excavated materials, within the same infrastructure work or for a new environmentalization of degraded sites external to the project, allows to significantly reduce the amount of waste to transfer to landfill. The protocol provides, in this regard, specific additional credits within the category of Resource Allocation that move the design towards more sustainable solutions of regeneration and reuse, which are fundamental for the practical implementation of the circular economy's paradigms.

The reuse of materials excavated within the infrastructural work becomes strategic to reduce CO_2 emissions linked to the transportation of land outside the building site.

The Protocol analyzes the possibility of achieving benefits in terms of greenhouse gases reductions, as required by credit CR 1.1 "Reduce Greenhouse Gas Emissions".

Italferr has developed a methodology, to be used during the design phase, for calculating the climate footprint of projects. This methodology, compliant with UNI ISO 14064-1:2006 and certified by a third party, allows determination of the environmental footprint of projects by calculating the quantity of greenhouse gases produced during construction. Furthermore, specific provisions are included in contracts, requiring construction companies to report on CO_2 emissions connected to the production of supplied materials (notably concrete and steel) and by their transportation to the production site.

3.2 Carbon footprint of railway infrastructures

Italferr is the first Engineering Company in the world to have acquired the Certificate of Conformity Regulations ISO 14064-1 for the methodology: "Calculation of greenhouse gas (GHG) emissions caused by the production of transport infrastructures" as part of an eco-sustainable strategy.

The methodology allows the measurement and reporting of GHG emissions produced in construction activities of railway infrastructures. Resorting to this new tool will allow a preventive energy assessment of the works to be performed, facilitating the project manager's interventions right from the first planning phases, necessary for the modification of any possible form of irrational consumption of resources.

Assessing the emissions generated by the construction of a single type of work or by each part composing the entire infrastructure, it will be possible to measure efficiently all forces that come into play in the assessment of the emissions. This system is effective in defining design solutions which deliver same performance standards with lower emissions.

A set of specific contractual regulations instruct the construction firms to procure their concrete and steel requirements from environmentally aware suppliers and to steer towards environmentally friendly means of transport, thus rewarding firms which actively collaborate and contribute to reduce CO_2 emissions into the atmosphere.

In this way, Contractors are assessed based on the environmental improvements originating from their 'environment-friendly' choices when procuring materials and selecting materials' transportation methods. They can choose low environmental impact products by buying them from suppliers who have gained "EPD" (Environmental Product Declaration) certification. As regards transportation, Contractors can privilege suppliers using environmentally compatible means of transport, such as trains instead of trucks.

Contractors, therefore, are expected to enter the information regarding the materials purchased into specific reports and to file the related registration documentation (TD, EPD, certificates of origin, etc.), while establishing and constantly updating a traceability register of the materials purchased. At the end of every semester, and anyhow by no later than January 30th after the end of the year of reference, the Reports envisaged by contract must be delivered to Italferr along with the documents that provide material traceability and the related certifications.

4 CONCLUSIONS

The new vision of infrastructure development cannot be separated from an overall assessment of the effective environmental, social and economic sustainability of the works. In the decision-making process behind sustainable infrastructure design, the question "are we doing the project right?" should necessarily be integrated with a new and more important question "are we doing the right project?"

In the transition to a model of economic development that targets not only profitability and profit, but also social progress and environmental protection, in order to outline more efficient regeneration and reuse systems, it becomes essential to concretely launch a circular economy capable of maximizing the usefulness and value over time of the planned infrastructures.

New Sustainability methods and protocols are operative tools to promote an innovative engineering concept that interprets each project as an opportunity to dialogue with the communities involved, focusing on local needs and environmental context, enhancing the reference territory and communicating to the collectivity, in a clear and transparent way, the benefits deriving from the realization of the works.

In this perspective, the Envision Protocol becomes the objective tool for measuring the sustainability of an infrastructure project also through an evaluation of the lifecycle of the work which aims at integrating the paradigms of the circular economy, which by its nature is regenerative, in the feasibility analysis of a specific project.

*Tunnels and Underground Cities: Engineering and Innovation meet Archaeology,
Architecture and Art, Volume 8: Public Communication and Awareness/Risk Management,
Contracts and Financial Aspects – Peila, Viggiani & Celestino (Eds)*

Best practice of communication for a complex transnational project – The Brenner Base Tunnel

K. Bergmeister
Brenner Base Tunnel BBT SE, Innsbruck, Austria

ABSTRACT: Communication is very important in complex, transnational infrastructural projects like the Brenner Base Tunnel. Here technical experts work together in the project society. They must be able to communicate in Italian, German and English. On the other hand, the construction sites are a workplace for people from 11 European countries. They speak different languages and come from different cultural backgrounds.

Within BBT SE, a 'multi-level approach' has been developed for communication. The aim is to inform all stakeholders, everyone from the local population to political decision-makers, students and scientists at the universities about this project for the longest tunnel ever built. Furthermore, a new method to determine the number of informed persons has been developed.

1 INTRODUCTION

Apart from the numerous technical tasks in areas such as geology, geotechnics, tunnelling and construction engineering, communication is one of the most important and challenging tasks when carrying out large infrastructural projects. At the Brenner Base Tunnel, communication takes place in many different ways and on many levels. Information is provided continuously through events in the townships along the tunnel railway line, through specific information sheets, technical lectures, internet, radio and TV reports as well as through articles in specific journals and daily newspapers. Every year, the project company organizes the "Open Tunnel Days". The population is invited to visit the construction sites and to gain insights into underground construction. Furthermore, information centres have been built in Innsbruck, Steinach and Fortezza.

Public relations are very important, especially for infrastructural and construction projects. A trustworthy leadership personality must be the direct communication partner (Spieker et al. 2017).

2 BRENNER CORRIDOR PLATFORM

The declared commitment to an expansion of the railway connection between Munich and Verona has been anchored by several treaties and contracts with interested partners in Germany, Austria and Italy, over more than 30 years. Examples are the "Memorandum of Udine" (1989), the "Memorandum of Montreux" (1994) and the "Agreement of the trilateral working group on measures relating to the construction of the Brenner Base Tunnel"(2002).

"We are not just building a tunnel, we are building an infrastructure project to create economic networks, promote sustainable rail freight transport and improve environmental-friendly mobility in our habitat"; this was the background idea of the author in 2007 to create a new Corridor Platform.

To avoid building an isolated tunnel project, the EU-coordinator Karel Van Miert (+06/ 2010), Herald Ruyters (at that time Assistant to Karel Van Miert, now Director of DG Move) and Konrad Bergmeister (BBT SE) founded the Brenner Corridor Platform. This cross-border platform has been directed by the EU Coordinator Pat Cox since 2010. The main task of the Brenner Base Tunnel will be the modal shift of goods from road to rail. Therefore, the three involved transport ministers ((Doris Bures – Austria, Wolfgang Tiefensee – Germany, Altero Matteoli – Italy), the five regions from Bavaria to Verona and representatives of ÖBB and RFI signed the Brenner Action Plan on May 18th, 2009. This Action Plan was revised in 2016 with the help of EU Coordinator Pat Cox and presented for signature on June 12th, 2017. The Action Plan defines provisions and deadlines regarding the access routes, railway logistics and terminals, cross-financing models, transport policy modal shift measures and environmental monitoring.

With a total length of 64 km, the Brenner Base Tunnel will be the longest underground railway connection in the world. The Brenner Base Tunnel runs under the Brenner Pass at 794 m above sea level. The Brenner Base Tunnel is an important connecting link in the high-capacity rail line of Trans-European Corridor No. 5 between Helsinki/Finland and La Valletta/Malta.

This cross-border tunnel will contribute to the strengthening of the European ideals both during the construction phase, and also when operative. For example: in the Tulfes-Pfons construction lot, people from 11 European cultures and languages work together. Communication has a central role in the success of this project.

In the 18th century, a journey in a horse-drawn carriage from Innsbruck to Bolzano took three days. With the dawn of the railway in the 19th century, travel time was about 9 hours. In the 20th century, the journey took 2.5 hours by rail and 1.5 hours by car. In the 21st century, the travel time through the Brenner Base Tunnel and on the southern access line from Fortezza to Ponte Gardena will be just an hour. The Brenner Corridor Platform was founded to realise this aim. The working groups shall support this development. Nevertheless, engineering sciences and communication have to make every effort not just to build a tunnel but to create a European, future-oriented rail infrastructure with – hopefully – one set of harmonized operating rules and one single language.

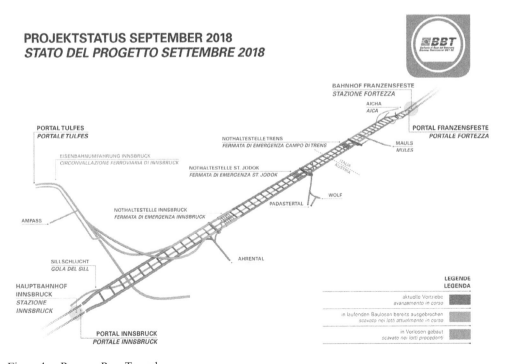

Figure 1.   Brenner Base Tunnel.

3   MULTI-LEVEL APPROACH

Communication takes place on various stages. The aim is to reach as many population groups as possible, from the population living within the project area to the experts from the worlds of science and politics. BBT SE places a high value on tailoring its communications materials to various target groups. Therefore, different information materials have been worked out for different age levels. As follows, the different ways to transfer information are presented in 4 levels. Certain numbers from this table can be found in the 2017 balance sheet report of BBT SE.

A theoretical model has been developed to estimate approximately the quantity of informed people, called "impact – I" on the basis of an exponential function "e", the number of direct contacted or informed persons "c" and the information level "i"

$$I = c \times e^i \tag{1}$$

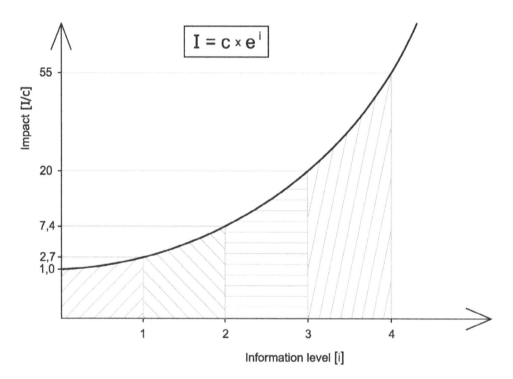

Figure 2.   Multi-level approach – theoretical estimate of informed persons.

Taking into account that in Europe an average family has 1.6 children (of which 1 over 12 years of age) and therefore includes 3.6 persons, with about 40% of households having only one person, the following calculation can be made.

Each informed person will report his/her knowledge to at least one child (over 12 years) and one partner. The following factor can be used for the calculation of the overall informed persons:

$$0.6 \times 3 + 0.4 \times 1 = 2.2 \tag{2}$$

The total number of persons who received information on the Brenner Base Tunnel can be estimated according to the sum of Table 1 to (calculated impact: 59.6 Mio x 2.2) approx. 131 Mio.

Table 1. Level of Information and informed people.

Information Level	number of informed persons	Media	Impact (theoretical model)	Counted informed person in 2017 of the BBT project	calculated impact
4	< 1,000,000	News channel	$I = c \times e^4$	>1,000,000 people	55 Mio
3	< 100,000	Internet, BBT Portal	$I = c \times e^3$	103,000 users	2 Mio
		International newspaper, radio, TV		100,000 people	2 Mio
2	< 10,000	Info-centres (Steinach, Fortezza)	$I = c \times e^2$	31,500 visitors	233,000
		Information sheets in the project area		25,000 people	185,000
		Open day's		7,800 visitors	58,000
		Fairs, exhibitions		4,000 visitors	30,000
		International scientific publications		3,000 experts	22,000
		Books (chapters)		2,000 experts	15,000
		Tunnel excursions		6,618 visitors	48,900
1	< 100	International project presentations	$I = c \times e^1$	2,000 people	5,000
		Local project presentation		350 people	1,000
		Master thesis, PhD thesis		100 experts	300

3.1 News channels

The biggest media response to an event concerning the Brenner Base Tunnel was a false fire alarm on December 11th, 2017, in the Ahrental-Pfons exploratory tunnel. A false report caused a fire alarm to be given. Within a few hours, an enormous information wave swept through the international media. This piece of "news" was mentioned 679 times in the USA, 358 times in Germany, 63 times in Canada, 59 times in Austria and 33 times in the United Kingdom. If a newspaper prints 50,000 copies, this news would reach more than 60 million people.

3.2 Internet portal – Internet appearance

Continuous, updated and transparent public relations work is of particular importance for all infrastructure and building projects. The internet portal is updated weekly and contains the most important information on the project and the actual construction phase.

At the end of October, 2018, about 90 km of the BBT system had been excavated (about 38%). All access and logistic tunnels have been built. More than 50% of the exploratory tunnel, with an overall length of 60 km, has been excavated so far. In 2017, there were more than 176,000 clicks and more than 103,000 visitors on the BBT SE homepage. In average, three sub-sections were opened by each visitor to the homepage. The most clicked sub-sections were "work progress" and "project overview". The average visitor stayed on the website for 2.55 minutes.

3.3 Information centres

At Fortezza, in the old fortress (built from 1833 to 1838 under Kaiser Ferdinand I and named after Kaiser Franz I of Austria) an information centre was opened in 2015. In 2017, more than

12,500 people visited the exhibition on the Brenner Base Tunnel. The information center "Tunnel World" at Steinach, in operation since 2016, had 18,992 visitors in 2017; 8,762 of them were young people and children. The exhibition concept at Steinach was created especially for families and children. It gives children and young adults the chance to gather information on the project. The visit to the information center was supported, at 97 elementary schools within the project area, by providing the schools with information material in order to give the children the possibility to get to know more about the themes related to the project implementation. School classes tour the "Tunnel World" information center. A questionnaire evaluates their knowledge of the project and – indirectly – the didactic quality of the exhibition. This information activity is supported by continuous updates of the information material and several short films about the project. Furthermore, experts provide information on a daily basis.

3.4 *Open Tunnel Days*

The involvement of the population has its high point every year in the "Open Tunnel Days", an event organized by BBT SE every year since 2007 on the construction sites in Italy and Austria. This event attracts thousands of people, even from outside the project area, and has become a real festival with many children and their families. The day usually starts with a Holy Mass and continues with construction site and tunnel tours. The tours are guided by BBT SE technicians and give an idea of the construction progress and the extent of the structures that have already been completed. Over the years, the "Open Tunnel Days" have attracted more and more people (more than 10,000 visitors in 2018).

Figure 3.   Impression of the Open Tunnel Days.

4   ADDED VALUE

The prescriptions made during the environmental compatibility assessment in Austria and Italy help to create high-quality structures and infrastructural improvements in the project area that would not have existed without the project.

Some examples are: the construction of noise barriers along the existing rail line, water lines, re-forestation and greening operations etc.

A very interesting example is the construction of the Wendelin chapel near Steinach on the basis of the Roman construction method "opus caementicium".

Underneath the largest disposal site of the Brenner Base Tunnel project, the Padaster Valley, there was a small wayside shrine with a picture of the Holy Mother Mary and St. Wendelin. The shrine had to be removed for the construction works.

Outside the disposal site, along the European hiking path, BBT SE built a chapel and gave the sacred images a new home. The chapel was built exclusively with tunnel spoil: material processing was optimized and adjusted so that Bündner schists could be used for this new construction method. The "Opus caementicium montium magistrale" method calls for broken-grain material (in this case, extracted from the surrounding mountains, hence "montium") to be poured into a formwork and grouted with self-compressing concrete (Bergmeister, 2014).

The chapel is visited by about 1,000 people each year and has become a small pilgrimage destination.

Figure 4.   St. Wendelin chapel in Roman concrete, "opus caementicium montium magistrale".

The stone fragments give the visible surface a particular aesthetic effect, thanks to the different minerals. Furthermore, this construction method saved 20% in costs and produced up to 40% less CO_2 as compared to traditional methodologies.

5   SUMMARY

Communication has an important role for all construction projects. Information levels have to be developed comprehensively and continuously in transnational, complex infrastructural projects. Over the past 12 years, BBT SE has done extensive work to develop communication levels, cultivating contacts with the local population and holding periodical "Open Tunnel Days" (at least once a year) as well as information events within the project area (several times a year).

Internet and digital media have an important role and must be curated regularly. Information must be transmitted in a transparent way.

REFERENCES

Spieker, A.; Wenzel, G.; Brettschneider, F. (2017): Bauprojekte visualisieren Leitfaden für die Bürgerbeteiligung ISBN 2366-1437, pp. 78

Bergmeister K. (2011): Brenner Base Tunnel – Der Tunnel kommt. Tappeinerverlag – Lana, pp. 263

Bergmeister K.: Opus cementicium montium magistrale – von der Idee zum Bau. In: Festschrift Prof. Hegger, Universität Aachen (Jubilee publication Prof. Hegger, Achen university) 2014

BBT SE: Balance sheet and financial report for 2017 by the Brenner Base Tunnel. Approved by the Supervisory Board on March 13th, 2018.

Tunnels and Underground Cities: Engineering and Innovation meet Archaeology, Architecture and Art, Volume 8: Public Communication and Awareness/Risk Management, Contracts and Financial Aspects – Peila, Viggiani & Celestino (Eds)
© 2020 Taylor & Francis Group, London, ISBN 978-0-367-46873-6

The Naples Line 6 underground metro system works: From the feasibility study to the construction

O. Carbone
ACaMIR | Regione Campania, Naples (Na), Italy

G. Molisso
Ansaldo STS |A HITACHI GROUP COMPANY, Naples (Na), Italy

S. Riccio
Direzione Centrale Infrastrutture, Lavori Pubblici e Mobilità | Comune di Napoli, Naples (Na), Italy

ABSTRACT: The Naples Line 6 underground Metro is part of an integrated network of railway transport with several interconnection nodes with other railway lines (metropolitan and regional railway systems).

The line is currently operating in the Mostra-Mergellina section with 4 stations. The section currently in construction, Mergellina-Municipio, crosses the central and most attractive areas of the city, behind the coastal strip. The line lies under the groundwater level along the whole length; the free water level for the most part is just a few meters below ground level.

The next step is the construction of the following: a. the tunnel from the Mostra station to the Campegna station, in the Bagnoli area; the final depot in the ex military arsenal area (below the Posillipo hill).

The design and construction of a metropolitan transportation system in a big city like Naples requires complex and in-depth analysis that was developed since the feasibility studies and in the end were considered in the municipal transport plan of the city of Naples.

This paper describes the transport analysis at the base of the conceptual design of the line, the environmental compatibility studies, the process for the stackholders involvement and communication, the urban integration of the system, the costs/benefits balance and the financial analysis issued for Naples Line 6 underground metro project.

1 INTRODUCTION

The Naples Line 6 is an important element of the public rail transportation network planned by the Municipal Transportation Plan (in Italian PCT) for the Naples Metropolitan Area.

The PCT aims at developing a balanced and integrated transportation system, i.e. a system articulated on networks that are strongly interconnected and structured so as to achieve a balanced distribution of mobility between the different modes of transport, using each mode within its own field of technical-economic validity and environmental compatibility. The urban and territorial configuration of Naples, the rich infrastructure of railway infrastructures, the intensity of the transport demand, the levels of traffic congestion, show that the backbone of the Naples transport system is an integrated and enhanced iron network.

The basic idea of the PCT is to build a "network of railways" by completing the projects started in the 80s and 90s and by transforming the terminal sections of the individual historical lines of the regional railways (Ferrovie dello Stato, Circumvesuviana, Circumflegrea, Cumana, Funicolari) into metropolitan railways, with new stations and new trunks and with more frequent services. The result is a network consisting of nine real metro lines or regional railways with subway characteristics for a total of 90 km, with 98 stations and 18 major inter-change nodes. The 9 lines form three interconnected rings, intersected by two transversal axes West-East at different points.

The Line 6 system is part of the multi-modal West-East coastal corridor (Fuorigrotta - Riviera di Chiaia - Centro - Porto) of the city of Naples, connecting the interchange nodes of Piazzale Tecchio, Mergellina and Piazza Municipio.

Line 6 is typical mass transit system, having the characteristics of "light rail metro", as defined by the UNI UNIFER 8379 regulation At the moment, it extends in the tunnel from the Mostra station, in Campi Flegrei node, to the node of Piazza Municipio, for a length of about 6,5 km with 8 stations and a Depot located in the dismissed military arsenal area of Coroglio (Figure 1).

In the following we present the methodology that has been followed in order to attain the European Commission approval and the authorization to use 2007/2013 ERDF Funding for the "Naples Line 6 Underground Metro System works – "Mergellina/Municipio functional Lot" Major Project. This analysis (and economic evaluations in general) focuses on the welfare impacts on society of the proposal submitted to the EC. These impacts are related to market and non-market costs and benefits, which in turn include the economic, social and environmental impacts determined by the above mentioned Major Project. The economic evaluation has therefore required to set out a two phase process to take into consideration:

- the relative impact of options at 'concept estimate' or 'developed concept estimate' -level of assessment;
- the investment impact based on preliminary design estimates of costs and benefits of the chosen investment solution of building the Naples Line 6 Metro infrastructure.

Cost-Benefit Analysis provides a robust method for evaluating the costs and benefits (including both market and non-market impacts) of a project or policy change in today's society as a whole. The estimated net benefits (total benefits minus total costs), and any significant impacts that cannot be valued, are used to help decision-makers rank and assess options, and to decide whether to implement them. In cost-benefit analysis (as in economic evaluations in general), the underlying core objective of government intervention is assumed to be focused in maximizing the society's welfare. Social welfare in this context is comparable to the level of

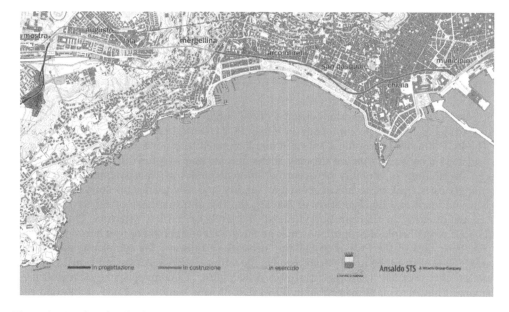

Figure 1.   Naples Line 6's chart.

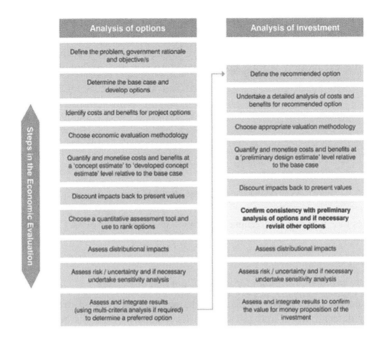

Figure 2. The economic evaluation process.

prosperity or living standards in an economy. This can be partly measured by standard economic measures such as Gross Domestic Product (GDP) and real income, but can also relate to less easily quantifiable factors that individuals value such as their own health, leisure time, benefits gained from enjoying environmental assets (and therefore the costs of pollution), and various social factors

The business case addresses a range of other objectives and factors that the government may wish to focus on in addition to the welfare maximization derived from the economic evaluation. Figure 2 shows the steps involved in an economic evaluation for a business case. The main issues that lead to the "Naples Line 6 Underground Metro System works" feasibility study can be found in the following pages and have been described according to the evaluation steps listed above.

2 NAPLES METRO LINE 6 – KEY ISSUES FOR THE ANALISYS

The urban transport functional lot "Mergellina/Municipio" concerns the costruction works of a new light railway system, as a completion of the line that had opened in September 2008 ("Mostra/Mergellina" connecting the Fuorigrotta district with Mergellina). At its completion, the Line 6 will continue to Piazza Municipio via the new Arco Mirelli (Piazza della Repubblica), San Pasquale and Chiaia stations. With its 5,5 km of lenght and 3 interchanges with other lines, the new line will create a crucial connection from Fuorigrotta to Piazza Municipio (the very heart of Naples Metropolitan Area).

The Major Proect was co-funded by the ERDF with a co-financing rate of 50% on the eligible cost of E. 173.051.488,23 (out of a total of E. 588.458.863) as required by the local government application for funding. According to the EC CBA guidelines, the investment consisted in:

- excavation works; tunnel (about 3,000 miles in length) and stations construction;
- catenary system and electricity supply networks signaling, spacing and traffic management systems and civil installations, including escalators and lifts.

4327

It is important to point out that some of the Stations included in the project are classified as "Art Stations" and required additional construction works (i.e.: pedestrian tunnels connecting to Municipio Station Archaeological Museum). Due to the massive archaeological evidence found during the excavation works, costs related to art and archaeology were necessarily included in the analysis.

PROJECT ITEM	NOMINAL VALUE (EUR)	TIME (to verify)
Mergellina/S. Pasquale	320.747.793,94	2007 2020
S. Pasquale/Municipio	267.711.069,25	2016 – 2020
TOTAL INVESTMENT	**588.458.863,19**	

Figure 3. Naples Line 6 Metro "Mergellina/Municipio" investment costs (VAT is excluded). The values reported are indicative.

2.1 Defining the problem, rationale for intervention and objectives

In Stage 1 (Conceptualise) of the lifecycle guidelines, the focus is on problem definition, and the preliminary determination and use of high-level benefits (and dis-benefits) to shape strategic options. As well as clearly identifying and providing evidence for the problem and the rationale for government intervention, the primary benefits or objectives have to be outlined. These benefits provide criteria to identify a preliminary strategic response to the problem. While Stage 1 involves looking at various strategic options from a high-level perspective, Stage 2 (Prove) involves undertaking a more detailed evaluation of the major options backed up by research, data and evidence, before determining the recommended option.

Benefits (and costs) which would be suitable for inclusion in economic evaluations are typically specific, tangible, able to be monetised, and can be linked unambiguously to an investment or activity. Under Stage 2, the identification and quantification of all costs and benefits to society is intended to capture the full range of welfare impacts of a given intervention. The Benefit-Cost Ratio (BCR) that is produced as part of an economic evaluation is one indicator of how successfully an intervention meets the objective of welfare maximisation.

Our case. By looking at mobility supply, in the mid-1990s, the rail transport network in Naples was quite articulated and wide. However, it suffered from a limited interconnection capacity between existing urban railway lines, and therefore was far from creating network synergies. Six different companies operated the existing lines in absence of an integrated vision of the urban rail transport system and in an un-coordinated manner. There were no interchange nodes between transport infrastructures and stations owned by individual companies, which handled lines designed as point-to point links, each with its own time schedule, frequency and tariff. In few words, many railway lines operating in the city were born as regional railways rather than urban ones, crossing many areas of the city with low frequency or without offering any stops and thus being inadequate to offer a reliable service during the whole day.

The public transport bus service also had its own weaknesses. It was born without an organic design and over time new lines overlapped old ones. The resulting system was a complex of buses lines and a network structure lacking of effective connections. Moreover, most of the bus routes covered long distances with limited frequency; in addition, the poor maintenance of buses continuously decimated the park of circulating vehicles.

On top of that, the urban road network was inadequate to allow a regular flow of passengers and goods vehicles for structural and functional reasons. 77% of urban roads had an actual width less than 6 meters, insufficient to absorb the traffic flows entering in the city. The network was also characterized by the absence of a functional hierarchy, i.e. a specialisation of roads in relation to the type of connection to be served (long/short trips within the city, commuting trips between the city and its metropolitan area) and the type of integration with the urban environment (trade, tourism, recreation, etc.). The consequence of this was a low average traffic speed during all the day, which further reduced during peak hours. At peak times, one-third of roads were at their maximum mass flow rate often exceeding the optimum density, traffic flow became unstable and even minor accidents resulted in persistent stop-and-go driving conditions. In addition, the level of air pollution from traffic was often above the legally admissible thresholds18 and interventions to completely stop the circulation of vehicles in the city was an option usually considered.

The main goals of the project, were therefore:

– Improving the accessibility to the city centre, flews coming from cruise traffic included.
– Reducing traffic congestion through a more effective system of connections with existing transport modes.
– Redeveloping and regenerating the surface areas and the surrondings in which the new stations related to the project would have been built.
– Increasing the quality of the urban transport service offered, through communication (i.e. rapid identification of a metro station in the city with appropriate signaling and real time and pre-trip user information along with clock-based and easy-to-memorize rail timetable); comfortable, clean, and modern stations, appropriate lighting and high-standards for security and safety.

Let's look at the financial management. It is worthwhile to note that this project and the Line 6 in general (along with its Art Stations) was conceived during a period of economic upturn, when public financial constraints were not as stringent as they are today. Therefore, on one hand the project has been negatively influenced by the financial crisis, an event unpredictable at the planning phase. On the other hand, the difficulty in guaranteeing the continuity of the necessary funds for the completion and the maintenance of the planned works since the beginning is undeniable, reflecting the poor managerial capacity to allocate the necessary funds by the stakeholders. This in turn has reflected in an implementation phase of the planned works that still proceeds with extremely longer lead time than expected.

Strictly linked to the previous point, there is the fact that archaeological explorations have accompanied the building work of the Naples metro stations and, of course, have had their own impact on the performance of the project both at infrastructural and service level. The assessment of the managerial capability to address archaeological issues produced mix results. Naples' experience represents a real integration of the worlds of transport engineering and heritage conservation, which have been considered as a mutual obstacle for centuries. From this point of view, Naples' experience represented a change in perspective, where opportunities from archaeology were maximised and to be replicated in other parts of the world. On the other hand, cost overruns and on-going construction delays mirror an archaeological risk not properly (ex-ante) mitigated. Although archaeology investigations were embedded in the metro project since its planning phase and permissions to build were obtained by the Superintendence of Cultural Heritage, archaeological related issues, such as unexpected discoveries and the low degree of "latitude" of the project are the main causes of the temporal misalignment (up to 8 years for some stations) between scheduled works deliveries and their actual state of play. Beyond increases in costs, delays generate two additional problems: the first one is the postponement of the stations opening, which negatively influence the performance of the whole metro line; the second one is that construction sites generate traffic congestion at surface.

Nevertheless, the positive effects towards the stated objectives are clearly stated in the analysis, although it is reasonable to consider that there is still large scope for further improvements. This will depend on the successful implementation of on-going investments (new trains), on the commitment by local institutions to implement the announced strategies.

As far as mechanisms are concerned, a favourable institutional, cultural and social context, positively contributed to the planning and construction phase of the project and, mainly, to its operating performance. Before the project, the mobility situation was so problematic that the investment yields a positive result in spite of many circumstances that has negatively influenced it, making the project itself underperforming. An appropriate selection process was essential in the planning and construction phase of the project that was delivered in time. The prioritisation mechanisms along with the involvement of the most important stakeholders was a well-chosen strategy for the definition of the project and to avoid delays in the completion works of this segment of the metro Line 6.

Similarly, the project design well mirrored the needs of city at the time of planning. The high quality and aesthetic standard of the stations greatly impacted on the city image, its urban quality and on the real estate market in Naples. Moreover, the design has undoubtedly influenced the well- being, the behaviour, and the quality of life of people that judge the new stations in a positive way. Having been fitted in a favourable economic cycle and because of the subsequent crisis and the lack of funds, today, these infrastructures are criticised for their high maintenance costs.

Experts from the European Commission pointed to an overestimation of benefits in the ex-ante CBA coming from two problems. The first one is related to the correct identification of the project. It is likely that the estimated benefits were computed by looking at the entire Regional Metro System, while costs only referred to the segment of the metro under assessment. The second one is that the analysis relied too much on the future availability of rolling stock. This supports the idea that the forecasting capacity that was probably affected by optimism bias.

Nevertheless, the process that has been successfully carried on in the evaluation of the Regional Metro System project as a whole (including Line 6) according to the process shown in Figure 4, is recognised as a best practice for having improved the quality of urban environment by designing eye-friendly transport infrastructures, which have strongly and positively impacted on the behaviour of people and their every-day life.

Today, the "Art Stations" (Line 6 ones included) represent a case study for architects and designers from all over the world, a valuable asset for the city, and an- extra attraction for tourists. However, developing such goal-oriented and costly infrastructures, also requires to ensure an optimal use of them, and specifically, to guarantee that the primary goal (i.e. the metro transport service) for which they were built is fully achieved.

Finally, our case study also proves that the ability to intercept the needs of the context in which the projects are to be implemented is a skill that pays off.

In the estamistions taken into account, a great effort was put in the process of identifying and incorporating stakeholder concerns, needs and values in the project decision-making process. In the same vein, a massive communication campaign to inform citizens about what the project would have look like was carried out. This also included the use of documents and posters as a useful way of presenting information to the general public conveying the main idea of the project progressing over time and the related events, among others, the opening of a new statiion.

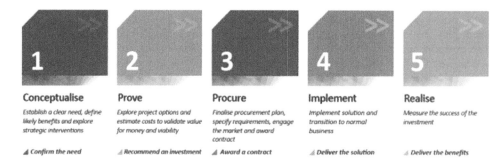

Figure 4.   Assessment process during designing Line 6 feasibility study.

3 INSIDE THE COST/BENEFIT ANALISYS (CBA)

A Cost-Benefit Analysis should capture, along with the financial analysis, allwelfare costs and benefits to society. The welfare impacts that should be captured are generally more specific and tangible than the benefits (and dis-benefits) referred to in the Stage 1 Conceptualise process. Not all benefits identified as part of an investment logic mapping process will be suitable for inclusion in an economic evaluation of an investment proposal.

Our case has therefore included the following.

3.1 Financial analysis overview

Operating costs (nearly 19,00 milion Euros per year) have been estimated according to Metronapoli S.p.A. balance sheets, the managing company running similar services at the time of the analysis. Additional estimations were made in order to properly consider the functioning stage. They include: a. labour costs (38% of the total yearly operating costs); b. supply costs (19%); c. costs for services (43%).

Revenues. By simulating the demand trends in the "intervention" scenario, a 30,215 milion passengers yearly demand (confirmed according to the new timescales of the completion, due for 2020) has been assested. The total revenues per year were estimated in 6,584 milion Euros (i.e.: (expected demand) x (net revenue per passenger of Euro 0,22 as provided by tickets+season tickets)).

As a result, the Financial Net Value of the Project (FNV/P) has been assested in 260,7 milion Euros (applying a 5% Discount Rate).

The Financial Net Value of the Investment (FNV/I), that considered national funding too, has been assested in 414,6 milion Euros (applying a 5% Discount Rate).

3.2 Economic and Social analysis

Among the direct benefits, according to the modal choise parametres on course in 2012, a special mention as to be given to the Valuel Of Time (Figure 5) related to the project comes out from the following formula:

$$V.O.T = \beta \, time / \beta \, cost \qquad (1)$$

whereas βtime e βcost are the parametres included in the modal choise model for time and cost.

Effects on time savings, energy consumptions reductions, air pollution decrease as well as rectuctions in accidents complete the list of the benefit generated by the project.

For this kind of analysis, accordign to the CBA European guidelines, the Residual Value of the project has been prudentially taken into consideration (25% of the investiment value) an has been estimated starting from the 35th year after the investment completion.

Effects on environmental sustainability are mainly related to traffic decongestion and, in turn, to the resulting air pollution.

Reason	Urban mobility (Euro/h)	Extra urban mobility (Euro/h)
Work	14,63	17,89
High School	4,69	4,69
University	13,55	15,37
Otheri	17,19	20,79

Figure 5. Value of Time.

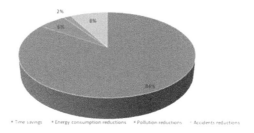

Figure 6. Economic and social benefits mix.

As for distributional impact, the high accessibility of metro stations and trains' platforms by all kind of users is ensured by appropriate signalling, escalator and lifts.

This positively contributes to social cohesion. Similarly, the Mergellina/Municipio extension of Line 6 will also increased the territorial cohesion making the access to the city centre easier with respect to the situation before the project implementation.

The baseline scenario of the CBA indicates that the project yields a positive socio-economic net present value (ENPV) equals to EUR 394 million, an economic internal rate of return (EIRR) of 8.57% and a benefits – costs ratio of 1.71 (Suggested Discont Rate: 3,5%). This proves its effectiveness and value for money from the society viewpoint.

3.3 Non measurable benefits

The high-architectural and technological solutions adopted for the metro terminals represents additional attributes in favour of the sustainability of the environment.

Energy saving systems have a limited impact on the emissions and consumption of natural resources, preserving them, as much as possible, for future generation.

Quantified benefits and their break down as a share of the total amount of indirect benefits generated by the project are shown in Figure 6.

It is however worth noting that these results are valid given the current state of play and knowledge about future scenarios, including the forecasts made on the future number of passengers. The expected increase in the number of passengers and the improvement in the transport service delivered as assumed in the CBA strongly rely on the extent to which the delivery of new trains and the construction works of the new extensions of Line 6 will respect the scheduled plan. The uncertainty related to these events is captured by the risk analysis, which, however, indicates that the project is currently characterized by a low volatility (i.e. the project's EIRR is likely to fall below its baseline value with a probability of 15% and below the forward social discount rate with a probability of 2%. Similarly, the ENPV has a 10% probability of falling under the baseline value). Future scenarios may range from a situation in which the project collapses to more successful perspectives should the main stakeholders put into practise the announced strategies. The CBA, which catches only quantifiable effects, could in future be supported by a qualitative assessment which is better suited to evaluate competing objectives, not measurable impacts as well as wider effects stemming from the project, which is a real must considering the peculiarity of the Naples Metro Line 6 Major Project.

4 THE PROJECT'S COMMUNICATION STRATEGY

The main stakeholders involved in the project design and in its execution first of all Naples Municipality and the General Contractor Ansaldo STS, shared a communication strategy based on citizens involvment from the starting stage of the project life cycle.

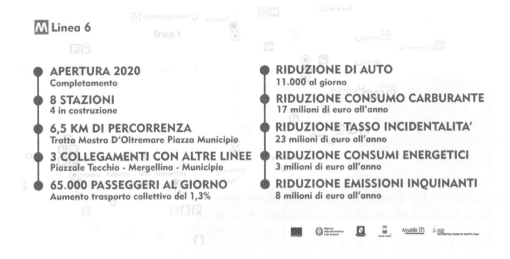

Figure 7. Line 6 to move better.

All the construction sites were externally equipped with particular rendering on the perimeter fences, rapresenting the design of the stations and the future superificial arrangement of urban areas after the works.

Furthermore, the project communication plan identified the channels through which to transmit to the public the vantages correlated to project realization, in order to mitigate the impact that citizens would have on construction sites.

A web site was issued for main informations concerning the schedule the social benefits and other advantages connected with the works, like the archeological retrievments found during the excavations (in the Figure 7 an example form the web site) is reported.

A general brochure of the project was published, now under review, and on the occasion of important project phase gates (new sites opening, institutional visits, inaugurations, etc.), Naples Municipality anticipate to mass-media a communication in order to correctly manage the information and to be sure the message is communicated to the citizens in a correct way.

This strategy achieved the main goal of the Municipality and General Contractor, to have transparency on the realization of an important public work and to delete the negative impact on the public opinion generated by the first stage of works in 90 years, when the fast tramway line (in Italian LTR) was abandoned.

REFERENCES

Piano Urbano della Mobilità Sostenibile (PUMS)
Dipartimento della funzione pubblica, Presidenza del Consiglio dei Ministri, La valutazione dei costi e dei benefici nell'analisi dell'impatto della regolazione, Italy.
Carli R., Paniccia M.R., (2003) Analisi della Domanda, Il Mulino, Bologna, Italy

Tunnels and Underground Cities: Engineering and Innovation meet Archaeology, Architecture and Art, Volume 8: Public Communication and Awareness/Risk Management, Contracts and Financial Aspects – Peila, Viggiani & Celestino (Eds)
© 2020 Taylor & Francis Group, London, ISBN 978-0-367-46873-6

TELT and the tunnel art work project

M. Virano, L. Beatrice, G. Dati & G. Avataneo
TELT-SAS, Turin, Italy

ABSTRACT: The Tunnel Art Work (T.A.W.) project, promoted by TELT – Tunnel Euralpin Lyon Turin –, and curated by art critic Luca Beatrice, brings together – the first case of its kind – three pieces of street art in the exploratory tunnel of Chiomonte, Susa Valley, Italy. Via the T.A.W. project, it has been possible, together with the Italian artists Simone Fugazzotto, Laurina Paperina and the French artist LUDO, to reach the depths of a mountain and create a new art space. Paperina has decorated the workers' convoy, while the other two artists have created original street art works at 2,800 meters from the tunnel entrance. Despite some international experience of dialogue between art and tunneling, such as the emblematic case of the Naples underground, there has always been a cultural and physical separation between the two fields. The T.A.W. project, therefore, aims to change this situation, promoting a reflection on underground creativity.

1 TUNNEL ART WORK

Playing on the double meaning of the word "gallery", which in Italian means also "tunnel", the Tunnel Art Work project, promoted by TELT – Tunnel Euralpin Lyon Turin –, and curated by art critic Luca Beatrice, brings – the first case of its kind – three pieces of street art in the exploratory tunnel of Chiomonte, Susa Valley.

Indeed, a gallery – intended as an art gallery – is the place where art is displayed, but "galleria" is also the Italian word for "tunnel." Going through a tunnel means traveling along a road, in the dark, seeing the light again only at the end of the path. There is always something magical, mysterious, and ancestral about these routes, a different sense of challenge compared to the overwhelming presence of white that is found in contemporary art galleries. This is the new, unknown, and growing Underground world.

Via the T.A.W. project, it has been possible to reach the depths of a mountain and create a new space, together with the Italian artists Simone Fugazzotto, Laurina Paperina and the French artist LUDO. Paperina has decorated the workers' convoy in her typical pop and comic-strip style, while the other two artists have created very original street art works 2,800 meters further away from the tunnel entrance.

Despite some crucial international experience of dialogue between art and tunneling, such as the well-known case of the Naples metro stations, there has always been a cultural, even more than physical separation between the two fields. The T.A.W. project, therefore, aims to change this situation, promoting a reflection on underground creativity and on the underground world. Its goal is to allow people to take back the underground environment, accompanying the technical development not only with the necessary communication about the work in progress, but also with a cultural operation of recreation and discovery with a novel definition of the hypogeum space.

2 BACK TO THE ORIGINS

Thanks to the T.A.W. project, Street Art goes back to where it belongs. Graffiti, this peculiar guerrilla-art form which became popular in the U.S. between the end of the Sixties and the beginning of the Seventies, was indeed born on urban walls and in the underground.

At the peak of his career, between 1980 and 1985, Keith Haring left more than five thousand drawings on the subway billboards. He used to draw some of his iconic subjects, such as the radiant baby and the barking dog, very quickly: just a few lines, drawn with some chalk, before moving to the following station.

His art was meant to be for everybody, thanks to his accessible and unique style, as well as to the choice of public spaces, where millions of people traveled every day.

The Graffiti movement was considered illegal for a long time. Its goal was to defy institutions. For writers, leaving a tag (the author's signature) in the most unreachable place, such as the top of a building or the depth of a tunnel, just beside the rail, became a way of self-representation, in front of their or other artistic communities. The tag is also used to mark the territory. However, it is also a way to communicate to the world outside. Hence, Street Art invades public spaces and it often entails a violation of private property.

It intends to provoke and to contest, going back to the Biblical expression "writing on the wall", which is often used to refer to an imminent danger and is mainly linked to an anarchist environment.

If compared to the vast urban areas in the U.S., or the most infamous New York neighborhoods, the beginning of the Graffiti movement in Italy in the Eighties found a much more complex environment, with significant stratification and a long cultural heritage.

From Bologna to Milan, the hip-hop culture brought the Graffiti art to Italy, as well as rap music and break dance, which both represent relevant elements of this avant-garde.

The turning point in our country was the exhibition "Arte di Frontiera. New York Graffiti", which took place in the Bologna Gallery of Modern Art in 1984. The curator was Francesca Alinovi, an out-of-the-box researcher who brought to Italy some great artists, such as Kenny Scharf, Keith Haring, and Jean-Michel Basquiat.

They all were part of New York's Old School that Alinovi discovered as she was exploring the Big Apple's neighborhoods, in search of avant-garde movements and new combinations of artistic languages.

"They are like geysers," she said, talking about the Graffiti movement's impact on traditional art. At that time, 10 to 15 years of ephemeral art had passed by in America: from performance to body-art, to conceptualism. Going back to the visual arts was, in a way, sort of inevitable. A crucial contribution came indeed from the Graffiti movement, which found an artistic environment ready to react to its challenges.

3 THE BIRTH OF STREET ART AND THE ROLE OF THE INTERNET

The Graffiti movement evolves according to its geographical peculiarities, adopting each time new characteristics and styles. In the Nineties in Paris, at a time when tags invaded the city, some groups of artists began exploring new ways of expression. They chose a much more pop and accessible language, leaving the tags behind and starting to draw symbols and paintings instead. Street Art was born. Starting with spray, artists then adopted a mix of different techniques.

The most practical one turned out to be the stencil, allowing to prepare drawings in the studio and to apply them on the walls later, in a few seconds. This technique is also the one often used by Banksy, the most well-known Street artist of all times, and it allows him to keep the secret on his identity.

A quick performance is also crucial for LUDO, the French artist who took part in the T.A.W. project, as we will see later.

As new technologies spread, performance became even more important than its result. The internet amplified this phenomenon. Starting with the first fanzines in the Nineties, which were exchanged between groups of people, it has allowed a potential public of millions to access every piece of work.

This process, as well as a much more universal style if compared to Graffiti, made Street Art famous and acceptable to society. It has even become a trend.

As it always happens to avant-gardes, the entrance of the Street Art in the official art system caused a loss in its original political intent. However, a significant exception originated around the Lyon-Turin opposition (the so-called "Movimento No Tav"), which found in Street Art a significant means of expression. In the Susa Valley, in 2017, the No Tav even organized an art festival, named "Wall Susa", which was a way to oppose the project and to make people aware to the No Tav's thesis and initiatives.

4 T.A.W.: ARTISTS AND ARTWORK

In the environment described above, the T.A.W. project brought together Italian and French artists in the first experiment of "Tunnel art" at an international level, willing to explore a new way to connect the TAV Project with the often antagonistic world of writers.

Simone Fugazzotto and LUDO were invited to express their creativity 2,800 m deep inside the exploratory tunnel of La Maddalena, Susa Valley. Laurina Paperina decorated the train that brings the workmen to the excavation face.

37-year-old Fugazzotto began his artistic career at the Brera Academy, in Milan, before moving to New York, where he took his inspiration mainly from the streets. He spent five years there, studying everything from painting to sculpting, from animation to Street Art.

Above all, he prefers painting on canvas, but he likes exploring different materials as well, from wood to plexiglass, from jute to concrete. Fugazzotto brings on canvas a reflection over contemporary human existence, which is lost somewhere between passion and vice.

In Chiomonte, the Italian artist presented "(Silence of a crossword)", a mural which brings attention to the enormous amount of information that overwhelms today's society. The monkey, which represents our era, finds itself in the middle of a crossword puzzle, where key-words about the Lyon-Turin project are displayed, e.g., speed, control, speculation, under-ground. All of these concepts have become so recurrent in the last decades that they have turned into catchphrases. The monkey, confused, tries to save itself from this word-bombing by seeking refuge under an umbrella.

Laurina Paperina was born in Rovereto, Trentino, where she lives and works as an artist. Her very peculiar style is based on irony and parody. Her natural environment is in the street, where she can draw and express her Graffiti attitude. By painting, drawing, working on instal-lations and amateur-video clips, she gives her reinterpretation of everyday reality, as well as the star system, the world of music, cinema and art, with their false myths and legends.

Her victims are rockers, artists, and cinema stars whom she often "kills" in parodistic cir-cumstances, or superheroes whose superpowers are represented in a satirical way. She uses almost everything in the process: canvas, re-used paper, post-its, paint-cans, picture frames bought at a flea market. She usually fills her world with fantastic characters who take after comic strips and cartoons. Paperina's inspiration comes from the Eighties and Nineties' imaginary that she discovered as a teenager, spending her afternoons in front of the TV or playing on the first computers.

In "Little Trains" Laurina Paperina decorated two wagons of the train that brings workmen and visitors into the exploratory tunnel. Here, she created a moving artwork, such as it used to be in the underground, where Graffiti was initially born in the Seventies.

LUDO, born in 1976 in Saint Germain en Laye, France, lives and works in Paris. Crucial to his artistic conception is the dialogue between very distant worlds: poetry and nature on the one side, and the harshness of modern life on the opposite one.

With a tagger profile, he took to paste-up in 2007. He usually draws large scenes which he prints with the help of a plotter. He then sticks the posters to the wall and finally paints on them, using just three colors, green, white and black. LUDO can imagine fantastic creatures, half living and half hyper-technological: birds, insects, flowers, and trees mix with war weapons, pistols, surveillance systems, and oil plants. A visual short circuit which can generate a reflection on our times, on human behavior towards his past and nature. However, there is no room for criticism, but rather for a trustful openness to the future.

In Chiomonte, LUDO prepared two artworks, one 12 metres long and the other 6 mentres long, bearing two Latin quotations, "Dulce Bellum Inespertis" and "Casus Belli," mixed one with roses and barbed wire and the other with a strange, robotic butterfly. More likely to display in a movie than on canvas, these science-fictional scenarios have a gothic taste, which offers an alarming point of view on the future.

5 CONCLUSIONS

Tunnel Art Work was inaugurated on October 10, 2016. Since then, dozens of articles have covered this project. Both traditional media and websites focused on this unprecedented cultural operation, contrasting it with the murals outside the worksite, in Chiomonte. Some of the most influential street artists in Italy, such as Blu, signed these paintings in order to express their solidarity to the No Tav movement.

The T.A.W. experience spread the debate also in the artistic community, where its impact was even more relevant than what TELT expected. "The art of controversy," titled the leading Italian newspaper, Corriere della Sera, which focused on "the new, cultural challenge between the No Tav movement and the promoters of the new railway."(Falcone 2016). La Stampa, another daily newspaper, published an interview with Paolo Damilano, former president of the Cinema Museum in Turin, who led the jury that awarded the contest for TELT's logo: "Culture can help to lower the tension and to release the spirits – he stated – Therefore, every initiative that goes in that direction is welcome." (Tropeano 2016).

In conclusion, we can say that if one of art's goals is to create hype, to open the debate and to make the public think, T.A.W. went right to the point. It helped rethink the underground environment, which often comes across with hostile and anxious feelings. Moreover, it opened the tunneling world to a non-professional public.

More than two thousand people have visited the exploratory tunnel since 2014. Most of them are technicians, researchers, geologists, and students. However, there are also groups of citizens among them, willing to discover the work in progress of a major European infrastructure. The T.A.W. opening made this participation increase, to the extent that a Japanese tour operator showed his interest in the project.

Another important influence on the T.A.W. project comes from the increasing trend of municipalities and other public bodies commissioning Street Art projects directly.

There are significant examples across Europe, sometimes linked to the underground environment. Just like in Naples with the "Metro dell'Arte" project, developed in '95 by the City in parallel with the renewal of its underground transport system. The stations, celebrated by the press around the world, host artworks by Paolini, Tatafiore, Paladino, Jodice, Alfano, De Maria and many others. In particular, the Toledo stop, designed by the Spanish architect Tusquets Blanca, won the Oscar for underground works in 2015, awarded by the International Tunneling Association. The Daily Telegraph and CNN also described it as the most beautiful metro station in Europe. The interior is enriched by two large mosaics and by an equestrian statue by William Kentridge, as well as by some photographs by Achille Cevoli and Oliviero Toscani. Bob Wilson instead, has turned the lowest part of the station into a "gallery of the sea."

At the opposite corner of Europe, in Stockholm, the Tunnelbana network has become the "longest art gallery in the world," with its 110 km of sculptures, mosaics, and installations, ranging from the pioneers of the 1950s to contemporary art experiments.

In the wake of these experiences, in the not too distant future the Mont-Cenis tunnel, which is currently under construction, may perhaps host exhibitions that lend themselves to still different fruition, starting with the enormous caverns where the Tunnel Boring Machines are assembled. Those may eventually become the "Sixtine Chapel" of underground art. In the background, remaining to be explored, is the idea of creating art works that can be enjoyed by spectators moving at high speed. This, however, is an entirely different story.

6 EXAMPLES OF T.A.W. ART WORKS

Figure 1.   The French artist LUDO in front of his "Casus Belli" artwork.

Figure 2.   "Dulce Bellum Inexpertis", LUDO's art work in Chiomonte.

Figure 3.   Simone Fugazzotto's "(Silenzio di un cruciverba)".

Figure 4.   The Italian artist Simone Fugazzotto in front of his mural in Chiomonte.

Figure 5.   Laurina Paperina's "Little Trains" in Chiomonte.

Figure 6.   "Little Trains", close up.

REFERENCES

Falcone, M. 2016. L'arte della polemica, *Corriere.it*, 26 October 2016 article.
Tropeano, M 2016. La street art è l'ultima sfida tra i Sì-Tav e il fronte del No, *Lastampa.it* 10 October 2016 article.

Risk management, contracts and financial aspects

*Tunnels and Underground Cities: Engineering and Innovation meet Archaeology,
Architecture and Art, Volume 8: Public Communication and Awareness/Risk Management,
Contracts and Financial Aspects – Peila, Viggiani & Celestino (Eds)
© 2020 Taylor & Francis Group, London, ISBN 978-0-367-46873-6*

Shallow TBM launch – alternative to cut and cover

W. Angerer
Jacobs, Dubai, UAE

A.M. Haimoni
Dubai Roads and Transport Authority, Dubai, UAE

A. Ozturk
Expolink Alliance, Dubai, UAE

V. Tellioglu
Jacobs, Dubai, UAE

ABSTRACT: This paper describes the decision-making process and implementation of a value engineering alternative that allowed a shallow launch of TBM on Route2020 Metro program in Dubai. This replaces a 300m long cut and cover tunnel included in the reference design. The TBM launch was originally targeted at the end of the cut and cover tunnel with 12m cover to the tunnel alignment. After award of the construction works, the client, contractor and designer investigated a value engineering alternative to launch the TBM from the open cut section, hence, replacing the cut and cover tunnel with a bored tunnel. The intent of the shallow launch was to de-risk the challenging construction program by reducing the construction works required prior to TBM delivery. Ultimately, the TBM was successfully launched with only 2.0m of cover. The paper will present an outline on the risk management, engineered mitigation measures together with a review of performance during construction.

1 INTRODUCTION

Dubai's award in 2013 of the World Exposition (Expo 2020) served as the catalyst for the Route 2020 project, a new metro route connecting the existing Dubai Red Line with the Expo site in the south of the city. Route 2020 is the first major rail extension to the existing Dubai Metro rail network. The project consists of approximately 15km of metro alignment which consist of 12km of viaduct and 3.2km of tunnel. It also features seven LEED Gold certified metro stations consisting of one interchange station, three elevated stations, two underground stations and a signature Iconic Station at the Expo Site as well as the supply of 50 trainsets, of which 15 are needed for the new line and 35 to increase capacity on the existing network. The new line will establish a direct metro connection between Dubai Airport, the City and the Expo Site. The number of passengers to use Route 2020 is expected to reach 125 thousand commuters per day in year 2020 and will rise to 275 thousand commuters by year 2030.

In June 2016, Dubai RTA (Roads and Transport Authority) awarded the design and build contract for the construction works to Expolink, an Alliance that consists of Alstom (France) in partnership with the Civil Works Joint venture of Acciona (Spain) and Gulermak (Turkey). The detailed design for the civil works was awarded to Jacobs Engineering by the Civil Works Joint Venture.

Tunnelling on the 2.4km bored tunnel between Discovery Gardens and Green Community was formally launched on October 24th 2017. Tunnelling was successfully completed on 11th

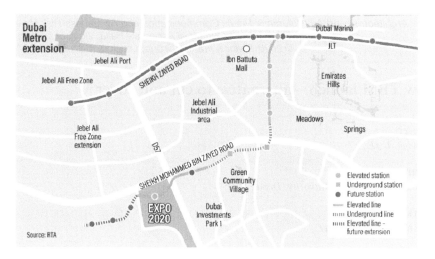

Figure 1.   Route alignment.

June 2018. This paper outlines some of the major decisions taken in relation to the shallow launch of the Tunnel Boring Machine (TBM) in Discovery Gardens.

2   PROJECT INITIATION

At the start of the project the Client, the Contractor and his Designer jointly reviewed the launch arrangement of the TBM in Discovery Gardens. Due to the tight construction timeline, a special focus centered on the reference design to identify if any improvements could be achieved to safeguard the critical activity of the TBM drive within the works construction program.

The reference design provisioned a classic cut and cover ramp approach leading to the TBM launch site. This consists of approximately 560m of below grade ramp section, of which 300m would be executed as cut and cover. Two TBM supply shafts were provisioned as shown in Figure 2.

This design had a significant number of interfaces with existing utilities and a complex construction program with numerous interdependencies. The Contractor studied this section with

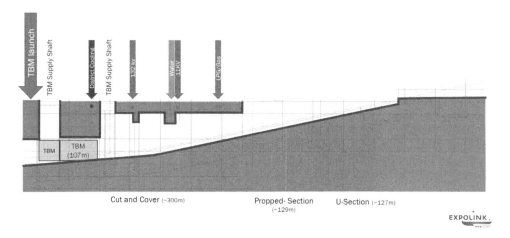

Figure 2.   Launch approach in reference design.

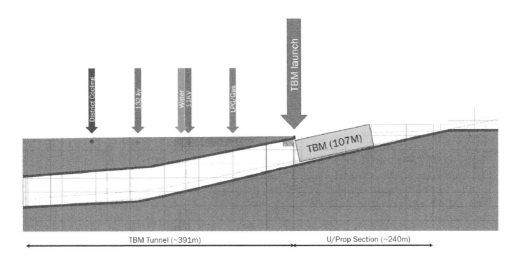

Figure 3. Alternative Launch of TBM.

different construction programs depending on the particular construction approaches and identified significant risks.

As such an alternative to the reference design consisting of a shallow launch of the TBM was brought into consideration.

The alternative design was aiming at reducing the works required ahead of the TBM delivery. In this approach the cut and cover section would be replaced with a bored tunnel that is constructed with minimum cover requirements. The utilities in this section would therefore remain largely unaffected and the critical activity on the construction program seemed improved. Furthermore, the utilization of the expensive TBM equipment is improved and the major utility corridor crossing the rail alignment would remain available for future usage of the local utility network providers. Therefore, this option was seen to offer many advantages and investigated in more detail.

3 RISKS AND OPPORTUNITIES

As a first task a risk and opportunities register was established for both options. The Contractor, as the owner of the project delivery risks, led the development. The below highlights the major considerations that were taken into account:

3.1 *Reference Design*

The design offered the following opportunities

- Construction ahead of TBM delivery possible although on the critical path
- Final structure provides more space for Mechanical, Electrical and Plant equipment (MEP)
- Reduced geotechnical risk profile

However, the design was perceived to pose the following risks

- Multiple stakeholder approvals required
- Three major utility diversions
- Stakeholder approval is on the critical path of the project
- Large construction area required close to residential buildings
- Complex construction works that lie on the critical path of the project
- Reduced ventilation performance
- Complex dewatering requirements

- Complex construction sequencing to account for the requirements of each utility
- Utility corridor largely compromised for future development (major utility corridor crossing alignment)
- Permanent diaphragm wall details are not seen as reliable as segmental lining

3.2 *Alternative Design*

The alternative design offered the following opportunities

- Major utility diversions can be avoided
- Reduced stakeholder approval risk
- Utility corridor remains available for future developments
- Ventilation performance will be improved, which is also an advantage for LEED certification
- Start of the open cut works and the provision of the required protection works for the TBM launch provides more float on critical path of the project
- Reduction of above ground works which will result in reduced noise pollution
- Reduction in dewatering requirements
- Lower Carbon Footprint due to reduced concrete quantity, which is also an advantage for LEED certification
- Reduced soil subsidence from tunnelling compared to cut and cover and decrease in influence zone of settlement

These opportunities are compared with the following risks

- Water mains running close to tunnel crown (require diversion)
- Increased geotechnical/tunnelling risk due to shallow cover
- Increase duration of TBM tunnelling works due to additional length of 390m to replace the cut and cover length present in the original design.

As can be seen in the above, the range of risks and opportunities affect all contractual partners on the project. However, all parties concluded that the Alternative Launch method represents a viable approach that may bring many benifets.

As such a working group to study the Alternative Launch was established which included members from the Client (and his representative), the Contractor and the Designer. The working group was tasked with the investigation and development of mitigation measures for the percieved risks related to the alternative approach prior to deciding to proceed.

4 MITIGATION MEASURES

4.1 *Structural and Geotechnical Design*

To mitigate against the risk of collapse the working group recommended a range of structural/geotechnical works that are required to be completed prior to TBM launch. These works are shown indicatively in Figure 4 and consist of the following

- Soil Improvement
- Construction of an inclined structural concrete slab – to remain as a permanent structure
- Diaphragm walls
- Geotechnical protection works – concrete slab – temporary works for protection purposes only

4.1.1 *Soil Improvement*
The ground investigation identified a layer of approximately 3m deep Aeolian sands present at the TBM launch location. This layer of sands is at the start of the TBM launch and has to

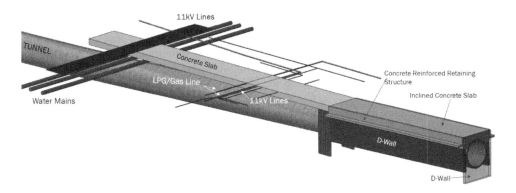

Figure 4.   3D Model of Alternative Launch (continued pile wall under concrete slab not shown for clarity).

be improved in order to allow the TBM to launch without completing a full EPB (earth pressure balance) chamber. It facilitates the build-up of face pressure.

A further soil improvement is implemented at the interface between inclined structural slab and protection slab. This interface is seen critical, as from this point on the geotechnical stability to the ground is solely reliant on the support provided by the TBM. The soil improvement is executed by the provision of soil replacement, with a target to give the material a temporary cohesion or cementation that allows excavation of the ground underneath the structural protection slab without any risk of localized instability during TBM passage. The geotechnical property targets were designed to achieve a near homogeneous material over the full face of the TBM cutterhead, to facilitate optimal steering control of the TBM. Low strength material is preferred compared to higher strength values for this reason.

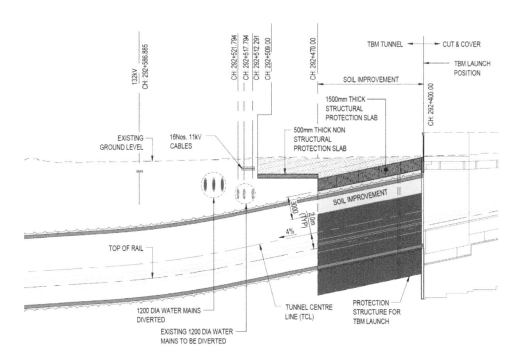

Figure 5.   Soil Improvement.

4.1.2 *Inclined concrete slab*

An approximately 60m long 1.5m thick inclined concrete slab is constructed above the tunnel crown in order to avoid buoyancy and provide structural support during the TBM launch. Some additional concrete reinforced retaining structures (contiguous piles) are constructed at each side of the slab in order to mitigate the risk of lateral geotechnical failure due to face pressure loss.

The structural protection has three functions:

* It addresses the geomechanical design in the area of shallow launch. The TBM excavates underneath the protection slab and ensures structural integrity of the material above the tunnel crown is maintained during TBM passage.
* It also creates the confinement stress required for the segmental lining to structurally function as a system according to its design.
* The structural protection slab is a permanent structure that will provide weight to counter buoyancy. It acts as a dead weigh rather than relying on the contiguous piles to act as tension members.

The inclined concrete slab is terminated at the location where a 2m of ground cover underneath the protection slab is reached. At this location, the TBM is judged to enter into homogenous material that allows adequate soil arch to form above the TBM to ensure stability and avoid local instabilities at the location of the transition.

4.1.3 *Concrete slab*

Following on from the inclined slab and for about 40m an unreinforced concrete slab 0.5m thick is constructed and is topped with engineering fill.

This slab is provided to prevent ground failure to the surface, to mitigate subsidence risks as well as to protect the utilities within the vicinity during the passage of the TBM. Moreover, it can offer a future protection to the tunnel lining by distributing concentrated loads after the completion of the reinstatement works at the surface should it be required.

The concrete slab is primarily a temporary measure. Its main purpose is to mitigate soil subsidence risks in the section of the drive. The slab reduces the risk of excessive settlement as well as differential settlement. The protection slab is constructed using a lean concrete mix with 5Mpa cylinder strength and without reinforcement. Reinforcement is avoided in order to allow the material to be easily excavated at a later date, should the need arise from any stakeholder for extension of their utility network.

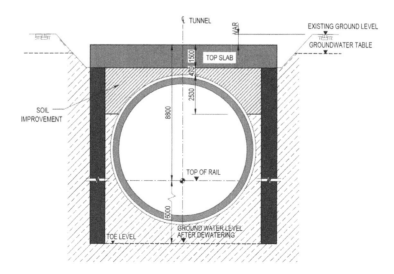

Figure 6.   Inclined concrete slab.

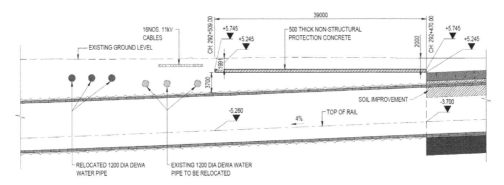

Figure 7.   Concrete slab.

4.1.4 *Major utility diversion*

Although one of the main aims of the shallow launch was to avoid major utility diversion, the levels of the water mains at the time of the decision-making process were estimated (based on the available information) to be higher than the actual levels encountered after undertaken a detailed investigation. The target of the design was to retain a minimum of 2m of cover between tunnel crown and main.

However, after execution of trial pits, it was identified that the water mains were existing at a lower level leaving a cover to the life water main of only approximately 1m. Hence, it was decided to divert the three water mains to allow adequate separation between the mains and the tunnel as shown in Figure 7, from the position shown in in light blue to the position shown in dark blue. The diversion was carried out using pipe jacking technique to avoid unnecessary disturbance to the ground that may affect the TBM drive.

4.2   *Schedule Analysis*

The Alternative Launch option affected the construction schedule of the TBM drive. The additional 391m of tunnel drive was planned to require additional 1.5 months on the TBM Tunnel construction schedule. In order to maintain the required float on the TBM construction program, meetings were held with the manufacturer and it was agreed to speed up the production and transportation of the TBM by 1.5 months. All TBM components were transported from Germany to Dubai by only one vessel, with the TBM arriving at the last week of June 2017 at Dubai port. The TBM installation was then started mid-July with completion achieved mid-September with the official launch being held end of October 2017. Tunnel excavation was planned according to below rates;

- First 353m @ 7.4m/day = 48 Calendar Days
- Next 646m @ 11.5m/day = 56 Calendar Days
- Next 709m @ 12.9m/day = 55 Calendar Days
- Last 680m @ 12.6m/day = 54 Calendar Days

The TBM arrived on the 1st day of June at the last receiving Station (R74 Station). The Intermediate Station (R73 Station) passage was completed in 3 weeks. Overall the executed tunnel excavation rates were in line with the planned. The maximum advance rate of the TBM achieved 36.8m in 24 hours.

5   RISK ASSESSMENT

Once all the mitigation measures were agreed, a final risk assessment comparing both options were carried out. The results of the assessment are shown in the figures below.

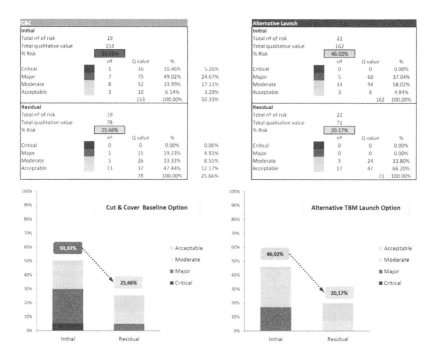

Figure 8.   Results of qualitative risk assessment.

Both options showed that a significant reduction in risks can be achieved if appropriate mitigation measures are implemented. Whereas the overall risk percentage is only slightly better for the shallow TBM launch option (25% vs 20%), it is noted that the shallow launch is considered to not include a major risk.

Although it may be perceived that using a new method, such as a shallow launch, is having a higher risk profile compared to an established construction method, such as cut and cover, the team implemented a very robust set of structural and geotechnical measures to ensure maximum mitigation of established risks. The actual calculations showed that a lesser amount of structural works would also have led to a code compliant and numerical safe approach. However, the team from the start based his decision-making process on qualitative risk criteria as well as design calculations, to ensure that the overall mix of measures leads to a successful execution.

6   EXECUTION

The construction was carried out under close monitoring and observations of the ground and TBM drive parameters through a daily management meeting that was attended by all concerned parties.

The whole section was successfully tunneled with subsidence never exceeding 2mm. This was achieved within the shallow launch area but also along the complete tunnel alignment including built environments such as an existing busy motorway, overhead power lines, crude oil and gas lines, and several multiple storey residential and commercial buildings.

6.1   Instrumentation & monitoring

Both in the shallow launch and in the remainder of tunnel boring works, the specification requirements on field instrumentation and monitoring were strictly followed. Both the structural and the non-structural-concrete protection slabs were equipped with sufficient number

Figure 9.   Tunnel Launch Location – as built.

of settlement bolts to monitor sagging/heaving movements on the surface across the tunnel cross-section as well as along the tunnel axis. Inside the tunnel, convergence prisms were installed to monitor the overall performance of the lining – no recordable ovalisation was recorded. Stepping and lipping were checked visually inside the tunnel with all rings remaining within the allowable range.

The first 15m length of shallow launch was contained within a dewatered area therefore the groundwater levels were observed closely before, during and after the passage of TBM.

The data from the field was assessed on daily basis by the management action team, who had multiple skill sets and experience, the findings and actions reported on weekly basis to the senior management of the client. The Contractor enabled the management action team to access the data remotely hence the team was able to communicate on a 24/7 basis, as required. No damage to built-environment was caused by TBM shallow launch and boring.

6.2   Face support pressures

The required cutting pressures, required minimum face support pressures, and maximum allowable to avoid blowout were estimated, cross-checked and monitored jointly by the designer and the builder.

Anagnostou-Kovari (1996) method was used to estimate the required minimum face support pressures in either dewatered or submerged state depending on site conditions. As the tunnel alignment descends rapidly at 4% gradient within the shallow launch area, multiple cross-sections had to be analyzed in face pressure estimation with varied ground cover and groundwater conditions. Special attention was paid to the boundary conditions such as the start-up, i.e. boring through the diaphragm wall, and interfaces with other structures, i.e. slurry walls, etc.

"DAUB recommendations for face support pressure calculations for shield tunnelling in soft ground (2016)" was used to estimate and control the maximum allowable face pressures to avoid blowout at the surface above the tunnel crown.

The estimated minimum face support pressures and maximum allowable pressures for the first 100m drive are shown in Figure 10 below.

It was clearly observed that the smooth operation of TBM with none or minimal changes in advance rate delivered good results in shallow launch. Good control on TBM parameters and coordination of the works amongst multiple disciplines primarily with instrumentation & monitoring require good management and technical skills to deliver a safe execution of shallow launching, which was successfully implemented on Route2020.

		Face support pressure [kPa]		
Chainage		Required min. support	Max. allowable pressure in tunnel crown against blowout	Remarks
from	to			
292+400	292+415	35-35	45-55	Dewatered. Confined dewatering within slurry walls. Under structural protection slab.
292+415	292+470	35-40	55-85	No dewatering. Under structural protection slab.
292+470	292+500	40-45	85-100	No dewatering. Under non-structural protection slab.
		(ref: Anagnostou-Kovari (1996)	*ref: DAUB recommendations for face support pressure calculations for shield tunnelling in soft ground (2016)*	

Figure 10.   Estimated face support pressures for the first 100m drive.

7   CONCLUSION

The rigorous risk management approach that was implemented led to a successful completion of the shallow TBM launch alternative on R2020 project. Structural and qualitative mitigation measures were designed and implemented in order to ensure a fully controlled construction environment. All decisions were carried out in a highly collaborative manner between the Client, the Contractor and the Designer. All parties were mindful of the drivers of each other and everybody in the management team had a similar understanding of the risks. From the beginning and during the whole tunnel drive everybody agreed on the measures to be implemented before instructions were passed to the operatives.

This success demonstrates that with a collaborative work model, innovative solutions can be executed in a structured and safe manner that can lead to benefits for the Client and the Contractor alike.

Tunnels and Underground Cities: Engineering and Innovation meet Archaeology,
Architecture and Art, Volume 8: Public Communication and Awareness/Risk Management,
Contracts and Financial Aspects – Peila, Viggiani & Celestino (Eds)
© 2020 Taylor & Francis Group, London, ISBN 978-0-367-46873-6

The potential for use of the observational method in tunnel lumpsum contracts

A. Antiga
ÅF-Consult Italy S.r.l., Milan, Italy

M. Chiorboli
ICMQ, S.p.A., Milan, Italy

U. De Luca
ANAS, Roma, Italy

ABSTRACT: Tunneling works are strongly associated with significant uncertainties and risks. Change orders occur for various reasons and almost all tunnel projects tend to change as they progress. Many of the reported problems are associated with risks related to "geo" conditions that are not managed sufficiently in the design, contractual and construction phases of the projects. In order to achieve a cost-effective and a more predictable outcome of the projects to a cost that can be estimated before the projects begin, it is essential to manage the geotechnical risks in an adequate manner.

The Observational Method is an example of effective reaction to this hazard and its application is common in "re-measurement contracts".

This paper will provide some reflections on the potentials of the Observational Method in underground works also with reference to its use into projects according to a "lump sum" logic.

1   HAZARDS AND RISKS IN TUNNELLING

Doubtless, tunnel design involves much higher uncertainties and hazards than any other civil engineering work. Many tunnel projects reported significant cost overruns and Contractors claims in percentage much greater than other civil construction works. There are many types of hazards and risks in tunnel works and the consequence of failing to manage these risks can be severe in many projects. These risks can adversely influence the cost and time schedule, health and safety, quality and the environment.

The ITA AITES guidelines for tunneling management (ITA 2004) groups hazards into general and specific hazards.

General hazards:

1. Contractual disputes,
2. Insolvency and institutional problems,
3. Authorities interference,
4. Third party interference,
5. Labour disputes

Specific hazards:

6. Accidental occurrences,
7. Unforeseen adverse conditions,
8. Inadequate designs, specifications and programmes,

9. Failure of major equipment, and
10. Substandard, slow or out-of-tolerance works.

There is broad consensus that ground conditions account for the largest element of technical and financial risk in tunneling projects.

This is because to be underground means that the context is difficult to know. Even in a comprehensive exploration investigation, we can test only a small part of the total ground and we can recover a miniscule drill core volume that is less than 0.0002 % of the volume of rock mass that one has to consider in tunnel design. For concrete works the controls are nearly 0.02% of the volume of a structures. This means that the ratio of knowledge between underground structures and concrete structures is nearly 1 to 100.

The second issue that do tunnel design very demanding is model uncertainty. The ground surrounding a tunnel is an integral part of the final structure and plays a pivotal role in its stability. Even if a tunnel structure needs supports made up of concrete and steel, the ground has both a supporting and a loading role and it is the major part of the structure. In tunnel designs, one has to work in a pre-existing equilibrium and proceed in some way to a "planned disturbance" of it in conditions that are only known in part.

Moreover, the excavation cost represents a part from 60% to 80% of the total construction cost.

However, we must remember that there are other types of hazards as ITA AITES guidelines highlight.

There is an interesting study (Avestedt 2012) that analyses many Swedish and American tunnels, built in the last 20 years, indicating how the most important hazards are "unforeseen ground conditions" (nearly 25 %) and "contractual disputes" (nearly 35 %).

On the other hand, it is showed that contractual disputes arise, in large part, by unforeseen ground conditions.

Ultimately, we can state that we must concentrate our efforts first and foremost on reducing risks deriving from hazards due to "geo" (i.e. geological, geotechnical, geomechanical) and contractual aspects.

2 LEVELS OF KNOWLEDGE ABOUT "GEO" UNCERTAINITIES

In the development of a project of an underground work we can distinguish various levels of knowledge about hazards and uncertainties and consequently about a risk event's occurrence.
We can have:

1) uncertainty associated to the tunnel section in which specific situations, nevertheless certain, will occur; for example, it is not known the actual distribution of rock types along tunnel alignment, the extent of fault areas, the exact area of sudden water in flows, etc. but it is certain that them will happen;
2) uncertainty associated to the precise intensity with which certain situations, expected by type (and as maximum value), will occur; for example, having to go through a certain rock mass, there are uncertainties about the maximum magnitude of the acting loads and of the consequent convergences, even though it is possible to assume a priori the extreme situations of reference;
3) uncertainty related to the possible occurrence of completely unforeseen situations. In this category we must include unfortunate geological feature, the presence of which was missed during the design stage and requires partial or total redesign of the project; we have also to include a type of geomechanical behavior that was not predicted in the design (Lunardi 2008).

We can define the first two uncertainties as "expected tunneling hazard"; we can prepare the tools for dealing with them during the design phase and therefore they are manageable and lead to the modification of quantities within a specific standard solution already identified in the project. The third is an uncertainty that can make the typical solutions imagined in the project no longer valid.

When we deal with ground conditions the "unknown" usually exist in inverse proportion to the amount and quality of the geotechnical investigations.

The first obvious, but not always respected, rule to mitigate the "geo" uncertainties is to prepare a plan of geognostic investigation in an appropriate manner.

With an adequate geological and geotechnical knowledge, it is possible to better refine the project in the context of the uncertainties of type 1) and 2) and reduce as much as possible the uncertainties of type 3).

The most effective strategy to reduce geotechnical risk is to minimize the unforeseen and site investigations are the basis on which the risks associated with the project are identified; on the other side without sufficient information from site investigations the inherent risks in construction would be unacceptably high. It would be obvious to conclude that a lot of time and economic effort need to be concentrated in site investigations. The greater the completeness of the site investigation program, the better the estimation of quantities on the plan and design stage and, the better the correspondence with actual quantities used for construction. Hence, an in-depth site investigation program would ensure that the project sees fewer change; resources dedicated to site investigations should be considered as an investment.

It is always important to remember that an "expected tunneling hazard" is a controlled hazard and small unforeseen hazards can be worse than big foreseen hazards. Therefore, an effective site investigation program is of major significance to obtain the necessary information essential for construction; we should increase our knowledge about possible hazard to be ready to treat them in a right way.

Nevertheless, a balance needs to be identified in determining the extent of site investigation. Soil investigations can be carried only to the extent that is reasonably possible, compatibly with time and budget available. It is necessary to establish a balance between the resources spent through site investigations and the losses prevented by them. Considering the increasing emphasis on cost cutting measures and considering practical considerations, site investigation is bound to be an area of compromise.

A good investigative program has been helpful in reducing costs of works in percentage much higher than the cost of the additional exploration (Figure 1).

We should remember that residual uncertainties can lead to excessively conservative design approaches, higher provisions for risks and higher exposure to the risk of contractual claims and any apparent economy in terms of cost could result in bigger residual geotechnical risks.

It is experienced that "geo" residual uncertainties tend to prevail despite robust site investigations and that other uncertainties are ineradicable, first of all model uncertainty.

In 1945 Terzaghi stated: "*In the engineering for such works as large foundations, tunnels . . . . many variables . . . remain unknown. Therefore, the results of computations are not more than working hypotheses, subject to confirmation or modification during construction*".

Despite the enormous developments in geotechnical theories and computational engineering since 1945, the variability of the soil and the model uncertainty will never vanish; tunnel engineers will always suffer from a lack of information. Thus, even the result of the highest sophisticated numerical model will be more or less a crude estimate.

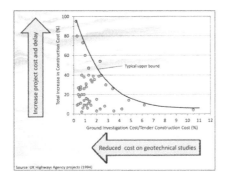

Figure 1.   UK Highways Agency – Increase in construction cost vs ground investigation cost.

3 RISK MANAGEMENT IN THE DESIGN PHASE

In the design phase, to deal with the uncertainties (first ground conditions and design model), the typical approach, known as 'predefined design', is to attempt to eliminate the uncertainty by assuming a conservative condition, derived from the data available prior to construction.

Various possible methods, inherited by design of structures, are used to account for uncertainties and to ensure adequate safety: the safety factor method, semi-probabilistic method (partial factoring), the probability-based calculation method.

In all three cases, the final design is determined in advance of construction, based on ground parameters that take account of uncertainties inherent in natural soil/rock.

The safety factor method is sustained by a vast amount of experiences. It introduces a safety factor that is defined as a ratio between total resistance and total force and should be kept a value more than 1 given by experience. It has some weaknesses. The main is that the safety factor is not an absolute measure of the safety of a structure and gives only a partial representation of the true margin of safety that is available. Through standards or tradition, the same value of factor of safety is applied to situations that involve widely changing degrees of uncertainty. In theory, large uncertainties, arising from data and model, require large safety factors. However, the safety factors established in design codes are not calibrated to consider different ranks of uncertainty; thus, equal safety factors do not necessarily imply equal safety level. Therefore, a large safety factor might be applied without necessity, even if the level of uncertainty is low, or, worse, a too low safety factor can be applied in a case with great uncertainties.

Many design codes in the world today are in the process of revision from traditional "safety factor method" to Limit State Design (LSD).

The uncertainties involved in basic variables concerning actions, material properties, shape and size, etc. are taken care by factors which are determined based on reliability analyses. In this case, the necessary safety margin is preserved by applying partial factors to characteristic values of basic variables. Then all the calculations are done deterministically; the designers, users of these factors, need not to have knowledge in reliability theory.

For geotechnical structures, this approach has two main weaknesses (Honjo 2009):

- in modeling the uncertainties of forces and resistances, it is the tail part of the Gaussian distribution that is really affecting the reliability of a structure. However, it is very difficult to accurately model the tail part from limited data.
- the design calculation method is always a simplification and idealization of real phenomena. It is rarely possible to evaluate the model uncertainty involved in this kind of method. Thus, the calculated failure probability cannot be an accurate indicator of the failure event.

At first sight, a probability-based calculation method appears an objective way to measure the distance to failure. Assuming that soil parameters can be modelled in the probabilistic framework, and provided that the probability distribution functions of load and resistance, as well as the correlation functions of the shear parameters are known, the failure probability could be a good measure for the distance to failure. However, there are too many unknowns and crude estimations in such calculations, so that the failure probability is just another qualitative indicator for safety. Moreover, this demands a high statistical expertise and it is very difficult for one person to have a sufficient grasp of both disciplines so to combine them sensibly (Fellin et al. 2005).

Usually this leads to "defensive design procedures" that uses the worst-case scenarios, both on the mechanical parameters of the rock mass (in the selection phase of the characteristic parameter) and in terms of the overall safety factor.

A monitoring system is usually used, also extensively, to verify the validity of the design assumptions and to calibrate the application of stabilization interventions according to the patterns defined in the design phase. It plays a passive role because it does not allow to improve the safety coefficients that generally remain those used in the predefined design without taking into account the greater knowledge acquired in progress. Basically, the use of monitoring produces an increase in safety, bringing it to levels higher than that defined in the design phase.

The truth is that one cannot tackle the design of a tunnel with the methods and principle of structural engineering. A different approach should be taken also regarding safety factors.

There is an important aspect in tunneling from that we must take advantage. Since a tunnel is excavated "step by step", we obtain a lot of information on tunnel behavior during construction so we can improve the original design during excavation proceeding with an observational approach. The approach is based on a design that is improved step by step during excavation by means of observations and data collection on soil structure interaction that allow to acquire useful elements for a better understanding of the geological and geotechnical reference model.

An Observational Method design assumes that the ground conditions will be near to a "reference value" of the predicted conditions, but alternative designs or contingency measures are prepared in case conditions should turn out to be worse than predicted.

During the design implementation a rigorous monitoring and observation strategy is used to check and confirm the actual conditions found during construction. Performance indicators are selected for monitoring, related to the critical risks. Essentially, it is a method for reducing the risk in the construction phase based on preventive design analysis.

This approach identifies that uncertainty has two components, risk and opportunity: the ground conditions could turn out to be better than expected or worse and therefore brings enormous potential in terms of reducing risks and optimizing the costs of the works.

In addition, hidden opportunities of better ground behavior than expected, which are revealed during construction, may be used (Van Staveren 2006).

The choice of the parameters of calculation for the "reference design" is one of the most debated item.

This choice conditions the possibility that the Observational Method is cheaper, with the same security conditions, compared to a traditional approach.

There are, essentially, two extreme approaches (Figure 2):

- based on 'most probable' conditions. Contingency measures are prepared before construction and are implemented if observed behaviors exceed critical limits;
- based on a "most unfavorable" set of parameters. Observations during construction are used to actively optimize the design.

There are, obviously, intermediate approaches.

Peck (1969) adopted the "most probable" design and then reduced the design to "moderately conservative" parameters, where triggers were exceeded. CIRIA (1989) considers a "safer" approach to design by adopting a "progressive modification" of the design starting with the design based on "moderately conservative parameters", and then reverting to most probable conditions through field observations. CIRIA uses the terms "most probable" and "most unfavorable" to describe the range of soil conditions as illustrated in Figure 2.

The "most probable" is a set of parameters that represent the probabilistic mean of all the data, although a degree of engineering judgment must be used in assessing this to take account of the quality of the data.

The "most unfavorable" parameter represents the 0.1% fractile (it represents the worst value that the designer believes might occur in practice).

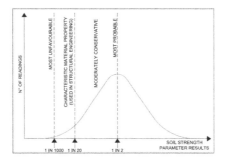

Figure 2.   Gaussian distribution.

The "moderately conservative parameter" (CIRIA 1999) or "characteristic value" of geo-technical parameters (Eurocode 7) represents an "cautious estimate of the value affecting the occurrence of the limit state". It should ideally result in prediction of the upper 5% fractile.

The "moderately conservative parameter" is not a precisely defined value.

It is a cautious estimate of a parameter, worse than the probabilistic mean but not as severe as the most unfavorable. In assessing these parameters, the designer should carefully consider the quality of the site investigation data and assess its suitability.

In the observational approach, the suitable limits of behavior, is a "serviceability" calculation. These provide the predictions that can be monitored and reviewed.

4 CONTRACTUAL ALLOCATION OF GEO-RISKS

Noted that residual "geo" uncertainties and cconsequently "geo" risks tend to prevail despite robust site investigations and advanced calculation models, a contract should incorporate a mechanism to determine how to deal with residual risks and should give a clear allocation of risks for the case they occur.

It is also clear that uncertainties of types 1, 2 and 3 must be treated in different manner and the choice of the contractual form implies a different allocation of risk to the Owner or Contractor. In general, in simplified terms, we can identify two most common forms of contract:

– unit prices contract
– lump sum contract.

The use of one or the other contractual form involves a different way of managing uncertainties and any unforeseen events (Klee 2015).

In unit prices contract it is easy to deal with large variations of quantities in a reasonable manner, as regulation mechanisms are built into the contract.

The experience shows that unit prices contracts are suitable to deal also with 'unexpected geological conditions', as long as the 'unexpected' element results only in variations in the quantities of work activities, that is for uncertainties of type 1) and 2).

This means that all necessary work activities must have quantities and preferably also 'standard capacities' for regulation of the construction time.

If "truly unforeseen" geological features (that is uncertainties of type 3)), for which there are no methods and quantities available in the contract, occur the unit prices contract must be supplemented by special agreement. Unit price contract allocates all or most of the risk for the ground conditions to the Owner.

For this reason, for underground construction where the impact of changed ground conditions can be very severe, Owners in many cases are reluctant to shoulder all the geological risk.

For purposes of financial planning, many Owners prefer awarding lump sum (i.e fixed price) contracts.

Fixed price contracts, with all risk for ground conditions allocated to the Contractor, may have an apparent predictability of cost, which may be attractive to the Owner.

However, this type of contract imposes risks on the Contractor that may at best be difficult to quantify, at worst disastrous if the "truly unforeseen" occurs (that is uncertainties of type 3). Such risks may become an Owner problem, no matter the contract text, if the Contractor is not able to bear the loss and complete the project.

Therefore, also lump sum contracts for underground projects may not provide the intended predictable cost.

'Adjustable fixed price' contracts, combining unit rate and fixed price, may prove to be more suitable than fully fixed price contracts, and easier to handle than unit prices contracts.

According to the FIDIC ITA Task Group 10, risk allocation and risk dependent costs can be shown in a simple way as in the Figure 3.

It suggests that a fair risk allocation is likely to produce the lowest construction cost (FIDIC–ITA 2014).

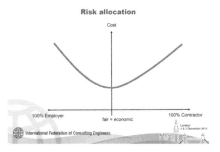

Figure 3. FIDIC Task Group 10. Lowest construction cost.

It is evident that, from a contractual point of view, the types of uncertainties described above must be addressed in a different way.

In case of uncertainties of type 3) one must think a subdivision of the risk that involves in a decisive manner the Owner. The idea that risk sharing must exist for this case is now well established. History shows that the attempt to transfer the total geotechnical risk to Contractor (by a lump sum contract) does not liberate the Owner of his final responsibilities. In the event of a high level of losses, the Contractor will (almost certainly) look for all possible ways to terminate the contract in advance; this can impact on the total costs, quality, safety of the entire project. Fair risk sharing between the Owner and the Contractor helps to reduce potential claims and therefore the total project costs.

We can say that for the uncertainties of type 1) and 2) we cannot speak of unforeseen or unforeseeable. In other words, there is the certainty of not knowing exactly anything, but moving in known areas for which the actors (Owner and Contractor) can evaluate and quantify the existing risk. In any case, these uncertainties do not allow to precisely define the cost of the work.

Therefore, it is necessary to find contractual mechanisms that distribute the risk of this non-definition on Owner and Contractor in a different way, also attributing a "risk premium" to the greater risk that one of the two is willing to charge. The theoretical ideal allocation of risk in a contract is that point where the sum of the risk premiums is minimized, the point where each party takes on the risk for which it is best placed to control and manage those risks.

For the uncertainties of type 1) and 2) a lump sum contract could be assumed.

The question to be asked must therefore be; how do we arrive at a fair risk sharing and the lowest construction cost for any given project, considering the uniqueness risk aspects?

It is necessary to define the risk assignment limit to establish what conditions were expressly or impliedly "foreseen" within the contract.

This can be done by the Geotechnical Baseline Report (GBR).

The Geotechnical Baseline Report is a document where contractual statements describe the geotechnical conditions predicted (or to be assumed) to be met during construction. With reference to the level of knowledge referred to in the previous paragraphs, we can say that it has to be viewed as a tool for defining those conditions that should be considered 'reasonably foreseeable' (i.e. . uncertainties of type 1) and 2)).

It has been developed by the American Society of Civil Engineers since 1970 (ASCE 2007) that published the guideline to be used for drafting this contractual document.

The contractual statements are referred to as "baselines" and they should be considered as the reference point for risk evaluation and pricing. In fact, the tenderer needs to know 'with a sufficient degree of certainty' the risk that he is going to price, while the employer needs to know what he is going to pay for. Both parties have to evaluate the cost of known conditions and to estimate the contingencies that should be included either in the contract price or in the Owners' budget.

The "baselines" serve as contractual references to establish where conditions encountered during construction that are significantly more adverse, onerous and time consuming, may be considered as 'unforeseen' (i.e type 3).

Risks associated with conditions consistent with or less adverse than the "baselines" are allocated to the Contractor, and those more adverse than the baselines are accepted by the Owner (Figure 4).

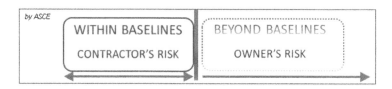

Figure 4. Baseline.

It is important to describe the possible range of property values or ground behavior. The recommended approach is to indicate the expected range of conditions and uncertainty, but then state a specific baseline (upon which bidders may rely) that has been established for contractual purposes. The baseline may be expressed as a maximum value, a minimum value, an average value, a histogram distribution of values, or a combination of them.

The Owner may decide to allocate certain risks and costs to the Contractor by baselines that are more adverse (higher "risk premium"). This will usually result in an increased bid price. Alternatively, the Owner may choose to share the risks and costs through less adverse baselines and utilization of either alternative payment provisions or a change order process, if the more adverse conditions materialize. To be useful for risk allocation, a geotechnical baseline must clearly establish the boundaries of those conditions that are expected and define how they are measured. The GBR has come into use in some countries and there are encouraging benefits of its use. The British Tunnelling Society and the Association of British Insurers, which represent insurers and re-insurers on the London-based insurance market, issued their the "Code of Practice for Risk Management" and the GBR concept is adopted in this code (BTS ABI 2003).

5 THE POTENTIAL OF THE OBSERVATIONAL METHOD

As seen in chapter 3, the use of the Observational Method allows, in general terms, a reduction of the risk present in the construction of an underground work.

If we compare an observational approach with that based on "predefined design", we must highlight that less conservative ground parameters must be applied. In a "predefined design" we proceed using the "most unfavorable" (worst scenario) set of parameters in the case of the Observational Method we should proceed using a less unfavorable set of parameters ("most probable" according to Peck, "moderately conservative" for other authors). Basically, we can work with safety factors lower than those used in a predefined design without reducing the real safety of the project. This is made possible by the knowledge acquired gradually during the work. To clarify this concept, we analyze the Figures 5. Assuming that we have defined the Gaussian distribution of load and resistance. If we design based on the "most unfavorable" set of parameters and apply a similar factor of safety that we would for the most probable condition, we will fall in the first category described by Peck (1969), i.e. a wasteful or over-conservative design (Figure 2). On the other hand, designing for the "most probable" decreases the degree of conservatism and would be more economical (Figure 5). However, as Peck (1969) stated, it is, to some extent, a gamble. If not there would be no need for the word "probable".

By adopting an observational approach and a working hypothesis of the most probable conditions, one could say that it is at least a risk-controlled "gamble". If we properly monitor

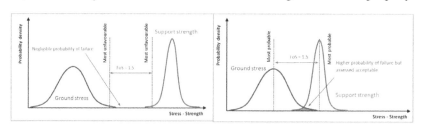

Figure 5. "most probable" and "most unfavorable" set of parameters.

the ground behavior during excavation, contingencies may then be put in place if we realize that our "gamble" has failed (Oliveira 2015).

Opportunities and risks could be found from modifications of the construction method on site because of the information obtained during excavation. They can result in additional or less works (considering the baseline construction plan that was based on the baseline scenario for the expected ground conditions).

The establishment of the initial design based on the "most probable" ground conditions drew most concerns. Muir Wood (2000) how sometimes the concept of "most probable" situation is mistakenly associated with the situation that will be overcome with little probability; in reality most probable is equivalent to 50% probability. Almost all subsequent authors, suggest starting with an initial "moderately conservative design", to be relaxed to a "most probable" condition during construction, should the observed behavior warrant it.

Anyway, it is important to highlight that the objective of an observational approach is to deal with uncertainties, and not reduce factors of safety.

6 CONCLUSION AND CLOSING REMARKS

In summary, adoption of Observational Methods helps in mitigating the geotechnical risks.

It is therefore established that by proceeding with an observational approach two important advantages can be obtained:

1) verification, during construction, of the solutions defined during the design phase
2) calibration of the input parameters and the consequent safety factors in consideration of the knowledge acquired during the works.

In general, only the advantage 1) is exploited, i.e. the predefined design solutions are refined.

Therefore, its potential, above all in terms of possible reduction of the overall costs of the work, is generally not fully exploited in the planning and construction phase.

It is standard practice to limit its use to refine the application of the stabilization interventions and construction methods defined by a "predefined" design.

In other words, a "basic" design is developed based on the available data and then in the construction phase this design is adapted to the real conditions found on site without modifying the safety factors as it would be possible by virtue of the greater information that are gradually acquired. Generally, it is applied, in the context of a unit prices contract with all risks are accepted by the Owner, without fully exploiting the opportunities provided by the method.

Therefore, there is a large "not exploited" reservoir of potential optimization of the total cost of the work. It is often suggested that cost uncertainty should be allocated to the party best able to anticipate and control that uncertainty, usually replacing the term 'uncertainty' by 'risk' although variability, ambiguity and systemic uncertainty may also be involved. On this basis, a tentative conclusion is that lump sum contracts are appropriate when sources of uncertainty are controllable by the Contractor, i.e. types 1) and 2) that are within baseline. An interesting analysis of Ward and Chapman (2011) concludes that a lump sum contract is usually risk efficient in allocating Contractor controllable uncertainty. An important conclusion is that even where Owner and Contractor share similar perceptions of project cost uncertainty, fixed price contract may be inefficient for the Owner, if the Contractor is more risk averse than the Owner. In this situation the Contractor will require a higher premium to bear the risk than the Owner would be prepared to pay for avoiding the risk. It is equally true that in the case of lump sum contracts all the uncertainties, even within the baseline, are the responsibility of the Contractor and consequently the Contractor may be driven to maintain a high 'risk premium' and then do not convenient offer for the Owner. For this reason, this type of risk should be mitigated by appropriate contractual architecture. On the other hand, under a fixed price contract, the Contractor is motivated to manage project costs downwards to formulate a competitive offer; it should be able to generate, an 'opportunity discount' for example, by increasing efficiency or using the most cost-effective approaches. With the use of the Observational Method one could exploit precisely this enhancement linked to the good work of the Contractor.

The definition of aggressive baselines involves high risks for the Contractor and the consequent attribution of a higher "risk premium" (i.e. expensive offer for the Owner).

This premium could be balanced by allowing the Contractor to use an Observational Method with optimization of the interventions. By exploiting the potential of the Observational Method, the Contractor could be able to manage all the risk (within the baseline) and could generate benefits for both Owner and Contractor. For that purpose, it is strategic to build a sufficient flexibility into the contract so that design can be adapted during construction according to "geo" properties encountered within an observational approach. With the use of the Observational Method the Owner could use a more aggressive baseline because thanks to the possibility of refining the "safety factors" (without reducing the overall safety of the work) there is a possible premium for the Contractor. This could allow for more aggressive lump sum bids and therefore an overall advantage for the Owner.

Ultimately, we want to highlight how moving within the limits of the uncertainties of type 1) and 2) then definable by the Geotechnical Baseline Report and controllable, the contract could also be a lump sum thus attributing all the risk to the Contractor.

Within the baseline, the Contractor would have the possibilities to evaluate and quantify the risk and considering a possible refining of the "safety factors" (made possible by the greater knowledge acquired during construction and therefore without reducing the overall safety of the work).

It must be emphasized that for a "right" use of the Observational Method, that wants to enhance all the potentials of the same, it is necessary that Owner and Contractor provide a team of expert designers able to support the mutual interests. The processes of design and construction need to be integrated and close co-operation is required between all those involved in the project.

Definitely, increasing investments in engineering design would allow a control of cost and time overrun and a certain reduction of the overall cost of the project.

REFERENCES

ASCE, Essex R., 2007. *Geotechnical Baseline Reports for Construction: Suggested Guidelines*. ASCE.

Avestedt, L. 2012. *Comparison of Risk Assessments for Underground Construction Projects* - Master of Science Thesis Royal Institute of Technology - Stockholm

BTS, ABI, 2003. *The Joint Code of Practice for Risk Management of Tunnel Works in the UK*. British Tunnelling Society (BTS) and the Association of British Insurers (ABI).

Ciria 1999. Report 185 - *Observational Method*

Eurocode 7 (BS EN 1997-1:2004) "Geotechnical design - Part 1: General rules

Fellin, W. et al. 2005) *Analyzing Uncertainty in Civil Engineering* - Springer - Verlag Berlin Heidelberg

FIDIC 1999. *The Silver Book: Conditions of Contract for EPC/Turnkey Projects*.

FIDIC – ITA Task Group 10 2014. Presentation of Task group 10, *Contract Form for Tunnelling and Underground Works*. London 2014.

Honjo 2009. *Code calibration in reliability-based design level I verification format for geotechnical structures*. Geotechnical Risk and Safety – Taylor & Francis Group, London

International Tunnelling Association 2004. *Guidelines for tunnelling risk management*, Working Group No. 2. Tunnelling and Underground Space Technology 19 (2004) 217–237

Klee, L. 2015. *International Construction Contract Law*. Wiley & Sons.

Lunardi, P. 2008. *Design and construction of tunnels – Analysis of controlled deformation in rocks and soils*. SPRINGER, Berlin Heidelberg.

Muir Wood, A. 2000. *Tunnelling: Management by design*. E & FN Spon, London.

Nicholson, D., Tse, C.M., Penny, C. (1999). *The Observational Method in ground engineering: principles and applications*. Report 185. CIRIA.

Norwegian Tunnelling Society 2017. *The principles of Norwegian tunnelling*. Publication 26

Oliveira, D. 2015. Post Linkedin

Patel, D. et al. 2007. *The Observational Method in Geotechnics*. Proceedings of the 14th ECSMGE: Madrid, Spain. Vol. 2, 365–370.

Peck, R. B. 1969. *Advantages and limitations of the Observational Method in applied soil mechanics*. Geotechnique, 19: 171–187.

Van Staveren, M. 2006. *Uncertainty and Ground Conditions: A Risk Management Approach* - Elsevier Ltd.

Ward, S. & Chapman C. 2011. *How to Manage Project Opportunity and Risk* – John Wiley and Son Ltd.

*Tunnels and Underground Cities: Engineering and Innovation meet Archaeology,
Architecture and Art, Volume 8: Public Communication and Awareness/Risk Management,
Contracts and Financial Aspects – Peila, Viggiani & Celestino (Eds)
© 2020 Taylor & Francis Group, London, ISBN 978-0-367-46873-6*

Extension of Milan's metro line M1: More than 10 meters of water table increase and two fails of the contracting companies: How to manage this contract

A. Antonelli, G. Galeazzi & A. Bortolussi
MM Spa – Milan, Italy

ABSTRACT: Extension of line M1 to the N-E outskirts of Milan (€ 85.135.081,82 for 1.9 km gallery) was found to be extremely complex in the management of numerous problems that were difficult to foresee. The project based for tender was carried out taking into account a reference groundwater level, on average stable, which, since design for tender until the award of the works, underwent an extremely high increase, of more than 5 m in 18 months, and has continued to increase by another 5 m in the 4 years following the assignment of the works. The unexpected increase of water table has made the execution and management of the project since its assignment extremely complex, implying the need to develop a variant project, which has required the definition of additional interventions for the execution of the provisional civil works, especially the consolidation for the traditional excavation of tunnel. During the course of the works, the principal company of the JV, for matters not connected to this contract, declared its failure. The execution of the works was entrusted to the 2nd principal company of the JV, which did not substantially fulfill the obligations envisaged leading to delays and failures with consequent termination of the contract. Now, work has been entrusted to the second classified company in the call for tenders, which has resumed work, albeit with the problems related to the restart of a partially completed and flooded/abandoned building site for over 2 years.

1 EXTENSION OF METRO LINE M1 TO MONZA (NORTH OF MILAN)

1.1 *Metro line M1 and its extension to Monza*

The first of November 1964, in the presence of the mayor of Milan and some authority of the time, line 1 of the Milan Metro was inaugurated. The "red" metro was born. A track of 11.8 km including 21 stations to connect the Lotto square (west suburbs of Milan) to Sesto Marelli (north-east suburbs).The first construction site of the M1 line, contracted on 24 April 1957, was opened in Monte Rosa Avenue on the following 4 May.

The absolute novelty developed by MM technicians, after experiments and studies carried out in collaboration with the Polytechnic laboratories, provided that the bulkheads could fulfill a function of containment of the thrusts of the looming buildings and a supporting function of the tunnel structures, through the reinforcement of concrete with improved adhering bars. The new system, innovative for the time, was then used for the construction of, for example, the Toronto subway, with the name of "Milan Method".

Architectural firm of Franco Albini and Franca Helg designed the stations of the new line, while Bob Noorda designed the signage. Both Albini-Helg and Noorda were awarded the prestigious Golden Compass for the Milan subway layout and signage project.

The demographic development of the territory of the municipality of Milan and the hinterland has made the continued extension of the red line (M1) in both directions, which to date has reached a 27 km route including 38 stations.

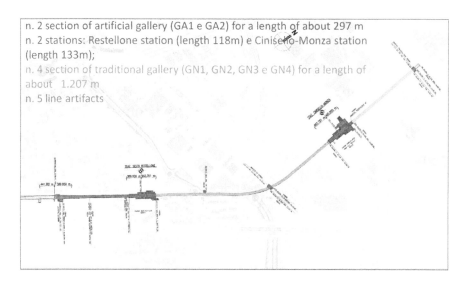

n. 2 section of artificial gallery (GA1 e GA2) for a length of about 297 m
n. 2 stations: Restellone station (length 118m) e Cinisello-Monza station (length 133m);
n. 4 section of traditional gallery (GN1, GN2, GN3 e GN4) for a length of about 1.207 m
n. 5 line artifacts

Figure 1.   Plan of metro M1 extention.

In 2011, specific Joint-venture of company was awarded the construction of the extension from the Sesto FS station, the current north-east terminus of line 1, to the Cinisello Monza station for a total contract award of around 61 millions of euros.

The intervention extends for a length of about 1800 meters and includes:

- n. 2 sections of open-air tunnel for a total length of 297m;
- n. 2 new stations respectively 118m and 133m
- n. 4 sections of tunnel bored with traditional method for a total length of 1207m
- n. 5 line shafts

1.2   Tender project

The methods envisaged for the realization of the extension to the North of the M1 are, generally, the same ones already adopted for the construction of the existing line. The process for choosing the excavation method is a function of the layout and the pre-existence and is aimed at optimizing construction costs while minimizing the impact and occupation on the surface,

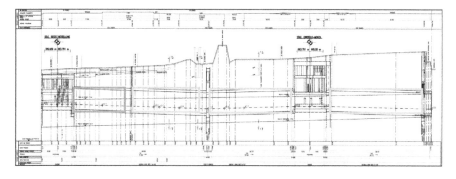

Figure 2.   longitudinal profile extension line - traditional method (yellow one) and cut-cover (blue one).

especially in the presence of significant traffic or pre-existences. For this reason, a significant part of the extension is bored with traditional method.

The stations, shafts and the initial stretch of the line up to the Sesto Restellone station are realize in cut-cover.

The initial part of the tunnel has in fact a significant width and a reduced coverage (railway at about 10.5 m from the road level) having to connect to the current three-track termination and still having to keep the three tracks to allow to differentiate the frequency of trains on the main line and the extension.

High width of the works and reduced coverage have determined the necessary use of the cut-cover method to build the first 297m of the artificial gallery (highlighted in blue in the figure).

For the remaining sections of the tunnel, with development of about 1,210 ml, the traditional excavation (highlighted in yellow in the figure) is envisaged.

1.3   Design solution for the excavation with traditional method

In tender project, the excavation technique provided for the construction of the line tunnel, from the Sesto Restellone station to the end of the lot, is the traditional one with "partialized section" with a first phase involving the excavation of the top section from the crown to the center followed by the digging down phase.

The excavation of the top section of the tunnel requires consolidation of the crown and the front by sub-horizontal jet-grouting column in advance, in order to obtain a band of consolidated material suitable to guarantee the formation of a discharge arc. After the excavation of the top section of the front, the pre-coating consists of metal ribs, arranged with a longitudinal pitch of 1.0 m and solidarized by means of reinforced or fiber-reinforced spritz-beton with a thickness of 0.20 m.

Once the shell has been excavated, jet grouting treatments to protect the piers, the subsequent downward excavation and the extension of the pre-covering are carried out from inside the cable.

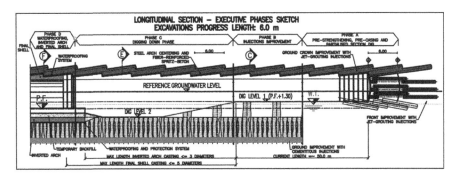

Figure 3.   longitudinal profile of the excavation phases.

Follow the jet of the inverted arch and the completion of the final covering after laying of the waterproofing. The lengths of the advancement fields, the intensity of the consolidations, the possible reinforcement of the columns in jet-grouting and the maximum distances from the front provided for the casting of the final coatings, are appropriately modulated according to the context and in relation to the presence of buildings that require attention to the deformation levels induced by the excavation.

As outlined above, the definitive tender project, drawn up in 2009 (PD 2009), did not provide the consolidation of the bottom of the tunnel (inverted arch), because from the monitoring of the piezometers in the areas of the work, the groundwater level was well below the excavation depth expected during the construction of the tunnel.

2 ADJUDICATION AND EXECTUTION OF WORKS

2.1 *Assegnment of work and excursions of the water table*

In February 2011, the realization of the M1 extension was awarded to a specific Joint Venture and in the following month of March the works were formally assigned, the total duration of which is set at 1460 days and 14 March 2015 is the general deadline for completion.

However, it was immediately found that, in the time between the drafting by MM of the final project for tender (October 2009) and the assignment of works (March 2011), the level of the water table has increased by an average of 4 m and the trend is a further increase, which is unforeseeable circumstance from the analysis of groundwater data used for the evaluation of the tender project (historical series and piezometric readings below).

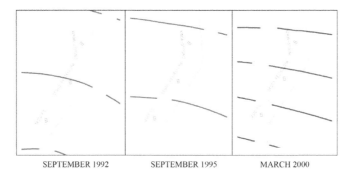

SEPTEMBER 1992   SEPTEMBER 1995   MARCH 2000

Figure 4.   Historical isopiezometric maps.

The Contractor, therefore, in the development of the executive design (PE 2011), in agreement with the Client, has revised the project assuming a reference water table consistent with the values recorded in the field increased by an appropriate amount. The excavation of tunnels is affected, for a significant part, by the new level of groundwater, so in this section the waterproofing of the excavation lower part of the tunnel must be provided with the ground consolidation.

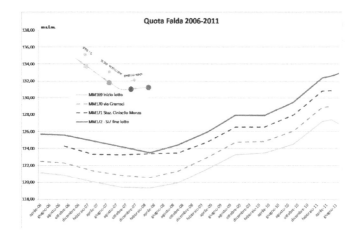

Figure 5.   Piezometric readings 2006-2011.

2.2 *Design solution developed by the contractor*

The new solution, developed by the Contractor in agreement with the Client, envisages, where the tunnel excavation intercepts the reference stratum, the execution of a partial section excavation. The tunnel excavation, with the height of the working plane above the level of the reference water table, is carried out with a consolidation of the crown by means of a jet-grouting treatment. Subsequently, from the inside of the cable is made the treatment of the lower part of the cable with injections of cementitious and chemical mixtures and, then, the excavation of the lowering block happens.

The cement and chemical injections will be carried out in a selective and repeated manner by means of manchette valves, with criteria of volumes and controlled pressures. The injected mixtures must permeate the ground present below the ground plane by consolidating and drastically reducing the permeability of the materials subject to the second drop, until the completion of the final design section.

The excavation of the tunnel intercepts the reference stratum for most of the route with a gradually rising waterway proceeding from south to north. Therefore, the maximum beating of the water is in the final section of the tunnel where the level of the reference stratum rises up to about +1.2 m from plan of railway.

The design configuration provides a shell of consolidated soil by injections in correspondence of the piers and the inverted arch where a useful thickness of the treatment of at least 3 m is obtained.

This thickness is technically suitable for:

- ensure a massive treatment of the volume potentially affected by filtration processes, statistically effective in obtaining a drastic reduction in the hydraulic conductivity and therefore in the flow of water to be grasped at the time of the excavation;
- generate a stabilizing arc effect during the excavation phase under the lifting action of the water exerted at the base of the consolidation, with a maximum hydrostatic head of about 5.5 m.

The solution, which provides for the waterproofing of the excavation by the invert consolidation, is positively evaluated by MM technicians and applied not only to traditional tunnels but also to the Cinisello Monza station and n.2 shaft (A4 and Tangenziale).

The quantification of the different intervention is been defined between the Contractor and the Client and included in an Administrative Act which, in addition to redefining the contractual amounts, updates the time for completion, bringing it to 25 May 2015.

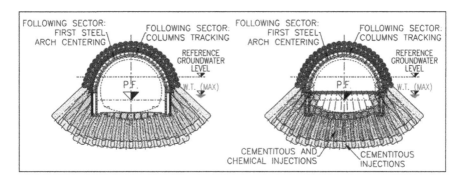

Figure 6. tunnel excavation sections – top heading excavation, bench-invert consolidation (right) and bench-invert excavation (left).

3 CONTRACTUAL PROBLEMS DURING EXECUTION OF THE WORK

3.1 *Failure of the mandatory company*

In December 2013, the Mandatory Company as part of a corporate restructuring plan aimed at safeguarding its assets, filed with the court of Florence, application for admission to the preventive arrangement. The second principal company in the JV, undertakes to complete the contractual works by proceeding to lease the business branch of the mandatory company.

3.2 *Takeover of the principal company*

Despite this new company takeover, production on site does not restart due to the considerable difficulties related to non-payment of suppliers and subcontractors. In order to cope with this situation, made even more critical by a further increase in the groundwater level, a Transacting Act is defined and formalized, with which the Client anticipates, subsequently restoring it as a percentage of the advance payment, to remedy suspended payments and ensure the resumption of work.

The drafting of the Act also includes the completion and commissioning of a section of the artificial gallery (first 150m) to meet the need of the metro service manager to park the trains during the night-time in Expo2015 event period.

Moreover, given the continuous increase in the groundwater level, Client and Contractor in according to the updated levels of groundwater shared the new methods of soil consolidation.

3.3 *Termination of the contract with the contractor*

The timetable attached to the Transacting Act provides for the definition of some milestones through which the Work Supervisor monitors the correct progress of the works including the opening of the first 150m of artificial gallery by 30th April 2015. Despite the achievement of the first milestone in the time schedule, between January and February 2015 the Work Supervisor has seen a worrying slowdown in the work until the total abandonment of the construction site, because of which the Client formalized, in March 2015, the termination of the contract.

4 NEW ASSIGNMENT OF WORK AND PROJECT REVISION

4.1 *Status of work and costruction site*

The termination of the contract is followed by the abandonment of the site before having finished the waterproofing of the excavations; this circumstance combined with the persistent increase of the water table inevitably produces the flooding of the various artefacts.

In spring 2015, the groundwater level reached its peak, with an increase of 12-13m compared to what was detected at the time of planning and put into the call of tender project (2009).

In the meantime, through the interpolation procedure pursuant to Italian Law for Contract (art. 140 D.lgs 163/2006), the MM, pending the refinancing of the work and having heard the financing stakeholders, is activated to stipulate a new contract by consulting the graduate company of companies that had at the time participated in the tender.

The company ranked in second place in the ranking of the tender, immediately manifests its willingness to enter into a new contract for the completion of the contract under the same contractual conditions agreed with the original contractor .

All the checks carried out by MM against the incoming company they give a positive result with reference to the possession of general and special requirements, of a

Figure 7. on the left, tunnel entrance GN3 from Tangenziale shaft and, on the right, Cinisello-Monza station and tunnel entrance GN2.

technical-economic nature, for the execution of the contract and, therefore, in June 2017 a new contract is signed, concerning the "completion of the contract integrated for the executive planning and for the realization of the rustic works, of the sub-services, of the road and surface arrangements, of the finishing, of the armament and of the systems of the extension of the Milan subway line 1".

4.2 Design revision that precede the new assignment

The period (2013-2015) between the difficulties of the JV awarded the first contract, the contract resolution and the selection of the new contractor was characterized by large fluctuations in the groundwater level that made it necessary both to develop, by MM, an adaptation of the Tender Project before reassigning the works (PD 2015 - raising water table dossier that takes into account the highest levels of groundwater level recorded in January 2015) and to contractually envisaging the redesign by the new contractor on groundwater levels further updated to the actual period start of work (because after the peak of 2015, the levels of groundwater recorded show a downward trend).

Therefore, to the complexity of the contractual and administrative management of the contract, there is also the need to adapt the project to the changing conditions at the boundary as well as to the different state of progress of the works along the section.

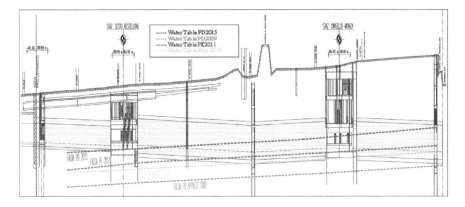

Figure 8. longitudinal profile of the tunnel with reference water table.

The profile is shown below with the identification of the groundwater levels used in the various design steps (PD 2009-PE 2011-PD 2015). It is clear how the maximum levels recorded in 2015, sometimes much higher than the center of the centers, are incompatible with the hypothesized advancement methods (traditional excavation with partial section and progress consolidation).

In addition to this, in the redesign of the tunnels for the reassignment of the works it was necessary to take into account the progress of the excavation work already undertaken along some sections.

By way of example, two sections of advancement of the PD2015 are reported, with groundwater level much higher than the floor of the centers, in the case of tunnel partially excavated and to be excavated ex novo.

In this section (case A in figure 9), the shell excavation has been carried out and the injections in the inverted arc have not yet been carried out prior to the downward phase and the completion of the definitive structure. The elevation of the water table is close to the shell. In this case, the tunnel must be filled by drilling from above and pumping granular material and expansive mixtures. Subsequently, since the groundwater quota makes it impossible to advance from the inside, the new open-air excavation between diaphragms is waterproofed by the ground improvement of the bottom with cementitious and chemical injections.

In this section (case B in figure 9) the tunnel to be excavated that underpasses the A4 motorway, is characterized by a level of high ground that does not allow you to work from the inside, also given the constraints of the surface can't be achieved even consolidating from above. Therefore, we opted for the construction of a tunnel in progress from which to perform a radial consolidation on the entire outline of the tunnel by means of integrated cement injections.

4.3 *Resumption of work and further oscillation of the water table*

The first activities carried out by the new contractor, in agreement with the Work

Supervisor, concerned the cleaning, the removal, the collapse of the equipment and devices abandoned on site and the securing of the building site areas, abandoned by the previous company in March 2015. This activity was preceded by the preparation of specific reports for assignment of the works, taking into account the state of consistency of the works carried out, the inventory of materials, machines and work, prepared pursuant to Italian Law for Contract (art. 138 D.lgs 163/2006)

The preliminary checks started after the writing of the aforementioned records essentially concerned the state of conservation of the works and materials that have remained exposed to various atmospheric agents for more than two years. For incomplete open-air structures, after in situ tests and careful analysis, it was decided to replace the waterproofing package of the vertical structures while the full functionality of the tie rods, although provisional, and exposed armor was confirmed. The state of preservation of the tunnels, where definitive structure have already been made, was obviously good, while some problems were detected for the waterproofing in the case of tunnels covered only at the inverted arch.

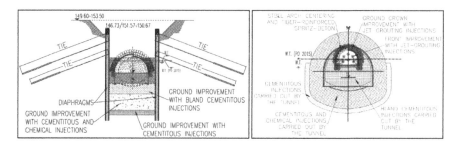

Figure 9.   on the left, case A- already partial excavated tunnel (PD2015 solution) and, on the right, case B – section to be excavated crossing the A4 motorway (PD2015 solution).

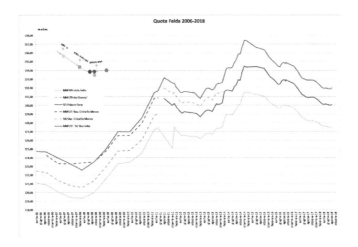

Figure 10.   water table trend (2006-2018).

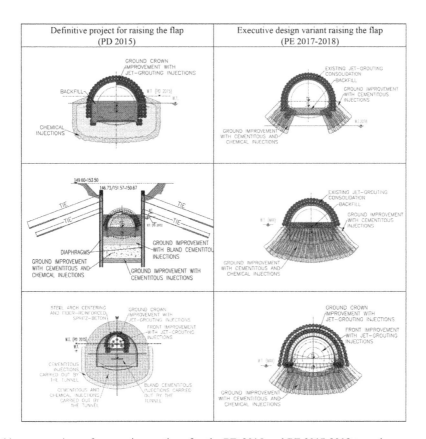

Figure 11.   comparison of excavation sections for the PD 2015 and PE 2017-2018 tunnel.

The water table, constantly monitored, showed a significant decrease compared to the peak of 2015 with a downward trend.

In the phase of executive development of the design solutions, in order to optimize technically and economically the same in relation to the favorable phreatic situation, various evolutionary scenarios of the same have been studied. In particular, for the traditional tunnel, more sensitive to groundwater oscillations, it was decided to proceed with observational method. Therefore, an abacus of typological excavation solutions was defined (variable for the groundwater level as well as for the progress of the works already carried out) agreed with the Client to be applied according to the piezometric surveys carried out in contradictory before the start of the works on a gallery section. The progress sections are evaluated on times of execution of the order of 3 months according to the sequence provided for by the time schedule of the works.

By way of example, some of the advancement sections hypothesized in the project PD 2015 (period in which the water table has reached the highest levels) are compared with the sections in the same section defined on the basis of the observational method in the executive project (PE) developed between 2017 and 2018 (period in which the water table trend is constantly reduced).

The application of the observational method allows to ensure optimization of the selected sections even in conditions of sudden ground oscillations; on the other hand, the methodology described requires an hard work by the Contractor and Work Supervisor to guarantee, in addition to the constant monitoring of the water table, the punctual drafting of reports for the agreement of the method of advancement and the accounting for the works.

5 CONCLUSION

The proposed case is emblematic of how, in complex contracts, the contingencies of an economic-contractual nature can intertwine and sometimes even generate technical-design problems and how the suspension of works can impact not only on costs and realization times of the work but even on the valence of the hypothesized design solutions.

In the case in question, in fact, the failure of Contractor produced a suspension of works of more than 2 years, linked not only to the technical completion of the new contract, but also to the planning and authorization process necessary in the changed conditions. Such suspension has affected the state of conservation of the works partially carried out and the time taken to recover the works necessary for the cleaning and the refolding of the equipment as well as for the tests in situ on the built structures.

Currently, after the assignment of works (January 30, 2018), the incoming company (the second in the ranking of tender) has restarted the activities on several fronts in order to optimize the management of the resources on the construction site that at present have reached approximately 80 daily attendance. The traditional tunnels with internal consolidations, more sensitive to the groundwater level excursion, are carried out on 3 shifts 7 days a week, while for tunnels with consolidation from the ground level, like the maneuvering tunnel, only two shifts are scheduled.

At the same time, we proceed to the constant monitoring of the groundwater level on the basis of which the choice of advancement in the subsequent excavation sections is calibrated, punctually recorded and recorded by the Work Supervisor to ensure the regular execution of the works that are proceeding in line with the programmed.

It is therefore evident that in complex situations, such as the one illustrated, the management of the contract requires the involvement and collaboration, at every stage, of all the figures in charge (Contractor, Engineer, Work Supervisor and Client) in order to intervene quickly in the resolution of unforeseen and unpredictable critical issues, considering the respective roles.

Tunnels and Underground Cities: Engineering and Innovation meet Archaeology,
Architecture and Art, Volume 8: Public Communication and Awareness/Risk Management,
Contracts and Financial Aspects – Peila, Viggiani & Celestino (Eds)
© 2020 Taylor & Francis Group, London, ISBN 978-0-367-46873-6

Evolution of risk management during an underground project's life cycle

A. P. F. Bourget
Amberg Engineering, Regensdorf, Switzerland

E. Chiriotti
Incas Partners, Paris, France

E. Patrinieri
CSD Ingénieurs SA, Lausanne, Switzerland

ABSTRACT: The paper provides practical guidance in transposing the latest ISO31000 and ISO31010 to underground works. The reporting technique of risk management varies through the project phases, and consistency in implementation throughout the project's life cycle is fundamental. After recalling background notions on risk management, the paper introduces the financial risk report (according to IFPS 15 – International Financial Reporting Standard – and the European Directive in force since January 2018) addressing each project phase: the design phase, concerned with the balance between cost, time and residual risk, in interaction with the owner's evolving risk tolerance; the procurement phase, concerned with the appropriate allocation of residual risks between the parties; the construction phase focused on the contractual perspective to understand if site conditions come under previously agreed risk allocations or generate new risks; the operations phase where risks relate to design life, structure's running costs or level of service availability.

1   INTRODUCTION

1.1   *Reference Standards*

The special importance of risk management in underground projects is well understood by all and needs no further justification. However, its implementation varies widely due to many factors. The publication of ISO 31000 in 2009, updated in 2018, has set a global standard which requires adaptations to existing corporate approaches, especially as the ISO GUIDE 73:2009 has changed many ISO GUIDE 73:2002 definitions, often leading to mis-comprehension between parties. The WTC's 2017 Bergen introductory talk by Professor Håkan P. Stille, confirmed the applicability of the ISO 31000 concepts to our field of endeavour.

Difficulties of interpretation can also arise between risks which can be managed by the parties to a contract, which relate to internal processes such as quality, health and safety, etc. (covered by ISO 9001, ISO 14000, etc..) and those which are outside the bounds of control by the parties, such as "Force majeure", geology or the built environment. The latter sources of risk are the main concern of the design process regarding construction risk reduction.

The industry in France has strived to produce guidelines on managing external sources of risk within tunnel construction contracts, with recommendations focusing on tender documents and Owner's requirements. The French government publishing an update to its Fascicule 69 in 2011 with guidance notes produced by the CETU (Center for Tunnelling Studies, a

French governmental institute) in December 2013. The French tunnelling association (AFTES) published its recommendations with regards to these risks through two main texts, GT32R2A1 in 2012 and GT32R3A1 in 2016, with a more general recommendations on contract in its GT25R3A1 guide in 2015.

1.2 Scope of the paper

The paper focuses on risk management from the owner's perspective with respect to a project, from inception to project delivery. As such, sources of risks derived from the external context – outside the control of parties to a contract, which are thus the owner's risk – are the prime focus of this paper. The internal risks of the consultants and the contractors are none-the-less discussed when these could become an owner's risk.

Operational risks, which influence design and are often managed through national or supra-national statutory requirements (in railway or road tunnels) with formats imposed by the national authorities are not addressed in the present article.

1.3 Regulatory obligations

The obligations imposed on the major companies of the European Union to adopt IFRS 15 (the International Financial Reporting Standard n°15: Revenue from Contracts with Customers) as the basis for their project financial reporting is a game changer and renders homogeneous reporting in our field of endeavour critical to our profession's credibility.

2 THE FUNDAMENTALS

2.1 Overview

The definition of risk is: *effect of uncertainty on objectives.* This definition is clarified in the ISO Guide 73:2009 with the following notes:

- a *effect* is a deviation from the expected — positive and/or negative;
- *objectives* can have different aspects (such as financial, health and safety, and environmental goals) and can apply at different levels (such as strategic, organization-wide, project, product and process);
- *uncertainty* is the state, even partial, of deficiency of information related to understanding or knowledge of an event, its consequences, or its likelihood.

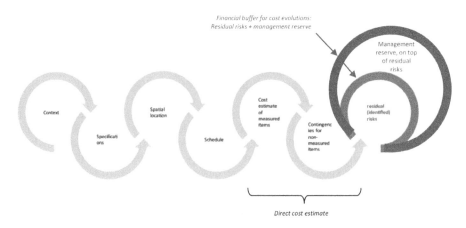

Figure 1.   Risk dependencies.

Thus, any risk assessment relates to potential deviations of any or all of the following:

- context, both external (such as statutory requirements or geology) and internal (such as owner's financial capacity),
- Specifications (inclusive of the risk management plan),
- Spatial localisation (alignement, dimensions, size, etc.),
- Schedule (base).
- Cost estimate of measured items
- Contingencies for none-measured items (lack of project definition), in accordance to AACEi's definition in its RP 10S-90.

Changes in any of the above aspects requires a re-evaluation of the risks based on the revised project referential. As risk is the deviation from the expected, a change in the expected implies a re-evaluation of the risks. The expected includes, thus are NOT risks:

- *Approximations*, e.g. a cost estimate of 1'000, with precision of +-20% indicates an estimate in which that the project's cost could be 1'200, as such the risks are those which increase this cost further. The notion of imprecision must be dissociated from that of risk (which are by definition identified).
- *Contingencies*. e.g. a contingency for non-measured items is nearly certain to be incurred. And is thus part of the direct cost estimate.

The cost to cover the occurrence of residual risks, plus the management reserve (for undefined or unexpected cost evolutions) constitutes the project's financial buffer. It is usual for the management reserve to decrease as the risks are better identified during the project's development.

2.2 *Objectives*

In underground works the term of *Risk Management Plan* is of common use and should follow the structure given in chapter 6 of the ISO31000:2018, As a minimum, it must include: a risk policy specific for the project; a clear and dedicated organisation; a systematic and iterative risk analysis (risk sheets, risk registers, risk treatment plans, etc.); the quantification of the provisions for identified risk; the format and process associated with risk treatment plans and the management of residual risk. Defining the objectives of a project from the Owner's perspective, for each phase of the project's development, is the first fundamental step within any Risk Management Plan.

ITA's WG2 recommendation identified 6 objectives, which can be extended with objectives relating to fulfilling the project's objectives (traffic capacity, power production, etc..). Each phase of a project can have its specific set of objectives, i.e. design activities rarely impact health and safety of third parties. Furthermore, the risk criteria associated with the objectives may be differentiated between project development phases, e.g. a 5 months delay may be of a greater importance to the owner during statutory procedures than during construction.

2.3 *Risk policy*

As part of a project's Risk Management Plan, it is the project owner's responsibility to define the level of risk which is acceptable for his project, and hence to take a major role in the definition of the scales of likelihood and impact to be used in the project and to finally validate the quantitative acceptability criteria, as well as their requirements in terms of degree of confidence to be applied by the project's actors. The Owner can should be assisted in this by his Consultants/Designers or an experienced external advisor.

2.4 *Risk source context and ownership*

Sources of risks that are of *internal context* – that is dependent on an organisation's management and processes – are, by definition, owned by that organisation.

Risk sources of *external contexts*, that are not under the control of the organisation belong to the Owner of a project, but can by transferred to another party, in totality or partially. In underground works, typical external risks include those related to geology, to meteorological conditions, to "Force Majeures", etc. As regards risks related to geology, and in line with ITA's and international best practice, fully transferring geological risks to contractors without carrying out proper investigation phases to reduce them is not a recommended approach.

2.5 Organisation for risk management

Risk management required a devoted organisation starting with the owner and ending to the contractor. It is fundamental that risk management be intended as a decision-making process fully integrated in the project, and not kept as a parallel and secondary activity.

2.6 Risk assessment

The terms and definitions are often repeated in articles dealing with risks, the aim here is to place each term in the general context of the risk management process (in bold the terms defined in ISO 31000 standard or ISO Guide 73):

The *Risk Assessment* is split into 3 activities:

Risk Identification which aims to provide a *Risk Description* – presented in light grey in Figure 2 – has 3 components. Uncertainty is given by the *Source of Risk* and is intrinsic to the *Context* of an organisation (*Internal*) or to site specific conditions (*External*). Thus, to proceed to the characterisation of an *Event* (deviation from the expected) it is required to make a baseline hypothesis. Events can combine in a cascade, resulting in distinct risks. *Consequences*, unlike Events, impact the *Objectives* directly in a quantifiable way, and each cascade of Consequences through knock-on effects or combinations become distinct risks.

i. *Risk Analysis* – presented in diagonal hash in Figure 2 – has two components. First, the estimation of the likelihood of any given cascade of Consequences, due to the relevant cascade of Events, on any specific Objective. Second, the quantification of its impact on each of the individual Objectives.

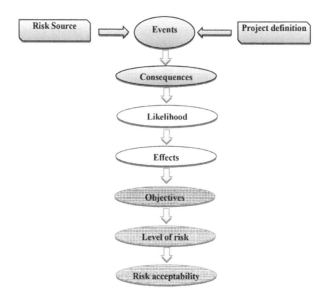

Figure 2.   Risk assessment process for a given scenario prior to risk treatment.

ii. *Risk Evaluation* – presented in grey dotted hash in Figure 2 – has two components. The global appreciation of the Level of Risk with its comparison with the Risk Acceptability, as defined in the Risk Matrix.

Furthermore, the ISO Standard requires that any *Risk Assessment* be accompanied with indications on, notably:

– the limitations of knowledge, in association with an evaluation of information reliability;
– the sensitivity and confidence level, relating to the Risk Analysis itself.

The implications to existing risk management cultures within organisations can be far reaching, as no risks assessment of any kind is possible without a minimum of information, no risks can be assessed on objectives which are undefined (such as "Other Objectives") nor can risks be identified without defining an initial baseline hypothesis (such as project location, time scale, alignment, geology, geotechnical hypothesis, etc..).

2.7 Risk treatment

Risk treatment is initiated when the *Level of Risk* is required to be lowered. As such, the risk treatment – although its impact on the risk reduction can be uncertain – is by its very nature an action with an expected outcome.

Actions such as obtaining further knowledge or better quality of information during the design phases, or monitoring risks during construction are not risk treatment measures, as these are not actions to reduce or eliminate risks or otherwise increase opportunities by modifying one or more of the steps of Figure 2.

It is to be borne in mind that *Risk Treatment Plan* may impact many risks by removing a source of risk (e.g., changing the project's alignment to avoid a geological fault) or a possible specific consequence existing in many different risks (e.g., using TBM to improve safety of the workers, to increase the excavation speed, to reduce settlement, etc.) or any or multiple combinations of each of the risk's aspects (source, consequence, effect, likelihood).

Risk treatment plans can be classified into three levels:

– *Level 1*: Those actions which modify the project's environment or redefine the project itself in order to *influence the possible events and/or consequences* (in light grey in 2b). Sources of risk relate to the project's environment. Examples are: legal, general geological, hydrogeological, geotechnical or climatic contexts. Project definition relates to alignment adaptations, alternative excavation methodology, etc.
– *Level 2*: Those specific actions which modify a specific or a limited number of risks by *altering their likelihood or their effect* on one or more of the objectives of the owner (in dashed oblique lines in Figure 2. Examples are: reinforcing a building; locally improving the characteristics of the ground (if they are too wide spread this can be equated to change a project's environment → level 1) or risk transfer to insurers, consultants or contractor - Beware only those risks identified and analysed can be subject to transfer.
– *Level 3*: Those actions that *revise the risk process itself* by adjusting the risk criteria or acceptability, as the cost of limiting the risks may become prohibitive (in dotted hash in éb). Examples are to increase the settlement or water inflow thresholds, etc.

Level 1 and 2 are hierarchically linked, while Level 3 is the iterative result of having studied and compared scenarios and associated costs and having given the Owner the opportunity to revise his policy accordingly.

The tracing of risk improvement for Level 1 and 3 requires the use of distinct scenarios for each proposed design each of them associated with their own risk profiles. Level 2 enables risk optimisations within pre-established design scenarios.

Before implementing risk treatment an appreciation of the risk analysis's confidence level must be carried out. The cost implications of a proposed treatment may justify the further investigations (e.g., site investigations, etc..) prior to any decision to proceed with it.

2.8 *Residual risks*

Residual Risks are those risks for which no further treatment action is sought, as such they are accepted by the owner. However this term is commonly also used to qualify the remaining risks at any point in the project's development, especially to distinguish an initial risk of a scenario from those remaining after implementing a Level 2 risk treatment plan, even if the level of risk is still too high and further studies are required further down the project's development cycle.

It must be noted that implementing a treatment of Level 2 will also generate a new scenario with the associated costs, schedule, etc. Therefore, the following should apply:

– any cost review must compare the initial cost, time and risk, globally with the cost, time and residual risk after treatment;
– any modification of part of the project during construction (e.g., adaptation of construction methods, implementation of a variant solution, etc.) which modifies costs, time or potential events affecting the project must be accompanied by a reconsideration of the risk analysis.

2.9 *Rigour in differentiating Risk Sources, Events and Consequences*

With the use of risk registers within contractual documents the imperative of coherence and consistency in risk assessment has become crucial, and precise rules are required established to determine what is what. As such:

Risk sources reflect *intrinsic uncertainties* (i.e. those of the project's global environment, independent from project definition). When described, the sentence should start with "Uncertainty in...". Typically for:

– External contexts: source linked to uncertainties in the political constraints, statutory procedures (legal/standards/administrative), macro-economic, geology, meteorology, existing structures, social/cultural, interfaces with third parties.
– Internal contexts: source linked to uncertainties in management capability, organization adequacy, process robustness, staff competencies, equipment reliability, materials quality.

Events can only be described when compared to a *baseline assumption*. Typically, their description will refer to:

– Variations from the expected, in terms such as: "less rigid than expected"; "harder than expected"; "larger than expected"; "more than expected" (e.g. "more settlement than expected").
– Sudden changes, phenomena or other disruptions in such terms as "collapse of", "flooding", "failure of"; "spillage of".
– Quality issues in terms such as "ovalisation"; "departure from alignment"; "out of tolerance".
– Availability or accessibility, in such terms as: "interruption", "inaccessible for"; "unavailability of"; "war"; "social unrest"; etc.
– Location, such as: "above", "below", "in-front", "behind", etc.

Events by themselves have no effect on the objectives. For instance, "greater settlement than expected" does not impact the cost objective of the client.

Events may cascade, for instance: "Collapse of the header leading to greater settlement than expected". The use of words such as "leading to"; "entailing"; etc. can be used to describe cascading events. The deviation from the baseline assumption has to be quantitative and measurable.

Consequences are generated when the deviation of an event from the baseline assumption generates an impact on the project's objectives. Typically, these relate to:

– Quantities due to changes in dimensions or to extra-works to mitigate a more impacting risk, in terms such as "more" (e.g., more injections than budgeted), "heavier than" (e.g., heavier steel sets than planned), etc.

- Damages to the project itself, to third parties or to the environment, in terms such as "repairing damages to" (e.g., repairing structural damage to "building no.21"); "loss of revenue"; "compensation to"; "relocation of"; "deaths"; "injuries", "pollution of", etc.
- Productivity, in terms such as "slower than" (e.g., slower progress);
- Prices, in terms such as "increased cost of" (e.g., increased cost of expropriation); "inflation";
- Performance, in terms such as "reduction in" (e.g., reduction in transport capacity); "reduced speed"; etc.

As with Events, Consequences can cascade, e.g., "repairing the structurally damaged building leads to the temporary relocation of the lodgers". Cascading events are to be considered as additional events to their component parts, e.g., "Repairing the structurally damaged building" (i.e. without relocation of lodgers) is a different risk.

2.10 *Interactions between likelihood and the effect on the objective.*

Likelihood is dependent on the importance of an effect. As an example, the likelihood of inducing damages on existing buildings can change according to the importance of such damage (i.e., aesthetic, functional or structural damage).

Depending on the effect it may thus be necessary to define "tranches" and to evaluate their respective likelihoods, leading to tranches of risk (risk of aesthetic damage, risk of functional damage and risk of structural damage in the example mentioned before). However, for many risks, adopting a mean damage estimate for which the associated likelihood can be ascertained is enough for design purposes and for the definition of the associated risk provisions. Although, for contractual risk allocations risk tranches may be required.

2.11 *Active period*

A risk must always be located both in space and in time. For tunnelling works this means identifying the risks' occurrence along the tunnel profile and on the schedule.

Within the general risks review process the risk location must also be reassessed and the active period may require increased monitoring and surveillance in relation to the identified risk.

3 RISK USAGE DURING PROJECT'S DEVELOPMENT

3.1 *Risk reporting formats*

Recording and reporting is a requirement of the ISO 31000:2018 and is the practical foundation on which the usefulness and the proof of application rests. The following chapters aim to provide guidance on setting up supports for recording and reporting the risks in the various phases of a project's development.

3.1.1 *Overview*
The risk reporting to stakeholders will need to consider differing objectives, thus a unified, single representation or format in the risk reporting can be difficult to impose. However, these fall into three global categories to provide:

i. Detailed documentation, principally the *risks sheets* describing each of them and their treatment in detail (oblique lines Figure 3) or the *risk treatment plans*, with the reasons why a particular treatment is to be implemented.
ii. *Risk registers* (light grey in Figure 3), provide a list of all risks and their implemented risk treatments in order to provide for risk management decision support, both to the current project phase and to future phases (notably construction, when in a design phase). Risk registers can be the main source of registering the risks or consolidate the information from

the detailed documentation. These take various forms depending on the objectives, but provide:

a. Easy comparison between scenarios through sorting risks by importance with a view to identify priority issues.

b. Overview of the effectiveness of level 2 risk treatment measures, comparing the initial (prior to the level 2 treatment measure) and residual risks of a given scenario.

c. Check the risks allocation between parties to a contract (services and construction), with the projects' residual risks, the detail of how this risk is distributed between the parties and the Owner's residual risk after transfer.

iii. Decision support to justify *risk provisions* for risk the Owner's organisation's accounts, usually referred to as the *financial risk report* (dashed oblique lines Figure 3). During any project phase, the financial risk report shall be updated regularly as risks become out-dated or occur (no more a risk), even if the risk register itself may not evolve. It must also be stressed that financial risk reports are produced for the project's current development phase, for which the organisation has been committed or is about to be committed to.

3.1.2 *Rigour of the process*

At the inception of a project it is common to seek to establish a financial risk report without documenting the risk appreciation process. This shortcoming renders, at later stages in a project, the task of understanding risk evolution impossible. Furthermore, the proposed likelihoods often reflect the lack of confidence in the risk analysis rather than any real likelihood of the risk occurring, which can forestall projects in the early stages.

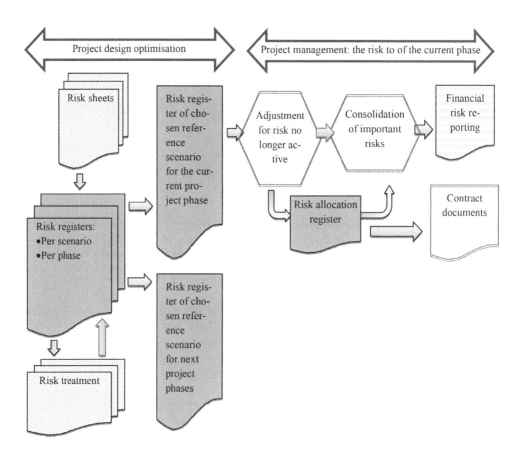

Figure 3.   Risk Management through a project and associated documentation.

Thus, at all stages in the project's life cycle, a full risk register for each scenario must be kept. However, risk sheets and treatment plans may be too detailed for the purpose and thus the risk register can directly include information usually only present in the risk sheets in order to avoid the multiplication of documents.

3.1.3 *Document management*

As can be understood, a robust identification and versioning system is fundamental to proper risk management, thus all documents require a header to identify its type, the phase, the scenario, its version and the date of its last update.

As the risk process can evolve the format revision must be indicated to identify the required steps in order to update the document to the organisation's current standard.

3.2 *Detailed documentation*

All aspects of the risk assessment process must be recorded, using risk sheets and risk treatment plans to record decisions. However, these are not discussed in the present article.

3.3 *Risk registers*

The risk register summarises in a formal manner information about identified risks (from the initial risk to the residual risk, through the mitigation and corrective measures for both) and thus it constitutes the fundamental document in being able to manage the project's risks within a contractual framework.

Each risk can then be developed in a dedicated risk sheet.

- Initial design phases: it is usual at this stage to concentrate on level 1 or 3 of the risk treatment plans. These stages tend to manage risks through the management of scenarios.
- Detailed design phase: it is usual at this stage to concentrate on level 2 risk treatment plans. As such, the use of initial and residual risk analysis becomes important.
- Procurement phase: during the procurement phase it is deemed that the project's risk cannot be further reduced through design, but to a certain extent it could be reduced with respect to the owner's objectives by risk transfer.
- Construction phase: the main focus of this register is to record new risks (i.e., unforeseen at procurement) or are risks identified in the contract, but which must have their likelihood or consequences need to be reassessed or to indicate they have occurred.
- Reception: at reception there maybe remain certain risks to the structure as such.

3.4 *Financial risk reporting*

3.4.1 *Issues*

As from project inception it is required to identify the risks, the financial risk reporting has become mandatory, both in public and private corporations, and a statutory obligation under IFRS 15, with a European Directive in force since the 1st January 2018. However, the pitfalls are:

Risk source			Risk Identification			Initial Risk assessment			Objectives						Risk surveyance	Risk transfert				Paiment conditions					Risk assessment after risk transfert			Objectives					
Category and sub category	Location, stratigraphy or structure	Description of the uncertainty	Event(s)	Consequence(s)	Risk reference n°	Likelihood	Cost	Delai	Corporal integrity	natural environment	built environment	Political - image	Performance	Level of risk	Measures (implemented in the contract to monitor the risk)	Description of transferred risk	Risk owner	Threshold of transfer	Capped risk amount	Deemed included in BoQ unit rates of contract	Penalties	Extra quantities on existing bill items	Bill items on works relating only to potential risks	Not part of the contract	Likelihood	Cost	Delai	Corporal integrity	natural environment	built environment	Political - image	Performance	Level of risk

Figure 4.   Example of risk register for procurement purposes.

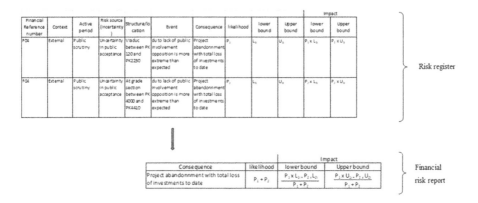

Figure 5.   Example of consolidating risks in the financial risk reporting.

– The financial reporting format of risks is usually based solely on a list of consequences, often with very simplistic linear likelihood levels. In addition, usually the format is usually independent of project size or complexity.
– The common use of stochastic estimations (e.g., consolidated unit rates, such as cost per meter of tunnel, and general progress rates, derived from previous projects), especially at the early stages of a project's design, most likely integrates risks which occurred on those projects.

As the financial risk reporting can be the sole constant in the project's life cycle it is important not only to ensure the traceability of the underlying assumptions, but just as importantly to define the methodologies used in:

– summing the likelihoods of identical consequences deriving from distinct events or risk sources;
– estimating the likelihood corresponding to the mean financial effect of a given consequence and if the effect has a too wide a standard deviation then the rules in defining "tranches" of effects;
– defining the cost estimate's range and confidence level above which any consequence impacting the costs are deemed a risk, not within the assumed tolerance;
– ensuring the list of risks are distinct and "independent"

Further difficulties are induced by the format of this type of report are:

– its general nature, which precludes too much information;
– its "consequence centric" approach, too simplistic for underground projects leading to inconsistencies in the terms used in the risk titles. These often refer not to a consequence, but to a risk sources, an event or even an objective. As such, the risks are not "independent". For example, a risk may be identified as "geological" (i.e a source of risk), another risk may be described as "building damage" or "project cancellation" (i.e. a consequence) and another simply describing a risk as "a cost risk" (an objective);
– its estimation as a single value, not a range of values, induces at best an over estimation of the effect of a risk, with the likelihood reduced to adjust for this; or at worst, the likelihood of the mean effect value being used, thus over estimating the likelihood a well;
– how to cumulate identical consequences but of different risks.

3.4.2 *Format*
Risk registers are too detailed for the purpose of constituting a financial risk report, and an organisation's financial reporting of risks may be too basic for a project with underground sections. Therefore, an intermediate format could be required in between the two mentioned documents, and this could consist of a table with the columns shown in the example given in Figure 6.

4384

						Sum of the squares	20	26
						Sum of the 2 most risky events	26	46
						Retained maximum risk	41	

| Reference number | Context | Active period | Risk source (incertainty) | Consequence | Percentage weighting | Effet on costs | | Weighted effect on costs | | Confidence level |
						lower bound	Upper bound	lower bound	Upper bound	
F01	Internal	Dossier preparation	Uncertainty in available competencies	Delay in project submittal	15%	50.00	150.00	7.50	22.50	2
F02	Internal	Dossier preparation	Uncertainty in corporate approval	Project abandonnment with total loss of investments to date	2%	75.00	95.00	1.50	1.90	3
F03	External	Public scrutiny	Uncertainty in public acceptance	Respond to oppositions	25%	5.00	7.00	1.25	1.75	3
F03	External	Public scrutiny	Uncertainty in public acceptance	Respond to oppositions and extra studies required entailling a delay in project implementation	15%	15.00	100.00	2.25	15.00	2
F04	External	Public scrutiny	Uncertainty in public acceptance	Project abandonnment with total loss of investments to date	25%	75.00	95.00	18.75	23.75	1

Figure 6. Example of an intermediate financial risk report format.

In the example, once public scrutiny is initiated, the two first risks (F01 and F02) can be eliminated as they are not active anymore.

4 CONCLUSION

The latest ISO31000 and ISO31010 transposed to underground works require a more rigorous dictionary of definitions which does not modify the substance of risk management as described by ITA WG2's recommendation but necessary to ensure risk tracking during a project's development, enabling the reporting structure to be adapted to each of a project's development stages. The paper addresses some common mistakes and imprecisions in developing an underground project's risk analysis and it gives advice on the reporting technique and on the importance of a consistent financial risk report (as required by IFPS 15 and the European Directive in force since January 2018) throughout the project's life cycle. An effort has still to be made by the international tunnelling community for the financial risk report to be considered as part of an underground project's Risk Management Plan as far as the risk policy, the risk register, etc. In this sense ITA could contribute to clarify this need and the associated format, e.g., as suggested in the paper.

REFERENCES

AACE®, 2018, *International Recommended Practice No. 10S-90 Cost Engineering Terminology*
ANSI EIA 748 *Standard on Earned Value Management*
Stille, H.P., 2017, Introductory Talk, WTC's 2017 Bergen
CETU (Center for tunnelling studies, a French governmental institute), 2013, *Guide to implementing Fascicule 69.*
AFTES guides GT32R2A1 (2012), GT32R3A1 (2016) and GT25R3A1 (2015).
Bourget, A.P.F., 2014, *Common mistakes in risk assessments in underground works*, Proceedings of the AFTES Congress, Lyon
Bourget A.P. F., 2015, *Using risk analysis in construction tendering processes*, WTC2015 World Tunnel Congress, Dubrovnik.
Bourget A.P.F., 2013, *Financial risk identification and tracking in underground projects*, WTC2013 World Tunnel Congress, Geneva.
ITA WG2, 2004, *Guidelines for Tunnelling Risk Management*, Tunnelling and Underground Space Technology, n.19, 217–237
European directive of 22 September 2016 amending Regulation (EC) No 1126/2008 adopting certain international accounting standards in accordance with Regulation (EC) No 1606/2002 of the European Parliament and of the Council published in the Official Journal on 29 October 2016 adopts IFRS 15 Revenue from Contracts with Customers issued by the IASB in May 20.

Tunnels and Underground Cities: Engineering and Innovation meet Archaeology,
Architecture and Art, Volume 8: Public Communication and Awareness/Risk Management,
Contracts and Financial Aspects – Peila, Viggiani & Celestino (Eds)
© 2020 Taylor & Francis Group, London, ISBN 978-0-367-46873-6

Hydropower tunnel failures: Risks and causes

D. Brox
Dean Brox Consulting, Canada

ABSTRACT: The hydropower industry has unfortunately experienced several tunnel failures over recent years. The risks and challenges for the successful design and construction of hydropower tunnels are much more elevated in comparison to other types of tunnels as hydropower tunnels are subjected to dynamic operating conditions that are not always fully known prior to operation. Many hydropower tunnels have been designed and constructed as unlined tunnels incorporating shotcrete lining only in limited areas based on technical assessments performed during and at the later stages of construction. The technical assessment can however be subject to limited access, limited data collection from mapping, concealed instability of TBM excavated tunnels, and the misunderstanding of dynamic hydraulic operations. The main risks associated with the design and construction of hydropower tunnels along with causes of recent collapses are presented and explained along with suggestions for improved industry practice.

1   INTRODUCTION

The design and construction of hydropower tunnels is based on significantly different design criteria and practices in comparison to other tunnels for civil and mining infrastructure. Hydropower tunnels are subjected to internal and dynamic loading conditions during operations and the performance of a tunnel can only be regularly monitored by pressure instrumentation. Physical inspections of hydropower tunnels to confirm their structural integrity require an outage of operations with dewatering of the tunnel that results in the loss of generation and associated revenue, and may also cause an impact to the integrity of the tunnel.

A number of hydropower tunnel failures occurred in the 1950's and 1960's including those at Kemano and Snowy Mountains (Jacobs, 1975). These same types of failures continue to plague recent and current practice in the industry. An alarming number of failures of hydropower tunnels have occurred in recent years, many of which occurred shortly after commissioning, and the reasons for which have some similarities. Figure 1 presents the post-collapse conditions of the 10 m, 5 km Rio-Esti hydropower tunnel in Panama. The risks associated with the design and construction of hydropower tunnels are presented along with causes of collapses.

2   HYDROPOWER TUNNEL PLANNING

The success of hydropower tunnel planning and design depends significantly on adequate geotechnical information as it has been demonstrated that cost and schedule overruns have been caused by a lack thereof (Hoek and Palmieri, 1998). Brox (2017a) presents the typical requirements for geotechnical investigations for hydropower tunnels. The planning and design for hydropower tunnels should adopt the well-established design principles and criteria including the prevention of aeration and leakage with adequate in situ confinement from the surrounding rock, the inclusion of a surge facility, adequate initial tunnel supports and final linings subject to the anticipated long-term behaviour of the geological conditions, provision of a rock trap for predominantly unlined tunnels, and steel lining sections where inadequate in situ stresses are present (EPRI, 1987).

Figure 1.   Multiple Collapses in Rio Esti Tunnel.

Typical project layouts including high elevation low gradient tunnels with connection shafts or inclined tunnels to underground powerhouses are presented by Benson (1989) and important geological and geotechnical considerations for the design of pressure tunnels including design criteria for the identification of lining types are presented by Merritt (1999) that should be fully respected without design shortcuts.

Hydropower tunnels planned to be located within young and active geological environments including mixed and disturbed volcanic bedrock of the Andes, and sedimentary bedrock of the Himalayas, should adopt special considerations for the understanding of the long-term behaviour and durability of the rock units. Other noted geological risks include the presence of karst, gypsum and/or anhydrite, as well as zeolites/laumontite within basalts, hydrothermal alteration within intrusive bedrock and thermal springs with gases.

The acceptability of an unlined hydropower tunnel is a major design decision that, while is associated with lower construction costs, warrants a very careful and thorough evaluation of the anticipated geological conditions, the long-term durability behaviour, as well as the risks associated with the most applicable and safe construction methodology.

3   HYDROPOWER TUNNEL DESIGN

The design of a hydropower tunnel should be thoroughly based on the consideration of the proposed or possibly worst-case future operating conditions. Reservoir project schemes typically maintain constant pressures implied to tunnels whereas run-of-river scheme are typically subjected to seasonal fluctuations of run-off with associated significant pressure variations. Finally, projects that are designed based on multiple intakes within a large watershed area with variable hydrology can be expected to result in significant pressure oscillations within a tunnel.

For predominantly unlined hydropower tunnels, the final design of the tunnel is only completed after a thorough evaluation of the encountered conditions and the design of a series of final linings to be constructed at each relevant section. It is therefore important to recognize that hydropower tunnels cannot be completely designed before construction and project developers should expect to incur additional costs associated with design modifications and scope increases during construction.

Hydraulic tunnels require special support designs for geological faults and other zones of weakness or moisture sensitivity. Standard designs of shotcrete for the final lining is not appropriate as shotcrete is of variable quality and permeability and allows rapid saturation of the surrounding rock resulting in deterioration, possibly with swelling, and increased loading upon the shotcrete lining that causes collapse. Only concrete linings should be considered for the long-term support at geological faults. The length of special tunnel linings should be carefully evaluated and designed to be constructed over an extended length of the intersection or exposure of the zones of weakness.

Diversion tunnels are required for the construction of dams and some large weirs and their performance and integrity must be ensured during the entire construction duration for all possible circumstances. As such, it is common practice that diversion tunnels are fully concrete lined to prevent any impact that may occur during the construction of a dam such as increased flooding upstream and associated pressure flow through the diversion tunnels. Shotcrete lining for diversion tunnels should not be considered as a schedule or cost saving measure particularly for diversion tunnels sited in fair or poor quality rock conditions which are common at shallow depths along the lower slopes of major valleys where they are typically constructed.

4 PROJECT DELIVERY METHODS

Engineering-procurement-construction (EPC) or Design-build (DB) project delivery methods are not considered to be appropriate for hydraulic tunnels. EPC commonly results in the contractor placing pressure on the design team to develop aggressive and non-conservative designs that do not follow good industry practice and minimize construction quality that can result in defects.

EPC may be desired by clients thinking that there are project schedule savings to be realized when they themselves are under time pressure to generate and deliver power based on their power purchase agreement with other third parties. The client can simply pass on the schedule demands to the EPC contractor which can also force the designer to adopt an inappropriate shotcrete tunnel lining design if or when the rock conditions are not suitable or favorable such as occurred at the Rio Esti project in Panama whereby a shotcrete lining design was used for a large span tunnel in low strength young volcanic rock units that were subjected to first time saturation and multiple collapses after 9 years of operations in 2011.

EPC is typically adopted by clients who wish to shed, rather than share, geotechnical risks and also require fixed prices even when limited pre-construction investigations have been completed. The often-used phrase "design as you go" does not lend itself for good application for EPC fixed price contracts as the Contractor prefers to build a tunnel lining design based only on shotcrete that is easy and faster to complete as he is under pressure to complete the project and not face penalties.

Design-build hydropower tunnel projects have commonly been executed to attempt to achieve schedule savings in order to start power generation as soon as possible. Design-build approaches have however commonly been based on very limited geotechnical investigations that require the successful DB team to perform their own additional investigations within a very limited time period in order to submit an acceptable design. But most of these projects do not allocate adequate time to perform comprehensive geotechnical investigations and rather proceed to construction and experience significant unexpected site conditions. In some cases, the actual site conditions have been so different from what was originally expected that the construction methodologies are no longer suitable, or at least, will not achieve the expected production rates as expected in the project schedule and typically results in disputes between the DB team and the project developer.

A design-build approach that separates the design liability between the initial and final tunnel support is also considered to be unwise as the design responsibility and liability can be uncertain both during construction and future operations. This approach incentivizes a contractor to minimize the amount of tunnel support to be installed in order to maximize production, thereby jeopardizing the overall safety of the works, which is not considered to be prudent.

Design-build can also result in a contractor not following the instructions of his designer in terms of tunnel support and take a chance on minimizing support in order to attempt to improve production when efficiency is lacking from the construction team. Overall, the EPC or design-build is not considered a good approach for the construction of hydropower tunnels. The reason that EPC design-build with allocated geotechnical risk to the tunnel contractor has been used in the past is due to the requirements of financiers for cost certainty. However, it is claimed that much better cost certainty can be achieved through comprehensive geotechnical investigations, characterization and risk identification based on competent industry practice.

Finally, the selection of the design-build project delivery method should not rely on project wide insurance to address shortcomings in executing good industry practice as expected by the International Tunnel Insurance Group (2012).

In comparison, Design-bid-build (DBB) is considered to be the most appropriate method of project delivery for hydropower tunnels that allows for design modifications during construction without significant cost increases when unit rate prices are included in the contract. With this project delivery method the client retains total control over the design and changes or modifications during construction in order to construct a low risk tunnel with appropriate tunnel linings.

5 HYDROPOWER TUNNEL CONSTRUCTION

Drill and blast excavation is typically adopted for the construction of hydropower tunnels when:

- Length generally short (< 4 km);
- Multiple access adits possible for long tunnels (> 4 km);
- Good drill and blast experience available from local labour, and;
- Moderate to high risk geological conditions are anticipated over an appreciable amount of the tunnel alignment, typically >20%.

In some cases it may be appropriate to consider high speed drill and blast methods including the Rowa Hanging Conveyor System or the Jacob's Sliding floor (Petrofsky, 1987) for long tunnels if the budget for construction is limited to attempted to be optimized or there is insufficient funds available prior to project approval for the procurement of major equipment (eg. TBM).

Long and deep hydropower tunnels with homogenous and low risk geological conditions typically warrant the use of a tunnel boring machine (TBM). If fair to good quality rock conditions are anticipated along most of the tunnel alignment with limited sections of poor quality rock conditions an open gripper TBM able to install rock support with follow up shotcrete final lining may be appropriate. However, since 1995 the hydropower industry has enjoyed over 300 km of successful construction of long, low-pressure tunnels using mainly double shield TBMs in conjunction with pre-cast concrete segmental linings with no reported problems to date (Brox, 2017b). Where there exists the risk of squeezing rock conditions it may be more appropriate to adopt a single shield TBM. These approaches have achieved very good rates of production as well as provides greater safety against risks associated with variable and difficult geological conditions including fault zones as well as overstressing including rockbursts for deep tunnels.

Correct and total backfilling of pre-cast concrete segments is critical for the long-term success and operations of these tunnel linings. The most appropriate type of TBM to use should be based on a thorough risk evaluation considering adequate geotechnical information. It is important to recognize that TBM viability is just not the ability to complete the construction of a tunnel, but to complete it efficiently achieving and sustaining good rates of production, which is only possibly if there is limited amount of poor quality conditions. Factory acceptance testing of a TBM must be entirely completed prior to approval by the project developer for transport to the project site. It is however important that no matter how extensive any geotechnical investigations have been performed, the project developer should refrain from specifying the means and methods of tunnel construction and leave this decision and associated risks to the tunnel contractor.

Wherever possible the design of long hydropower tunnels should be planned to include multiple access locations to provide future access for inspections and maintenance. Intermediate access adits used for drill and blast construction serve this function. For long tunnels constructed using TBMs it is important to incorporate practical access facilities such as adequately sized openings within the downstream surge shaft and the upstream gate slot in order to facilitate possible future inspection using remote operated vehicles (ROVs).

6 HYDROPOWER TUNNEL OPERATIONS

Clients must include all relevant information about the planned modes of future operations especially for design-build project delivery methods so all modes of operation can be fully understand and considered and incorporated into the final design of tunnel linings.

The internal operating pressure is an important aspect to consider for the design of pressure tunnels. Static and in particular dynamic internal pressures should be carefully evaluated and understood by the design team.

Hydraulic transient analyses should be performed by the hydraulic designers to confirm the magnitude and variation in the internal operating pressures to be expected during normal and non-normal operating periods.

The Norwegian cover criteria can be used for the preliminary design of the minimum length of steel lining sections typically required along the downstream end of tunnels. However, it is important to consider more detailed analyses especially if a tunnel is aligned parallel to a major valley with limited side confinement or there exists other irregular topography along the tunnel alignment where leakage could occur. Geological faults are commonly associated with gullies and topographic depressions that typically represent zones of potential leakage that should be evaluated.

Finally, instrumentation monitoring with multiple points of data collection along the tunnel should form part of the permanent operations for pressure tunnels. The monitoring of tunnel pressure provides critical information to confirm acceptable operations or detect headlosses that may occur over the life of a tunnel.

Rock traps are an essential component of predominantly unlined or shotcrete lined tunnels to prevent minor rock debris and fragments from damaging the mechanical equipment. Brox (2016) presents the design and functional requirements for rock trap for hydropower tunnels. The inclusion of a rock trap, including its over design capacity, should not be considered as an alternative to completing adequate shotcrete and/or concrete linings for all sections of poor quality rock conditions since rock traps eventually fill up and must be cleaned. A full trap that has not been cleaned poses a serious risk to ongoing operations. The fill volume of rock traps should be inspected at any outage opportunity during operations and cleaned accordingly.

7 HISTORICAL AND RECENT COLLAPSES

7.1 Historical Collapses

Notable historical hydropower tunnel collapses occurred in the 1950's and 1960's at the Kemano project in Canada, at the Snowy Mountain Project in Australia, and at the Lemonthyme Project in Tasmania respectively. The collapse at Kemano occurred after 3 years of operations and included a large 25,000 m³ collapse. A total of 3 collapses occurred at Snowy Mountain after 4 years of operations and included a total volume of about 3,000 m³. The collapse at the Lemonthyme tunnel occurred after 5 months of operations involved about 1,500 m³.

Each of these collapses were similar in nature in that they occurred at geological faults or fracture/shear zones that were not supported as part of the original construction and serve to confirm the difficulty in the design of tunnel support and linings for weak zones. All of these tunnels were mainly supported with thin (19 mm) layers of gunite as was the practice at the time. It is interestingly noted that the repairs for each of these cases did not include bypass tunnels but rather re-excavation through the collapse zones and re-support of the tunnels.

7.2 Recent Tunnel Collapses

The following hydropower tunnel collapses have occurred within the last decade:

Table 1. Hydropower Tunnel Collapses.

Project	Year	Size, m	Length, km	Max Internal Pressure, m
Glendoe	2009	5	8	670
Rio Esti	2010	10	5	122
La Higuera	2011	6	16	15
Shuakhevi	2017	6	18	530

Insurance investigations were performed and legal proceedings resulted for most of the tunnel collapses owing to large costs involved with the repairs that were required and the loss of power generation revenue.

7.3 Glendoe, Scotland

The 100 MW Glendoe hydropower station is located above the southwest side of Loch Ness and includes an 8 km, 5 m power tunnel that was excavated using an open gripper TBM. The total internal operating pressure is 600 m and flow capacity is about 18.6 m^3/s. The tunnel was constructed at an upward gradient of 12%.

The project was constructed under a design-build contract with a design life of 75 years. The hydropower plant was to be operated intermittently as a stop-re-start system subject to reservoir water supply availability thus causing highly variable internal pressures to the tunnel.

A major regional geological fault, the Conaglean Fault, was present and intersected by the tunnel along the upper third section of the tunnel. Geology along the tunnel mainly comprised quartz mica schist. An intake shaft was originally planned to be sited within the fault. A geotechnical drillhole completed into this area prior to construction confirmed highly fractured conditions including clay infillings. In addition, and most relevant to serve as a red flag, was that a raisebore pilot hole was unsuccessfully completed into the fault during construction and cancelling of the intake shaft at this location. This critically important geotechnical information appears to be have been overlooked by the project teams.

The fault was considered to be stable following TBM excavation and therefore no high capacity rock support was considered necessary to be installed for hydraulic operations. In hindsight the stability of regional geological fault zone was masked or concealed by TBM excavation with no disturbance during excavation. Drill and blast excavation through the fault would have likely resulted in collapse during excavation and then a correct design for this section for long term operations. The headrace tunnel operated for about 8 months prior to collapse and shutdown.

An approximately 600 m long access as well as a 600 m long bypass tunnel around the fault zone were constructed along with additional shotcrete lining of newly identified non-durable rock sections over about 50% of the tunnel alignment. The total outage for the tunnel repairs was about 36 months. Figure 2 illustrates the failed rock fragments within the tunnel after the collapse that generated a volume of about 20,000 m^3 (Palmstrom and Broch, 2017).

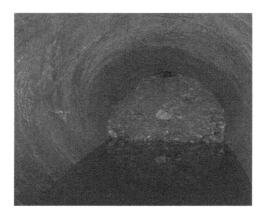

Figure 2.   TBM Tunnel Collapse.

7.4   *Rio Esti, Panama*

The 120 MW Rio Esti hydropower project is located in Chiriqui Province of western Panama about 30 km north of the city of David at an elevation of about 220 m. The scheme comprises the Chiriqui and Barrigon Dams and reservoirs, a 6.5 km canal, a 4.8 km, 10 m size headrace tunnel, and a surface powerhouse. The total head is 112 m with a flow capacity of 118 m3/s.

The project was required to be constructed under an extremely compressed design-build contract within 33 months to meet the deadline for a power purchase agreement for generation. The headrace tunnel was designed and constructed as a near-fully shotcrete lined tunnel since there was inadequate time in the project schedule for concrete lining even from via multiple access adits. The geology along the tunnel comprised horizontally bedded volcanic sedimentary rock including mudstones, andesites, tuffs, and agglomerates with a depressed groundwater table below the tunnel alignment. The vertical alignment of the tunnel was varied as much as possible to avoid intersection of the tuffs.

The headrace tunnel started operations in 2003 and continued adequately until early 2010 before a headloss was detected. The hydropower plant was continued to be operated until mid-2011 when the headloss increased further and the scheme was finally shutdown for an unwatered inspection of the tunnel. Numerous collapses were observed along the entire tunnel alignment upon dewatering of the entire tunnel. Figure 3 presents a major collapse of nearly total blockage of the tunnel that occurred near the surge shaft with an estimated volume of 14,000 m^3.

The failure of this hydropower tunnel is believed to be due to the first time saturation of the weak volcanic sedimentary rocks that increased the loading conditions on the shotcrete lining at various locations. The acceptable operation of the hydropower tunnel of nearly 7 years is remarkable given this design. The repair of the tunnel comprised re-profiling with the installation of steel ribs and 100% concrete lining for the entire 4.8 km. The total outage for the repairs was about 23 months.

7.5   *La Higuera, Chile*

The 155 MW La Higuera hydropower project is a run-of-river scheme located within the Tinquirirca valley about 150 km south of Santiago. The headrace tunnel has a length of 16 km and forms part of the downstream component of a three project cascade scheme. The tunnel was constructed by under an EPC contract using drill and blast methods from multiple access adits at a relatively low gradient and included shotcrete final linings throughout most of the tunnel.

A prominent geological fault was intersected along the most downstream section of the tunnel where nominal tunnel support comprising rock bolts, mesh, and shotcrete were installed as final support. An inferred weak geological formation/layer can be identified on surface using Google Earth near the location of the collapse. The headrace tunnel operated for about 9 months prior to collapse.

Figure 3. Major Collapse of Partial Blockage.

Figure 4 presents the weathered rock associated with the geological fault at the collapse location with a volume of about 12,000 m^3 (Palmstrom and Broch, 2017). Tunnel repairs comprised the construction of a 240 m long bypass tunnel. The total outage for tunnel repairs as about 21 months.

7.6 *Shuakhevi, Georgia*

The 181 MW Shuakhevi Hydropower project is a run-of-the-river plant located in southwest Georgia and includes over 35 kms of tunnels, 2 dams and 2 surface powerhouses. Construction of the tunnels commenced in 2013 using drill and blast methods with multiple intermediate access adits. Georgia is associated with complex and disturbed geology due to its location

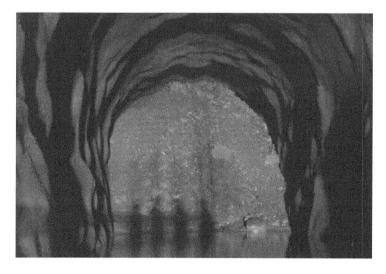

Figure 4. Weathered Rock at Collapse.

between the Eurasian and Africa-Arabian tectonic plates with continental collision. The rock types noted along the tunnels comprised highly disturbed basaltic andesites, conglomerates and volcanoclastic sandstones as part of a major geological syncline oriented sub-parallel to the downstream headrace tunnel.

Operations commenced in mid-2017 and after about 2 months a series of multiple collapses and other damage occurred in the main 19 km downstream tunnel. Most the multiple collapses were reported to have fully blocked the tunnel. The total volume of the multiple collapses is estimated to be about 15,000 m^3.

The repairs of the tunnel were planned to include the construction of a 250 long bypass tunnel at one of the major collapse areas and shotcrete repairs at the other smaller collapse areas. The total outage for the repairs was believed to be about 18 months.

8   ROOT CAUSES OF COLLAPSES

The design life of hydropower tunnels is subject to whether they have been maintained and repaired. The majority of hydropower tunnels are not inspected and maintained as per other hydropower components and therefore their design lives are limited. Brox (2017b) presents historical failures of hydropower tunnels from which it can be inferred a typical design life of less than 30-40 years. A number of past examples exist where failure occurred sooner but can be attributed to inadequate designs and lack of repairs. The common misconception in the hydropower industry is that hydraulic tunnels are not recognized and respected as engineering structures but rather as geological anomalies.

Accordingly, the hydropower industry has often assigned the responsibility, or relied upon the judgement of geologists to make the final design decisions with regards to the final lining designs to be constructed to address any poor quality and weak geological zones along a tunnel without regards to the nature of future hydraulic operations and imposed loadings for the tunnel. In addition, as per other significant and critical engineering structures like bridges and important buildings, there appears to be no independent checking of the final tunnel lining designs by another engineering practitioner.

The occurrences of collapses in hydraulic tunnels has typically been due to the incomplete identification of weak geological zones such as medium to large faults as well as minor fractures zones containing dissolvable minerals including clays, anhydrite and gypsum, the communication of these important conditions to the tunnel design team, misunderstanding of, or non-recognition of cyclic loading conditions from oscillating internal pressures, and correct high capacity tunnel lining designs to effectively protect and stabilize these conditions during future hydraulic operations.

Some project schedule demands from owners may limit the design and construction approaches that can be adopted for a project for a given geological environment such as occurred at the Rio Esti project.

9   TYPICAL REPAIR OF TUNNEL COLLAPSES

The repair of a tunnel collapse by re-excavation and support is a high-risk work process with greater risks than the original tunnel excavation. The tunnel collapse area is unstable and material typically has to be removed in order to install new rock support. The work has to be planned carefully and may require the use of remote operated equipment for scaling and mucking out of the failed material as well as for the installation of new rock support.

The most common and safe method for the repair of a tunnel collapse of limited extent is with the construction of a bypass tunnel and plugging off the original tunnel between the bypass. The design and construction of a bypass tunnel requires careful planning for the safe excavation and support of passing through the same collapse area. Bypass tunnel excavation typically involves extensive probe drilling to clearly define the geometry of the unstable zone and possibly a small size pilot drift to excavate through the unstable area safely without resulting in a further collapse.

The repair of a tunnel collapse either in the form of a geological fault of limited extent or multiple collapse areas commonly requires the installation of a new concrete lining in order to provide adequate support for long term operations. Concrete plugs of large volumes are also required for the sealing off of the original tunnel section in which the collapsed occurred.

10 NEW TUNNEL SUPPORT AND FINAL LINING DESIGN STANDARDS

The use of shotcrete for the long-term stability and support of major geological faults within hydropower tunnels is not considered to be adequate, particularly for peaking power plants and where highly variable internal pressure oscillations are normal that can result in preferential scour, saturation, and deterioration of such final linings. New tunnel support design standards comprising the use of concrete linings are considered necessary for the long-term stability and support of major geological faults within hydropower tunnels.

11 CONCLUSIONS

The following conclusions are considered to be applicable to the occurrence of the recent hydropower tunnel failures:

- Hydropower project developers should take note of the root causes and lessons learned from the recent series of tunnel collapses in order to prevent future occurrences;
- Root causes of failure are associated with inadequate tunnel support for the prevailing geological conditions commonly comprising a geological fault or non-durable rocks;
- Adequate geotechnical information should form the basis for a coherent design and for contract information;
- Realistic project schedules should be established that are site-specific for constructability and not subjected to power generation agreement deadlines;
- EPC or design build approaches should be avoided to allow total design control by the client to prevent any compromising of construction quality;
- Appropriate tunnel mapping should be performed with detailed descriptions of suspect weak zones;
- Final tunnel support and linings should not be based on the application of rock mass classifications;
- New tunnel support design standards comprising the use of concrete linings are considered appropriate for the long-term stability and support of major geological faults within hydropower tunnels;
- Appropriate tunnel linings should be carefully designed for all suspect weak zones;
- Hydropower tunnel operations should be curtailed at the onset of any detectable headloss to prevent damage, and;
- Independent external technical reviews should form part of the design and construction of hydropower tunnels as per other critical infrastructure.

REFERENCES

Palmstrom, A. and Broch, E. 2017. The design of unlined hydropower tunnels and shafts: 100 years of Norwegian experience. International Journal of Hydropower and Dams. Issue 3.
Benson, R. 1989. Design of Unlined and Lined Pressure Tunnels. Tunneling and Underground Space Technology. Vol. 4. No. 2. Pp. 155-170.
Merritt, A. 1999. Geologic and Geotechnical Considerations for Pressure Tunnel Design. ASCE.
Brox, D. 2017a. Risk Reduction for Underground Hydropower Project, World Tunnel Congress, Bergen, Norway.

Brox, D. 2017b. Practical Guide to Rock Tunneling, Taylor-Francis, pgs. 248.

Rosen, S. 2005. Geotechnical Risk Assessment and Management for Maintenance of Water Conveyance Tunnels in South Eastern Australia. AGS-AUCTA Mini-Symposium: Geotechnical Aspects of Tunneling for Infrastructure Projects.

Jacobs, D. 1975. Some Tunnel Failures and What they have Taught. Hazards in Tunneling and Falsework. Institute of Civil Engineers.

Petrofsky, A. M. 1987. The Jacobs Sliding Floor: Current Competitive Applications. 6th Australian Tunneling Conference, Melbourne.

Code of Practice for Risk Management of Tunnel Works, 2012. International Tunnel Insurance Group. London.

Fippen, R., Sketchley, J., Redhorse, T., and Tsztoo, D. 2018. In-Depth Inspection of a Century-Old San Francisco Water Tunnel, North American Tunnel Conference, Washington, DC.

Brox, D. 2016. Design and Functional Requirements for Rock Traps in Hydropower Pressure Tunnels, International Journal of Hydropower and Dams, Issue 1.

*Tunnels and Underground Cities: Engineering and Innovation meet Archaeology,
Architecture and Art, Volume 8: Public Communication and Awareness/Risk Management,
Contracts and Financial Aspects – Peila, Viggiani & Celestino (Eds)
© 2020 Taylor & Francis Group, London, ISBN 978-0-367-46873-6*

A general risk management approach: The result of integration between the semi-quantitative and quantitative methods

P. Caiazzo, M. Ciarniello, M. Ponte & M. Foresta
Italferr S.p.A., Rome, Italy

ABSTRACT: The long lasting experience gained in the application of quantitative risk analysis for Italian railway tunnels and semi-quantitative methods, employed for several complex projects abroad in the last decade, has provided a wide risk ranking framework and feedback about design choices. The integration between these methods and outcomes allowed widening the field of analysis, framing the tunnels within a wider infrastructure context and in all life cycle stages, identifying all key-players in each phase, with roles and responsibilities. The integrated approach, based on the operations risks forecasting in the design stage ensures the continuous improvement of project processes, by managing all related risk preventing overhaul and changes (increased times and costs) and, eventually, infrastructure management issues in the construction/operation phases. As an example, a case study related to project risk management for a long tunnel is reported.

1 QUANTITATIVE AND SEMI-QUANTITATIVE ANALYSIS: INTEGRATION OF PROCESSES AND IMPROVEMENT

In the last decades several Italian tunnels have been investigated in compliance with Italian Rules (i.e. Italian Ministerial Decree, Safety in Railway tunnels, 28/10/2005), which identified risk analysis as testing tool to assess a railway project in the respect of safety targets, especially for the operations risks. The application of the quantitative method to the tunnels starts, in fact, from the analysis of accidental databases, classifying the main causes (Technical, Human and External) in relation to the different sources (Rolling Stock, Infrastructure, etc.), in order to identify applicable hazards, related risks, mitigation/prevention measures through a cost-benefits analysis for risk reduction (Procedures, Policies, etc.) and, in general, potential actions to be apportioned to the responsible owner (Designers, Infrastructure Managers, etc.).

The quantitative analysis helps to define all the necessary and sufficient measures to be integrated within the design envelope and their prioritization, by quantifying exactly the prevented fatalities, the avoided delays and/or operations criticalities and emergency, thanks to the application of each measure. In particular through the costs-benefits analysis, the economic benefits are compared with manufacturing/implementation/maintenance costs, so that each subsequent amendment or addition to the basic project requirements can be identified but also evaluated from the point of view of effectiveness and financial commitment. This approach, based on the operations risk forecasting in the design stage, guarantees the continuous improvement of project process, allowing the well-timed adjustments for safety needs and preventing successive overhaul and changes (increased times and costs) and, eventually, in the construction and operation phases, infrastructure management issues (with potential risks for workers and users). The large number of tunnel case studies, analyzed in terms of infra-structure configurations (single or double bore, including stations, etc.), rail traffic data (passengers/freight/hazardous freight services), different equipment and project solutions (firefighting systems, smoke estractors, exits and/or cross passageways, etc.), environmental context and local interferences (hydrogeological factors, overlapping and crossing with other infrastructures and/or anthropic influences) provides finally an articulated reference framework where it is possible identifying significant dangers and measures, owners, areas of responsibility and/or involved life cycle stages (Project, Operation, Maintenance) as function of basic data and parameters.

The application of semi-quantitative method for RAMS (Reliability, Availability, Maintainability, Safety) planning and demonstration, according to European standards (i.e. CEI EN 50126), has contributed to broad the field of analysis, by framing the tunnels within a wider infrastructure context and in all life cycle stages, identifying all key-players in each phase, with their roles and responsibilities. This structured approach, applied in several complex projects abroad in the last decade (underground and high speed railway tunnels, urban junctions, etc.), together with the long-standing experience in Italian risk analyses applications and the integration between the two methods and results can be considered more than a detailed framework where all main project risks may be identified, treaced, classified, customized, assessed and "solved".

The key to exploit the is interaction between the two methods is their integration. In this sense, in the context of project risks to be managed it is necessary to distinguish:

- Project risks arising from evaluation errors/underestimations about *operations* criticalities forecasting. In order to fulfil safety targets (and also the reliability), this kind of risks can lead to:
  - project overhaul in advanced stage/system upgrading, for risks reduction, if the design errors are noticed during the as-built stage or construction;
  - operation procedures/maintenance policies, for risks management, if the design errors are noticed during the monitoring stage.

- Intrinsic Project risks, arising from error in project management: lack of integration and compliance among different design aspects, disjointed project, etc. This kind of risks can lead to:
  - project significant changes, for risks reduction, if the management errors are noticed during early stage of design;
  - project overhaul in advanced stage/system upgrading, for risks management, if the design errors are noticed during the as-built stage or construction.

- Project risks arising from evaluation errors/underestimations about *maintenance* criticalities forecasting. In order to fulfil maintainability targets (and, then, consequently, the availability), this kind of risks can lead to:
  - project overhaul in advanced stage/system upgrading, for risks reduction, if the design errors are noticed during the as-built stage or construction;
  - operation procedures/maintenance policies, for risks management, if the design errors are noticed during the monitoring stage.

The following figure represents as a pattern what explained above:

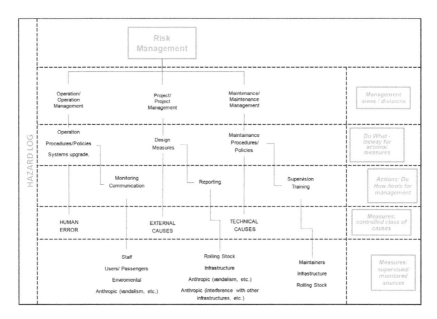

Figure 1. Risk Management process as result of integration between the semi-quantitative and quantitative methods.

The figure can be read top-down and viceversa. The bottom-up sense provides an idea of quantitative and semiquantitative risks analysis process, where the start point are the accidents/undesired events, the monitored sources generate the hazards precursors of main accidents together with the causes, grouped in classes. Then actions and procedures, as function of specific analyses, are identifiable to manage risks. The top-down sense represents a pattern of risk management process related in particular to project risks.

The main tool for risk management is the Hazard Log where all hazards, risk levels, mitigation actions, those responsible for these actions and their monitoring are recorded; this tool summarizes the information of the risk management plans.

2 TUNNEL PROJECT RISK MANAGEMENT

The experience gained in the integrated risk assessment, as a decision-making process for evaluation of mitigation measures needs and the effectiveness monitoring throughout their life cycle, represents fundamental support within the decision analysis and control process of company risks in the design, construction and management of railway infrastructure and tunnels in particular, in full compliance with ISO 31000. (Risk Management - Principles and Guidelines) and ISO 31010 (Risk Management - Risk assessment techniques).

This support for risk management activities is expressed in terms of the highly reliable technical contribution about projects quality, scheduling and costs, and also as management assistance of the tools and coordination of the actors involved in all phases of the process.

The following figure shows a schematic representation of the risk management process, where the main activities are highlighted and interconnected, as illustrated in ISO 31000.

The skills gained thanks to the RAMS support team activities facilitate and expedite the risk assessment in the several phases of risk identification, analysis and evaluation, ensuring a full risk mapping and the correct risk estimation, based on the detection of risks categories (i.e. safety and/or costs and/or scheduling related), types of damage (i.e. severity and/or increased costs and/or delays) and the definition of probabilities and consequences ranges, useful to risk level classification.

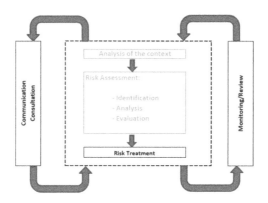

Figure 2.   Risk Management Process, based on ISO 31000.

In particular the statistical processing of on-field data stored, carried out for risks analyses, provided outcomes about the effectiveness of implemented equipment and adopted organizational measure, leading to the criteria for the risk acceptability definition, in terms of probability, severity and ranking. For example, the accidents monitoring and reporting, throughout the last twenty years, has shown a reduction in accidents due to human factor, thanks to the implementation of Automatic Train Protection and Control Systems, implying the systematic adoption of this technology and suggesting threshold value of residual risk according to ALARP principle (As Low As Reasonable Practicable).

In addition to risk assessment step, the technical contribution can be expressed even in the context analysis, consisting in defining scope and aims of the process, describing criteria of success and explaining the constraints and limitations.

In particular, the experience gained in foreign projects, requiring the local factors analysis and study (environment, rules and regulations, authorities and third parties, etc.). support to establish the context by using tools and specific know-how acquired in different countries from several point of view (geographic position, economics/politics, etc.). For example, specific environmental features as presence of sand, wind stress, significant thermal excursion, implied the adoption of further design, maintenance or procedural measures, customized in order to assure the project quality and targets.

Moreover, the studies carried out so far suggest when a risk should be treated or transferred, how to plan and define the action and the owner. In particular the added value in the management of such aspects arises directly from the collaboration with specialists/engineers/ managers implying an organizational chart and model, for the definition of competence areas and responsibility, as well as the identification of coordination roles and development of appropriate management tools.

The following figure represents an example of organizational chart, where it is marked the direct relationship between risk manager and risks specialists and all correlations among the different competence area managers and experts:

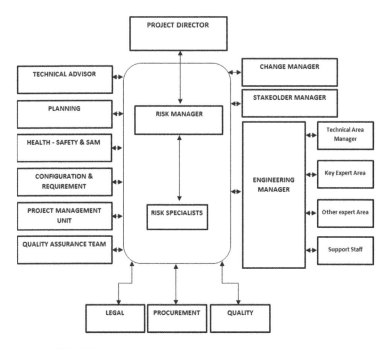

Figure 3. Example of Risk Management organizational chart.

As it can be noticed, the organizational chart reflects the schematization of the activity process, where the core of risk management is the context analysis and risk assessment and treatment and it corresponds to risk manager and risk specialists team, while communication/consultation and monitoring/review are activities correlated by two-way exchange flows. All information and input/output flow constitutes the contents of Risk Management Plan and Risk Register, fundamental tools for Risk Manager and also Customer.

The optimization key to merge the technical know-how and coordination skills, gained thanks to RAMS analyses, within the risk management process would be to share, among all competence areas, even techniques and tools functional to all the intermediate steps, guaranteeing a common "language" and uniformity of approach.

In this sense even aspects as legal, procurements, etc. should be treated as the strictly technical ones, by using same analysis methods of historical data, in order to guide the "information structure" and facilitate also the final implementation of Risk Register.

In the following table a pattern of main steps of context analysis and assessment process is represented in a general form, applicable to various areas of competence analogously to technical aspects:

Table 1. Structured Approach: steps, models and tools.

Structured Approach					
Methodological Steps			Model/Techniques	Input Support Tools	Output Support tools
Context Analysis		Issues /Criticalities Preliminary Analysis	Empirical Data Analysis	Reports, Projects, Contracts, On-field data, etc.	Lesson Learnt / Reporting
		Issues /Criticalities Mapping	Creative Techniques, as for example: HAZOP Brainstorming	Lesson Learnt / Reporting	Databases of issues and criticalities
Risk Assessment	Analysis and Evaluation	Identification	Creative Techniques, as for example: What if Analysis	Databases of issues and criticalities	Check Lists
		Probability	Data Processing	Check Lists	Probability Indexed Archives of data
		Severity			Consequences Classes
		Risk level			Risk ranking matrix, Risk indicators

4401

3 TUNNEL PROJECT RISK MANAGEMENT: A CASE STUDY

The case study below refers to a project risk management activity requested by a Stakeholder to evaluate the impact on the project of an overhaul with respect to the solution established to create a railway link. The alternative proposal compared to the contractual one had a smaller number of tunnels but a longer overall length and appeared advantageous for the customer in terms of initial economic investment and implementation phases. In order to establish the quality of the project and the effects in terms of scheduling and costs of the proposed alternative, the risks related to the different characteristics of the new project have been assess. Moreover, mitigation measures have been suggested for the risk treatment, thus providing the Customer with support in choosing the design solution most advantageous to him considering the overall life cycle of the railway infrastructure to be realized.

Therefore, the project risk management process has been developed as follows:

- Context analysis
- Classification of the probability of undesired events
- Classification of the severity of undesired events
- Risk classification and treatment
- Results of the analysis

3.1 *Context Analysis*

The first step of the process was the examination of the context aimed at understanding:

- scope and aims of the alternative solution proposed;
- the specific interests of the End-customer;
- system design choices related to specific technical-environmental criticalities;
- degrees of freedom in the definition of mitigating measures.

As a result of this analysis in the respect of the interest of the Client the study was focused only to possible events with negative consequences on the final objectives (opportunities was excluded) and the main risk categories to be investigated were defined (environmental, technical, organizational).

3.2 *Classification of the probability of undesired events*

Once the context was established, a risk assessment of the alternative solution was performed. So the overhaul risks were identified and evaluated in terms of Probability and Severity levels of related undesired events.

Probability classes have been defined considering that the event may occur during the construction phase as well as the operation of the railway system.

For tunnels, in particular, geotechnical aspects are crucial and therefore a lack/insufficent planning of the necessary geotechnical surveys can be the cause of many undesired events.

According to the category of risk, different criteria in probability class definition have been adopted. In particular for technical-environmental ones, the number of survey were identified as a critical element having a strong impact on the project (Costs, Time).

In the following, the probability classes adopted in the study are described and the related assigned score is reported.

Table 2. Probability classes for events versus geotechnical survey.

Probability Class	Description	Score
Very High	less than 40% of the necessary surveys are carried out	5
High	up to 40% of the necessary surveys are carried out	4
Medium	up to 60% of the necessary surveys shall be carried out	3
Low	more than 90% of the necessary surveys are conducted	2
Very Low	Not used	1

The "necessary surveys" is the number of surveys as per Best Practice for the specific type of terrain and the level of design analyzed. The reference for an Detail Design level is as follows:

- very variable ground at the considered depth: at least 1 survey every 0.5 km;
- ground of usual variability: at least 1 survey every 1 km;
- ground with a tendency to be homogeneous: at least 1 survey every 1.5 km.

The number of surveys per km takes into account the incidence of critical hot spots and the use of complementary investigation techniques.

With regard to the design quality, in particular for strictly technical risk category, the probability classification refers to the completeness of and compliance with consolidated standards.

Table 3. Probability classes for events related to the quality of the design.

Probability Class	Description	Score
Very High	Project with some undeveloped parts	5
High	Project with some parts not fully developed	4
Medium	Project developed without explicit references to consolidated standards	3
Low	Project developed with reference to consolidated technical and quality standards	2
Very Low	Project compliant with consolidated technical and quality standards	1

For events not included in the two previous categories, for example organizational risk category, the classification shown in the following was used.

Table 4. Probability classes for other events.

Probability Class	Description	Score
Very High	The event will almost certainly happen, if repeatable occurs often or continuously over a long period.	5
High	The event is likely to happen, if repeatable occurs occasionally several times over the long term.	4
Medium	There is complete uncertainty about the possibility of occurrence, it may occur during the technical life.	3
Low	The event is unlikely to happen, it is rarely present during the technical life.	2
Very Low	It is reasonable to neglect the unwanted event.	1

3.3 CLASSIFICATION OF THE SEVERITY OF UNDESIRABLE EVENTS

On the basis of the expected impact of the risk, different criteria in Severity class definition have been adopted, so the categories of consequences taken into account have been differentiated in terms of Costs, Times, and Safety.

For the "Costs" Risk Category, the thresholds of the Severity Classes have been defined as a percentage of the investment value.

For the "Time" Risk Category, the definition of the Severity Classes was differentiated for the Constructive (C) and Operation and Maintenance (O&M) phases.

The classification of the severity of undesired events and the related assigned score is presented in the Table 5.

3.4 Risk classification and treatment

The Risk of an adverse event is the combination of the probability that the event will occur and the severity of the consequences. This combination expresses the degree of undesirability of the event or, in an equivalent manner, a measure of the willingness to act to reduce the possibility of occurrence of the event and/or the consequent damage: the risk assigns a priority of importance and treatment to the event.

Generally, the representation of the probability and severity combination, and therefore Risk Level, is made by means of a matrix. In this case study, the matrix shown in Table 6 was used.

Table 5.   Severity Classes.

Severity Class	Cost	Time	Safety	Score
Catastrophic	> 15%	**C:** Irrecoverable delay of unsustainable magnitude **O&M:** the whole or part of the line is unavailable for more than 5 weeks	Affecting a large number of people and resulting in multiple fatalities	4
Critical	10 -:- 15%	**C:** irrecoverable but moderately long delay. **O&M:** the whole or part of the line is unavailable from 1 to 5 weeks.	Affecting a very small number of people and resulting in at lest one fatality	3
Marginal	1 -:- 10%	**C:** Larger recoverable delay **O&M:** all or part of the line is unavailable from 1 day to 1 week	No possibility of fatality, severe or minor injuries only	2
Insignificant	< 1%	**C:** Minor recoverable delay **O&M:** all or part of the line is unavailable for less than 1 day.	Possible minor injury	1

Table 6.   Risk Matrix.

	Severity	1	2	3	4
	Probability	Insignifican	Marginal	Critical	Catastrophic
1	Very Low	1	2	3	4
2	Low	2	4	6	8
3	Medium	3	6	9	12
4	High	4	8	12	16
5	Very High	5	10	15	20

The scores in the grid express the Risk Level and are obtained as a product of those associated with the Probability and Severity classes. The colors provides a clear indication of priority to be attributed to risk in treatment, from green -low priority- to red -very high priority-:

- da 1 a 3: Low priority;
- da 4 a 6: Medium priority;
- da 8 a 9: High priority;
- da 10 a 20: Very High priority.

On the basis of the Customer requirements and degrees of freedom in the definition of mitigating measures evaluated in the context analysis, potential risk reduction actions have been suggested and the residual risks have been assessed.

3.5   RESULTS OF THE ANALYSIS

The analysis showed that 33% of adverse events have at least one risk with a very high priority, in the cost, time or safety categories, based on the current project conditions. If the risks with very high priority are added to the ones with high priority, the percentage rises to 57%, which is reduced to 15% with a correct implementation of the proposed mitigation measures.

Approximately 39% of adverse events are due to design weaknesses such as poor product quality or completely undeveloped parts.

The following figures summarize the results of the analysis in terms of expected impacts on the project (Costs, Time, Safety), highlighting the effect of the mitigation measures.

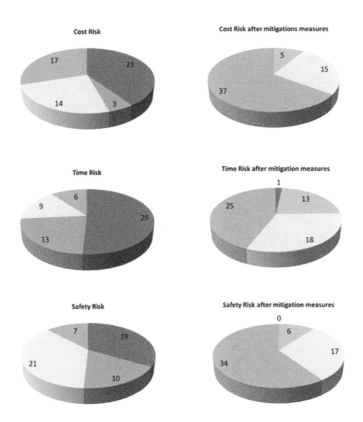

Figure 4.   Results of analysis.

All elements useful to risk management - such as undesired events, risk levels, mitigating measures - are reported in the Risk Register, as shown in the excerpt below reported (Table 7):

Table 7. Excerpt from Risk Register.

Risk ID	Event	Probability Description	Severity Costs	Severity Time	Severity Safety	Initial Risk Level Costs	Initial Risk Level Time	Initial Risk Level Safety	Mitigation measure	Residual Risk Level Costs	Residual Risk Level Time	Residual Risk Level Safety
C-T.E.1.1	Interception of karstic cavities in the excavation phase (traditional excavation)	Deep surveys carried out for galleries less than 40% of best practice. Very high	It is necessary to fill / close the cavity	The intervention exceeds 5 weeks	There are no particular critical points (safety procedures respected)	10	20	5	Perform deep surveys near the tunnels: at least 1 survey every 0.5 km- very variable terrain; at least 1 survey every 1 km - terrain of usual variability; at least 1 survey every 1.5 km - generally homogeneous ground	4	8	2
C-T.E.1.1.a	Interception of karst cavities in the excavation phase (mechanized excavation)	Deep surveys carried out for galleries less than 40% of best practice. Very high	It is necessary to free the front from the cutter (25% of the cost of the tunnels)	The intervention exceeds 5 weeks	There are no particular critical points (safety procedures respected)	20	20	5	Perform deep surveys near the tunnels: at least 1 survey every 0.5 km- very variable terrain; at least 1 survey every 1 km - terrain of usual variability; at least 1 survey every 1.5 km - generally homogeneous ground	8	8	2
C-T.E.1.2	Intercepting layers / underground water courses during excavation (traditional excavation)	Deep surveys carried out for galleries less than 40% of best practice. Very high	The excavation face is reactivated with a suitable design solution	The intervention exceeds 5 weeks	A major water coming can cause significant safety problems (even with safety procedures respected)	10	20	15	Perform deep surveys near the tunnels: at least 1 survey every 0.5 km- very variable terrain; at least 1 survey every 1 km - terrain of usual variability; at least 1 survey every 1.5 km -generally homogeneous ground	4	8	6
C-T.E.1.2.a	Intercepting layers/underground water courses during excavation (mechanized excavation)	Deep surveys carried out for galleries less than 40% of best practice. Very high	It is necessary to free the front from the cutter, (25% of the cost of the tunnels)	The intervention exceeds 5 weeks	Even in the case of compliance with the yard safety procedures, a major water coming can cause significant safety problems	20	20	15	Perform deep surveys near the tunnels: at least 1 survey every 0.5 km- very variable terrain; at least 1 survey every 1 km - terrain of usual variability; at least 1 survey every 1.5 km -generally homogeneous ground	8	8	6

4 CONCLUSIONS

The generalization of the approach adopted in the last decade in the safety assessment of Italian railway tunnels and RAMS studies for railway infrastructure in general enabled to provide a reliable support in project risk management process, both for technical and managerial aspects.

In fact, the variety of aspects to be treated have improved technical skills whereas the interaction with the different actors involved in the process (stakeholder/engineers/managers) have developed coordination skills.

As further improvement, the risk management performance would take advantage from the use, for all competence areas involved, of the same structured approach and related techniques/tools developed thanks to risk analyses applications guaranteeing common language and uniformity in the two-way information flows, by facilitating the coordination activities of all key-players.

REFERENCES

International Standard ISO 31000, Risk management - Principles and guidelines.
Foresta M., Caiazzo P., Ponte M. 2016. The evolution of risk analysis: from decision tool in railway line safety design to support for an integrated risk management approach, *11th World Congress on Railway Research (WCRR), Proc. intern. symp, Milan, 29 May - 2 June 2016.*

Tunnels and Underground Cities: Engineering and Innovation meet Archaeology,
Architecture and Art, Volume 8: Public Communication and Awareness/Risk Management,
Contracts and Financial Aspects – Peila, Viggiani & Celestino (Eds)
© 2020 Taylor & Francis Group, London, ISBN 978-0-367-46873-6

The Isarco River underpass at Brenner Base Tunnel: Main aspects of work supervision in a challenging design solution

B. Carta & A. Pigorini
Italferr S.p.a., Rome, Italy

A. Bertotti & R. Palla
hbpm Ingegneri S.r.l., Bressanone, Italy

G. Albani
Pini Swiss Engineers S.r.l., Lomazzo, Italy

ABSTRACT: The Isarco river underpass is the southern construction lot of Brenner Base Tunnel that will connect Italy and Austria with the longest railway tunnel in the world.

A solution with 4 tunnels underneath the Isarco river has been chosen to cross the fluvio-glacial loose deposit allowing the construction of 2 main tunnels and 2 interconnection tunnels. The design foresees heavy ground improvement works to increase both resistance and deformability parameters together with a strong reduction of permeability of the mainly gravel and sand with big cobbles soil; this has to be achieved coping with a high water table flow rate.

Works foresee vertical and horizontal jet-grouting, cement and chemical grouting, horizontal ground freezing. In order to achieve the design requirements the Site Supervision has been involved in controlling and analyzing specific tests and in situ trails field together with specific control procedures.

1 GENERAL DESCRIPTION OF THE WORK

1.1 *Main works*

The Isarco river underpass construction lot connects the Brenner Base Tunnel to the historic line North of the Fortezza station. This is a particularly complex lot, completely underground, which also includes the crossing underneath the Isarco river, the SS12 state road and the A22 motorway, with shallow overburden and characterised by the presence of loose water bearing deposits. For this reason too, it is essential to improve soil properties before excavation by using techniques such as freezing, jet-grouting and injections. In this lot, worth approximately 303 million euros and with a deadline for completion in December 2022, it is planned that the contracting joint venture (Salini Impregilo – Strabag – Consorzio Integra and Collini), will build two main tunnel for a total of 4.3 km and the two interconnection tunnels with the Fortezza station for a total of 1.5 km (Figure 1). The project includes several sections of excavation in both rock and fluvioglacial deposit .

Among the latter, we differentiate between the sections consolidated from ground level and those from tunnel face. From a technical point of view, the most challenging section consists of the Isarco river underpass, a short stretch of 60 m characterised by a cover of a few meters under the riverbed, accessible through the construction of 4 shafts near the river from which to start the ground treatment with cement/chemical injections and ground freezing through a system of freezing pipes with liquid nitrogen and brine.

In addition to the excavation, the tunnel waterproofing system is particularly innovative and complex. In fact, the sections under groundwater have a "full-round" subdivision system with

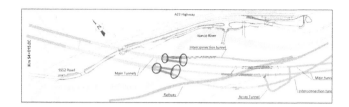

Figure 1. Key plan of the Isarco river underpass lot: TR (blue lines), GB (ciano lines), GN (orange lines), UT (red lines), GA (green lines), PO (blue circle), open cut (light pink lines).

pressure-tight characteristics, including a two water-proofing PVC layers and water-stop joints connected to the lining construction joints, that allow injections through a system of pipettes. Over time, this system will allow the repair of the hydraulic seal in case of water leaks during the service life of the works (200 years according to the max. value of Eurocodes standards).

Consequently the materials have been designed to meet high quality controls as to ensure the durability of the structures.

1.2 *Geological and geotechnical information*

The project area is an area of approximately 1 square km in the valley of the upper Valle Isarco, North of Fortezza. The valley has a rather narrow configuration and is bordered on both sides by steep mountain slopes. The valley shows an anthropic environment with important infrastructures on both sides of the river.

The A22 Brenner motorway and the SS12 state road are on the left riverside and the Brenner railway line, built on a few meters high embankment in the alluvial fan, is on the right riverside (Figure 1).

The area is characterised by the presence of intrusive vulcanic rock formation of granodiorite, known in literature as the Bressanone granite. This formation is covered by fluvioglacial landform with heterogeneous deposit of sand and gravel and sediments of fine-silty sands, with the presence of large cobbles and boulders.

This very thick deposit, with a shallow groundwater level, characterizes the Isarco Valley and affects the construction of the Isarco Underpass. It consists of fluvial deposits of the Isarco river and debris flow deposits fed by side ravines and slope coarse materials. Table 1 shows the different types of loose soil and rock formations present in the project area with their geotechnical parameters. The local hydrogeology is affected by the basin of Isarco river and by the Rio Vallaga and Rio Bianco creeks. The main aquifer, with a groundwater level between 4 and 10 m below the surface, flows in the alluvial deposits of the valley and the side alluvial fans. The aquifer flows in the same direction of the Isarco river although unconnected. The flow speed varied between 2.4 and 16 m/day, with the highest values recorded in the area of the underpass. (Lombardi & Perello 2017).

1.3 *Challenges and main requirements of the project*

The project involves the implementation of methodologies and technologies able to solve the critical issues arising from the complex context in which the works are located, while minimising the environmental impact in the areas affected by the works, reducing interference with the existing infrastructure, and ensuring durability of the works, consistent with their service life.

Table 1. Main geotechnical parameters.

	γ [kN/m³]	Φ' [°]	c' [kN/m²]	Es [MN/m²]	k [m/s]
Debris Flow Deposits	19 - 22	25 - 37	0 - 35	40 - 120	1×10^{-3} - 1×10^{-6}
Slope detritus	19 - 21	35 - 45	0	40 - 120	5×10^{-2} - 1×10^{-6}
Alluvium	19 - 22	33 - 39	0 - 5	25 - 80	1×10^{-3} - 1×10^{-5}
Bressanone Granite	26	30 - 62	300 - 3300	500 - 7000	1×10^{-4} - 1×10^{-8}

Namely, the most challenging aspects of the Contract concern:

- the Isarco river underpass through tunnels under shallow overburden, using the technique of ground freezing;
- the use of technologies, such as jet-grouting and cement/chemical injections, to reduce permeability and consolidate the ground, in combination with traditional excavation methods;
- the construction of 4 large shafts up to 25 m deep on the sides of the Isarco river;
- conventional full-section excavation, also in the fluvioglacial deposit sections under the groundwater level;
- the construction of a waterproofing system to protect the final lining (full-round system - "pressure-tight" tunnel sections);
- the use of concrete for the final lining that meets the requirements of durability, fire resistance, resistance to freeze/thaw cycles for the exceptional service life of the works, set at its maximum as provided for by the Eurocodes.

2 CONTROLPLAN FOR DIFFERENT TUNNEL STRETCHES AND COMPONENTS

In order to guarantee the challenging performance required in the project, the Works Management (carried out by the Temporary Grouping of Companies: Italferr s.p.a - Hbpm s.r.l. - Pini Swiss s.r.l.), verify, supervise and monitor, by means of specific Quality Control Plans, the control stages necessary in infrastructure projects that, in this specific case, are increased by the complexities associated with ground improvement that must guarantee stability of the excavations and the reduction in permeability.

The tunnel sections can be divided into the following categories (Figure 1):

- Tunnels in rock (2856 m) (TR);
- Tunnels in fluvioglacial deposit - ground improvement from ground level (997 m) (GB);
- Tunnels in fluvioglacial deposit - ground improvement from tunnel face (671 m) (GN);
- Isarco underpass tunnels (243 m) (UT);
- Cut and cover tunnels (972 m) (GA) and 4 shafts (PO).

2.1 *Tunnels in rock (TR)*

For the construction of tunnels in rock the excavation is carried out in full section by the use of drill and blast technique (to date approximately 85% of the total length has been excavated).

The types of rock supports, defined in accordance with the characteristics of the rockmass, consist of the use of rock bolts, reinforced shotcrete and steel arch, addition steel pipes umbrella for tunnels underpass and faults zone. A standard control plan is provided for these support operations.

2.2 *Tunnels in fluvioglacial deposit consolidated from surface (GB)*

Where the overburden vary between 10 m and 17-18 m, allowing for the ground improvement works to be carried out from ground level, the tunnels are excavated (in full section) once the ground improvement around the cavity has been fully formed . The soil improvement consist of jet-grouting columnar treatments with double-fluid technology, with sub-vertical geometries and with a diameter of 2 m. The project defines a quincunx mesh of 1.45 m x 1.65 m that provides for a theoretical minimum overlap of 10 cm (Figure 2). These tunnels are also divided into compartments by means of transversal partitions that consist of columns of consolidated ground. The design requirements for the shell ground improvement are as follows:

- Uniaxial compression strenght: 5 MPa (on core samples h/d= 2)
- Maximum permeability: 1×10^{-7} m/s (from in-situ Lefranc tests). For hydraulic sealing purposes, flow rates of less than 5 l/s per 1000 m^2 of draining surface are required

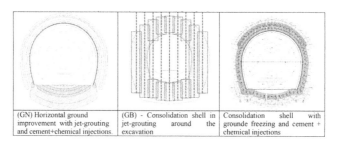

(GN) Horizontal ground improvement with jet-grouting and cement+chemical injections.	(GB) - Consolidation shell in jet-grouting around the excavation	Consolidation shell with grounde freezing and cement + chemical injections

Figure 2.   Various uses of ground improvement techniques.

- Lack of treatment maximum surface allowable: 0.5 m^2.

The control plan is designed to ensure that the performance mentioned above is achieved by means of the following verifications:

- verifications during works (positioning of drilling rigs, injection parameters, deviation of the drilling from the vertical plane);
- verification of the mechanical resistance (boreholes and laboratory tests on samples of improved soil);
- verification of the continuity of the treatment (definition of a 3D model of the jet-grouting columns carried out and identification of any lack of treatments);
- verification of the large scale hydraulic seal by pumping tests of contained volumes.

The measurement of the deviation of the drills from the vertical plane allows for the three-dimensional reconstruction of the jet-grouting columns, which can be represented with horizontal sections at the specific level (tunnel top arch, centre plane and invert arch). From these sections, considering the diameter of the nominal jet column at 2 m, it is possible to identify the areas in which the columns are not interpenetrated (nominal voids). In case the nominal voids have a surface bigger than the admissible value (0.5 m^2), additional columns must be carried out in order to restore the continuity of the treatment (Figure 3).

If the pumping test carried out after the implementation of additional columns does not show that the hydraulic seal requirements of the treatment have been met, further investigations (i.e. geophysical investigations) must be carried out in order to define further corrective actions.

2.3   Tunnels in fluvioglacial deposit consolidated from tunnel (GN)

In the transition areas from rock to soil, in fluvioglacial deposit areas and in particular in the underpass of the existing road infrastructures (SS12 state road and A22 motorway), excavation is made by using the full-section excavation method with ground treatment carried out from the tunnel face ahead the tunnel core, using cement/chemical injections and jet-grouting columns (to date approximately 36% of the total lenght has been excavated). The purpose of ground improvement is to create a shell around the cavity, with both static and low permeability functions, and a vertical shell at the front, so that the excavations can be carried out under

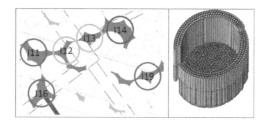

Figure 3.   3D reconstruction: nominal voids with area more than 0.5 m^2 (in red) additional treatments performed (coloured circles) and 3D plat (right).

4410

hydrostatic condition, namely without a significant drainage of the aquifer. The ground improvement works consist of a jet-grouting columns umbrella, made with single-fluid technology. In order to reduce the permeability of the fluvioglacial deposit and thus better guarantee the success of the jet-grouting treatment, an outline of cement and chemical injections is carried out in advance. At the same time, in order to reduce the shadow effects during jet-grouting injection due to the presence of large cobbles and boulders, an umbrella of steel pipes has been installed in the top arch. The works are completed by front and invert arch ground improvement, carried out with cement and chemical injections. In this area, a double ring of injections, both cement and chemical, is implemented by means of valved fiberglass reinforcement (Figure 2).

Since performed under the water table, all the drillings are carried out using the preventer in order to prevent the uncontrolled entry of flows of water carrying fine materials.

The design requirements for the ground improvement are as follows: $\gamma(kN/m^3) = 21$, fcd (MPa) =2.5, $\varphi'(°) = 35$, c'(kPa) = 650, E(MPa)= 1500.

The water inflow limit for each consolidated tunnel "chamber" (7 m in length) is about 3 l/s.

The jet-grouting operating parameters and the diameter of the column have been defined through test columns in the first excavation chamber.

The control plan implemented to ensure that the requirements are met provides for:

- the use of trial columns to verify the operating set parameters (in the trial fields);
- permeability tests in the drill hole (in the trial field);
- the verification of the drilling direction;
- the verification of the injection set parameters (pressures and volume injected).

2.4   Isarco underpass tunnels (UT)

The excavation of the tunnels in the underpass of the Isarco river is carried out, starting from previously constructed shafts (described below), with the use of an eco-compatible technology that consists of ground freezed shell around the entire perimeter of the excavation, previously treated by means of cement injections to reduce soil permeability. The tunnel face stability is increased by cement grouting pipes and fibreglass reinforcement (Palomba et al. 2017). The first ground improvement activities are scheduled to begin in March 2019, therefore the control results on these works are not the subject of this report.

2.5   Cut and cover tunnels (GA) and shafts (PO)

The excavation of cut and cover tunnels has been designed based on jet-grouting columnar treatment carried out on the walls and bottom of the excavation in order to create a thick wall of improved soil to withstand the high external hydraulic pressures and to reduce the soil permeability. The maximum depth of the excavation is of approximately 17 m and the maximum level of the groundwater level is 3 m below ground level. The average thickness of the lateral treatment varies between 3.5 and 5 m, while the thickness of the jet-grouting sealing slab varies between 7 and 10 m. Up to date the ground improvement works for two cut and cover tunnel sections have been completed.

The excavation of the tunnels for the Isarco river underpass is carried out from 4 shafts (2 north and 2 south) with an elliptical shape, an area variable from 1407 to 687 m² and a depth of 25 m. The maximum level of the groundwater is approximately 7 m below ground level. A jet-grouting shell made up of interpenetrated columns Ø 2 m (using the double-fluid technique) allows the excavation of the shaft using the top down technique. Walls made of improved soil have a thickness of 3 m (with an additional 1 m in thickness in the deepest sections) and a sealing slab with a thickness of 8 m (Figure 4). To date, all jet-grouting treatments for the 4 shafts have been completed. The design requirements and the control plan implemented for these tunnels and shafts are the same as those for the tunnels built with ground improvement works from ground level. In addition to these, thermal leakage detection tests were carried out on the two northern shafts to verify the quality of treatment in order to

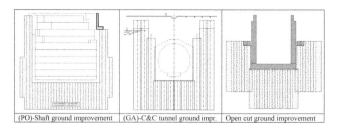

| (PO)-Shaft ground improvement | (GA)-C&C tunnel ground impr. | Open cut ground improvement |

Figure 4.   Jet-grouting works for shaft, cut and cover tunnels and open cuts.

guarantee the requirements for the hydraulic seal (chapter 3). In this specific case, the evaluation of the results obtained allowed for the definition of additional interventions.

For all the types of tunnels described, waterproofing package and final lining is foreseen.

2.6   *Waterproofing*

The project includes drained and pressure-tight sections. In the first case, the waterproofing package includes a layer of geotextile to protect the waterproofing membrane in adherence to the first shotcrete lining, a subsequent layer of draining geocomposite, and finally the 2 mm thick PVC membrane. In the second case, on top of the geotextile layer, a double PVC sheet is laid, consisting of a 3 mm main sheet and a 2 mm protective sheet, inside which the injection system for restoring the hydraulic properties is housed. Expanding waterstops and PVC-P sealing tapes at the construction joints complete the waterproofing structure. In order to guarantee the final requirements of the waterproof structure, verifications are carried out on the properties of the materials supplied, on the conformity of the surfaces to be waterproofed and on the quality of the installation.

The control plan implemented includes the verification of the density of the geotextile layer (at least 900 g/m^2), the resistance to perforation (at least 7 kN) and permeability under the operating conditions (at least 7 l/m*h). For the PVC membrane, the control plan includes the verification of the tensile strength with a minimum elongation of 330% in the two main directions, the verification of homogeneous behaviour in case of bending at cold temperatures (-20°C), the verification of maximum reduction of 20% of the tensile strength under ageing in water at 80°C, and finally the verification of the fire reaction class.

Works Management authorises the Contractor to lay the waterproof package after the development gradients of the convergences have reached values lower than 0.5 mm/day and after the laser scanner verification has returned an excavation profile free of any subprofiles. It is also checked that surfaces are free of protrusions and/or cavities with radii of curvature greater than 0.2 m and with a thunderbolt/rope ratio greater than 1/10, in order to avoid the onset of tensions in the membrane during concrete casting.

Upon completion of the waterproofing process, Works Management tests all the welds between the heat-sealed PVC sheets by means of a test with compressed air inflated in the double weld; a pressure of 2 bar has to be maintained for 10 min, verifying that the final loss of pressure does not exceed 10% of the initial value. Typically, the joints undergo different tests such as the one with the vacuum bell or the one with electrical voltage.

2.7   *Concrete for the final lining*

The final concrete lining has been designed for a service life of 200 years, therefore specific quality control plans have been necessary to guarantee the verification of durability. In fact, different classes of exposure (XA1, XC3, XF3 according to UNI EN 206) and therefore of concrete resistance are provided for the project in relation to the different conditions of environmental aggression, such as resistance to freeze/thaw cycles due to thin soil covers, resistance to sulphates due to the aggressiveness of the aquifer, carbonation resistance and fire resistance. Thus 8 cm of concrete cover are foreseen for reinforced segments and addition of polypropylene fibres (2 kg/m^3) and a wire mesh adequately spaced from the reinforcement are foreseen to limit the effects of concrete spalling in fire scenarious. Through a holistic approach that prefers the performance method to the compositional method, the production of concrete on site is subject to different levels of control ranging from self-control in the plant in accordance with the UNI EN 206 standard, to the on-site control of Works Management, at the various stages of casting and requires the systematic collection of samples to verify compliance with the requirements in the specifications. Namely in addition to the samples provided for regulations ("concrete resistance acceptance verifications"), Works Management verifies the properties of the fresh concrete: consistency class, air content (for exposure classex XF3), fibre content and the water/cement ratio. Samples are also taken for the frost/thaw resistance tests, water pressure penetration tests and hydraulic withdrawal tests.

Finally on the casted concrete non-destructive tests are carried out with ultrasonic techniques to double check the thicknesses of the lining and the concrete cover for reinforced segments.

3 TRAIL FIELDS RESULTS AND OUTCOME

Trial fields have been carried out in order to establish the technologies and the operating parameters to be used for ground improvement works and to collect evidence of achieving the design requirements. The first trial field was completed in 2015 during the design phase. All the ground improvement techniques provided for in the project were tested: jet-grouting, cement injections, chemical injections and ground freezing; this allowing for the final project to be calibrated.

Focusing on jet-grouting works, "single treatments" were tested first, where 10 single jet-grouting columns were used, varying the sets of operating parameters in relation to the 4 nominal diameters under assessment (0.65 – 1.5 – 1.8 – 2.0 m). The first test campaign was completed with the execution of 3 interlocking columns, and 2 additional single columns, aimed at verifying the diameter created under the most difficult conditions, applying different techniques for measuring the diameter (coloured rods and thermal tests).

The 2015 trial field confirmed the success of jet-grouting with a nominal diameter of 2 m up to a depth of approximately 21 m (Palla 2015).The mesh of jet-grouting columns has been also defined to increase the hydraulic seal efficiency. At depths greater than 21 m and in correspondence with a layer of soil with a finer grain size (silty sand), it was difficult to reach the nominal diameter. It was therefore considered essential to carry out, also in accordance with the contractual provisions, additional trial field to verify the whole treatment (and not only the single columns), thus to verify the group behaviour (massive) effect. A first "field test" was carried out during construction phase on a main tunnel sector, about 15 m long, to be excavated in conventional method after jet-grouting works carried out from ground level. The purpose of the field test was to check and verify the massive effect of jet-grouting treatments. It should be noted that, as detailed in the previous paragraphs, the jet-grouting technology is also used to create very thick walls of improved soil (thicknesses between 3.0 m and 4.5 m) and it is therefore logical to expect a "group effect" in which the treatment of a section of the ground can occur from more than one injection point, with a more effective sealing of the voids.

Two types of jet-grouting treatment have been tested and for this reason the sector has been divided so as to create two distinct chambers, in one the columns have been created with the standard double-fluid method, and in the other with the pre-washing system which consists of a preventive treatment of the ground by means of high-pressure water injection, so as to facilitate the desegregation of the soil and the injection of a cement mixture in the following step (Figure 5).

For all the columns, the drilling and injection parameters were continuously recorded and the deviation from the vertical plane was measured. The following tests were then carried out in both sectors to verify the success of the treatments:

- gradual pumping test by means of a well built in each chamber;
- boreholes for cores stratigraphy, collection of samples and Lefranc permeability tests;
- 3D seismic tomography and cross-hole Vp/Vs tests.

The purpose of the pumping tests is to verify the hydraulic seal of the compartment. Both tests gave flow rate values lower than the reference value of 5 l/s per 1000 m^2 of draining

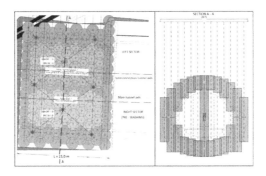

Figure 5. Plan and section of the implementation of the first "test field".

surface. It is possible, however, that they did not involve the complete emptying of the chambers, because a rapid rise of the piezometric at the end of the pumping process, probably due to a very tight drawdown hydraulic head at the end of pumping, has been observed.

The Lefranc permeability test values ($1{,}63 \times 10^{-6}$ m/s average value) show a remarkable reduction in permeability compared to the natural soil, but did not reach the reference values (1×10^{-7} m/s).

The seismic investigations help verify the continuity of the jet-grouting treatment by examining the speed of seismic waves (Vp) detected by cross-hole and down-hole tests, with a three-dimensional tomographic processing. Namely, 3D seismic tomography (Figure 6) shows the effects of jet-grouting treatment in terms of improving the seismic speed by 2 to 3 times compared to the natural soil (Vp value for fluvioglacial consolidated deposit > 1600 m/s).

All the compression tests carried out on samples extracted from the cores have returned values of strength and deformability significantly higher, sometimes by more than double of the minimum reference (Rc > 5 MPa e E > 1600 MPa). However, in some cases it was not possible to take the sample at the requested level because the material did not have characteristics for the collection of the sample.

In conclusion, the jet-grouting treatment, overall, has reached an adequate level of ground improvement, but the field test has not provided enough evidence that the entire expected performance has been achieved, for this reason it was decided to perform a second "test field". For the second test field a second sector of the main tunnel was chosen, which was investigated by means of a thermal leakage detection test, as well as with the usual pumping tests, boreholes and Lefranc permeability test. The thermal leakage detection test provides information on the continuity and hydraulic seal of the treatment. It consists of constantly measuring the temperature of the soil/groundwater inside the tunnel test field chamber during the pumping test. The piezometric gradient, produces a cooler flow of water from the outside to the inside and causes a lowering of temperatures, which is more pronounced in areas where the hydraulic seal is not guaranteed. The Figure 7 illustrates the outcome of the test performed.

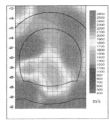

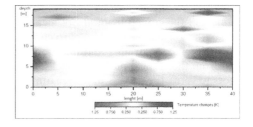

Figure 6. Trend of the Vp values around the tunnel.

Figure 7. Vertical section along the chamber perimeter, showing temperature variation during the thermal leakage detection test.

The coloured areas represent a temperature gradient: namely the blue areas identify a lowering of the temperature probably due to a water inlet from outside the chamber.

In conclusion, the 3D measurements of the deviations revealed a rather small presence of nominal voids. The thermal leakage detection test did not detect any significant infiltration. The gradual pumping test showed a flow rate value well below the flow rate limit. The core drilling, and the in situ and laboratory permeability tests showed that the jet-grouting treatment was massive and with good columns interpenetration. From the results of the "test fields" it is therefore possible to express an overall positive evaluation of the jet-grouting treatment carried out around the investigated sectors that have certainly benefited from the group effect induced by the massive treatment, thus confirming the possibility of obtaining the design requirements with 2 m diameter columns with a mesh of 1.45*1.65 m, performed with double-fluid technology and the tested operating parameters. The outcome of the field test also made it possible to provide a criterion for the ongoing verification of the success of the jet-grouting treatments carried out from ground level around the tunnel chambers and for the identification of corrective actions.

4 DATA ANALYSIS

In this paragraph we intend to represent part of the results of the verifications carried out by Works Supervision Management and by the Contractor on more than 9000 columns of consolidated ground relating to the jet-grouting interventions from ground level for the construction of natural and cut and cover tunnels and shafts, with particular reference to the performance required in the project. The design requirements, as mentioned above, concern the geometry of the treatment, namely the achievement of the massive effect, the treated ground impervious requirement or namely the lowering of the permeability value and the mechanics of the treatment, namely the mechanical properties of the soil improvement.

The first requirement was evaluated by analysing data on the verticality of the drillings and it has returned an average value of the deviations (measured at the bottom of the column in % with respect to its length), between 1.1 and 1.7%, lower than the maximum value required in the project specification (2%). It is interesting to note that the average deviation is approximately 1.3%, generating at the foot of the longer columns, deviations of 30-40 cm. In addition, the 3D graphic representation of the columns as actually implemented has allowed to identify the presence of nominal voids due to the lack of columns interpenetration and the consequent need to drill additional columns. Figure 8 shows the correlations between the average value of deviation of the columns and the number of additional columns implemented.

As it can be seen, the number of additional columns implemented for the various works varies with the highest values recorded for the shafts (PO) columns characterised by greater average treatment lengths (average length > 35 m) than the cut and cover tunnels columns (GA) and the tunnels (GB) columns (average length up to 20 m and 30 m). Thus the number of additional columns is higher for deeper columns: in fact, the values of deviations from the vertical, although similar for columns of different lengths, have created larger nominal voids.

To guarantee the strict requirement thus the lowering of the treated soil permeability, 10 pumping tests have been carried out on different works up to date (4 on the shafts, 4 on tunnel compartments and 2 on cut and cover tunnels). The results have constantly returned values of maximum drained water flow always below the limit flow value set at 5 l/s per 1000 m² of draining surface.

However, these were not always reliable because in some cases the pumping test failed to cover the entire chamber volume, thus returning partial data. However, together with the other tests carried out both on the overall volume of treatment (thermal leakage detection and geophysical tests) and on the individual columns (boreholes), the overall results confirm the achievement of a lower soil permeability. To date, 3 of the 10 sectors tested (2 sections of tunnel built with ground improvement works from ground level and 1 shaft) have been completely excavated. It has been found in 2 cases (tunnel chambers) that the ground improvement works actually met in full the design requirement (flow rates for each compartment much below the reference value of 4.7 l/s).

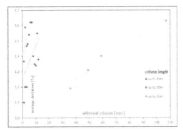

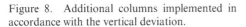

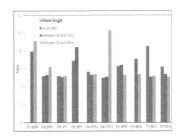

Figure 8. Additional columns implemented in accordance with the vertical deviation.

Figure 9. Minimal values for uniaxial compression strength.

On the other hand, in the first shaft, recently completely excavated, water inflows both concentrated and diffused from the consolidated walls have been observed. The drained water flow rates have reached values above those expected (100 l/s instead of 30 l/s), and this made it necessary to carry out injections with polyurethane resins. While not completely eliminating the concentrated flows, the resins injectios have nevertheless allowed the adjustment and completion of the excavation and the concrete lining of the shaft. From a mechanical point of view, at the request of Works Supervision Management more than 280 jet-grouting samples were taken for the determination of the jet-grouting compression strength and the elastic module. No critical results have been found, highlighting a certain homogeneity between the different tunnel sectors. The compression uniaxial tests showed that the minimum design value (5 MPa) was constantly reached, with a tendency to increase with the sampling depth (Figure 9). However, the average values of approximately 8-12 MPa show a wide variability of results, probably due to the heterogeneity of the core material, due to the presence of micro-lesions due to the disturbance of the coring action and of cobbles of different sizes. A similar trend is recorded with jet-grouting density data.

5 CONCLUSIONS

The paper wanted to highlight how the knowledge of the control process of works and materials is important to acquire awareness of the main aspects to be verified for the achievement of the project objectives. The controls and inspections carried out by Works Supervision Management have not only concerned the standard aspects of the construction works of a tunnel, but have also involved innovative methods, techniques and technologies adopted in the challenging project of under crossing of the Isarco River. Namely it has been possible to study in depth some test and control methods that have allowed Works Supervision Management to monitor more effectively the achievement of the design requirements for the huge ground improvement works carried out before tunnel escavation, to ensure both mechanical strength and hydraulic seal of the soil around the tunnels.

REFERENCES

Lombardi, A. & Perello, P. 2017. Il sottoattraversamento del fiume Isarco: la sfida idrogeologica nei terreni fluvio-glaciali nella valle Isarco. Proc. symp. *"Le grandi infrastrutture ferroviarie alpine in costruzione: le scelte progettuali e costruttive per la realizzazione di tunnel lunghi e profondi"*. Verona: SIG.
Palomba, A. & Cassani, G. & Gatti, M. 2017. Il sottoattraversamento del fiume Isarco in galleria: utilizzo della tecnica mista di iniezione-congelamento dei terreni. Proc. symp. *"Le grandi infrastrutture ferroviarie alpine in costruzione: le scelte progettuali e costruttive per la realizzazione di tunnel lunghi e profondi"*. Verona: SIG.
Palla, R. 2015. Relazione finale assistenza specialistica campo prova Sottoattraversamento Isarco. Unpubl..

Tunnels and Underground Cities: Engineering and Innovation meet Archaeology,
Architecture and Art, Volume 8: Public Communication and Awareness/Risk Management,
Contracts and Financial Aspects – Peila, Viggiani & Celestino (Eds)
© 2020 Taylor & Francis Group, London, ISBN 978-0-367-46873-6

Hydrogeological Excavation Code, a value added methodology for water safeguarding

M. Coli
Department of Earth Science, Florence University, Italy

L. Martelli & P. Staffini
JASPERS Joint Assistance to Support Projects in European Regions, Roads Division, European Investment Bank, Luxembourg.

ABSTRACT: Water inflows during tunneling can put safety at risk and cause irreversible damages to the natural water resources. Both the construction phases and the long-term operation of a tunnel can be disrupted, resulting in losses of time and reliability, delay and cost overruns. Assessing the hydrogeological situation at a very early stage in the project cycle is crucial. The use of a methodology called Hydrogeological Excavation Code (HEC) has demonstrated to help transport infrastructure authorities manage the EU environmental safeguards and value for money, EU legislation compliance and hydrogeological risks in major underground projects. HEC is compatible with FIDIC works contracts. Recently, HEC has been used in several tunnel projects in Europe addressing challenges created by estimated water inflows ranging from few liters/second to real underground streams. The feedback from tunneling projects confirms that HEC can be an effective tool to control costs, environmental impact, disruptions and, in general, the viability of the project.

DISCLAIMER: The comments and opinions expressed in this report reflect the views of the authors and do not necessarily state or reflect the views of JASPERS and its partners (European Commission and EIB). In particular, the views expressed herein cannot be taken to reflect the official opinion of the European Union.

1 INTRODUCTION

Today, the increasing number of long and basis tunnel projects requires a high-level investigation at the early stages of the project development. This allows to obtain realistic geological, hydrogeological and behavior models, and to mitigate the risk of design weaknesses, delays, costs increase, adverse working conditions and short/long-term impact on the natural water resources. For example, experiences in the last century show many examples of heavy impacts on the water resources in the territory surrounding the tunnel project, demonstrating that an accurate forecast of water inflows, based on well-addressed investigation and design approach, appears to be a key factor for the success of the underground project, and well worth the time required.

Impacts on the water resources depend on the position of the tunnel with respect to the water head, the regional water flow and the water recharge. These, in turn, are closely linked to the hydraulic conductivity inside the rock-mass. Therefore, knowledge of the rock-mass hydraulic parameters is important already in the preliminary phase of a project, and it requires a detailed analysis of the hydrogeological setting of the area in order to evaluate the hydraulic flow.

The hydrogeological properties of a rock-mass essentially depend on its geostructural setting, because water flow occurs through the networks of discrete and ubiquitous discontinuities, whose features strongly affect the hydraulic properties. Many authors have dealt with the permeability and the water flow capability of a rock-mass (see in Coli & Pinzani 2014 for *Bib.*). Some of the proposed models are not easily applicable in normal practice, where the need of practical results calls for adopting simplified methods based on statistical analysis of the available geostructural data obtained in the field.

2 WATER FLOW INTO TUNNEL

Water flow mainly occurs along the hard rock-mass discontinuity networks, given the much lower permeability of the intact rock block, which is usually neglected.

In tunneling, water inflow can occur at the front head, behind the front in the working segment and behind the primary or secondary reinforcement as a long-term flow (Fig.1).

Water inflows can be categorized into three groups:

- Diffuse water inflows from the network of the ubiquitous discontinuities - impact on the natural water resources and on the project is typically low or neglectable.
- Concentrated water inflows along main specific geological features (faults, thrusts, fracture bands) - risks for safety and depleting of local water resources and possibly high impact on project cost/time.
- Water and debris inrushes from large open/decompressed fault/fracture-band filled by incoherent granular material and with a high water head - high risks for safety on site, impact on the environment due to permanent losses on the water natural resources, and impact on project costs due to remediation measures - where possible.

Water flows through the network of discontinuities, according to either the gravity in the vadose-not saturated zone, and to the regional gradient in the saturated zone. The discontinuities controlling the water flow are the sets number, opening, spacing, persistence, roughness, filling, block size and shape. All these features concur to the definition of the rock-mass permeability and connectivity, which are to be defined for each RMZ (Rock Mass Zone, ISRM 1981) crossed by a tunnel.

The main challenge during the preliminary investigations is defining the permeability for each RMZ, since the working scale (Fig.2) is a factor in determining the investigations needed to define the right permeability and connectivity, controlling the water flows through the rock-mass and hence defining the appropriate working scale and REV (Representative Elementary Volume). Therefore defining the correct working scale is key:

1. *Very near field*: water flows through a single discontinuity, whose hydraulic parameters can be defined by in situ and lab tests and field measures.
2. *Near field*: the water flows through a few defined discontinuities, whose properties and attitude can be surveyed in detail in the field.
3. *Far field*: the water flows through the ubiquitous discontinuity network of a single RMZ, which behaves like a porous medium; the definition of the hydraulic parameters requires

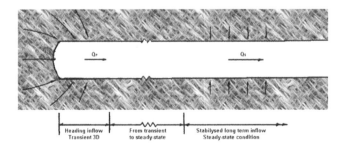

Figure 1. Different types of water inflow into a tunnel (from Coli & Pinzani 2014).

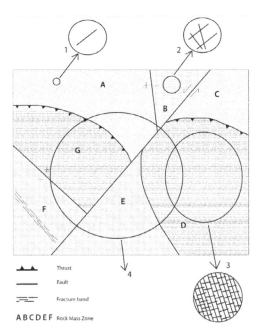

Figure 2. Sketch of the subdivision into RMZ, circles 1 to 4 resemble the four defined working scales for water flow (from Coli & Pinzani 2014).

the statistical analysis of the geo structural data for defining the main sets and features of the discontinuities.

4. *Very far field*: the water flows through the ubiquitous discontinuity network of RMZs and specific discontinuities related with the main tectonic lineaments such as faults, thrusts, and fracture bands; this requires basin scale hydrogeological analysis and balances plus the detailed analysis of the ubiquitous discontinuity.

Water flow along discontinuities, assuming a laminar flow in-between the discontinuity walls, follows the Darcy's law and is mainly controlled by the discontinuity aperture that influences the water flow rate at the cube (cubic low: Lomize 1951).

Rock-mass permeability can be defined by using the Snow (1969) or Kiraly (1969) equation. Barton et al. (1985) and Wei et al. (1992) equations allow to take into account the influence of discontinuity roughness simulating the water flow in a discontinuity as a laminar flow. Other authors incorporated the role of the in situ stress (Barton et al. 1985) and the increasing stress with depth (Wei & Hudson 1988) into the permeability equations.

All these approaches can be used to define the permeability value to be used into empirical abacus (Heuer 1995), analytic (Goodman et al. 1965, Perrochet 2005) or numerical analysis (FEM, DEM, - see a comparative critical analysis in Rutqvist et al. 2009a, 2009b) for evaluating the water flow by numerical analysis.

3 THE EC WATER FRAMEWORK DIRECTIVE

The Water Framework Directive 2000/60/EC (WFD) sets out environmental objectives that must be met for all water bodies and intends to protect and enhance their (both surface and underground water bodies):

"*Prior to receiving consent, new infrastructure project developments that have the potential to affect water bodies (or infrastructure projects, which lead to modifications of water bodies) should be assessed against the Directive's environmental objectives to determine whether they have the potential to prevent these objectives from being met*".

For underground works, it is necessary to evaluate the impacts and the compliance with WFD. The measures necessary to achieve the following main objectives should be implemented (the list is not exhaustive):

– prevent deterioration of the status of waters;
– protect, enhance and restore all bodies of surface waters and ground waters;
– ensure sufficient supply of water;
– protect the marine environment.

The European Court of Justice ruled in favor of a challenge against the WFD objective assessment process, and its findings, for a proposed dredging scheme on the River Weser in Germany, in July 2015 (Case C-461/13, Bund für Umwelt und Naturschutz Deutschland eV v Bundesrepublik Deutschland). The Court clarified how compliance with the Directive's key environmental objectives should be interpreted in the assessment of new developments and scheme proposals. It also established that decision makers have a duty to refuse development consent in cases where development would result in non-compliance with WFD standards, unless derogations were made out, in which case they must ensure consistency with the ruling.

The European Commission published detailed guidance on the application of WFD exemptions in 2009 via Technical Report 2009-027 "Common Implementation Strategy for the Water Framework Directive (2000/60/EC), Guidance Document No 20, and Guidance on Exemptions to the Environmental Objectives".

Article 4.7 of the WFD sets out the objectives that any underground project should be assessed against to ensure the protection of the status of all water bodies (in our case it will be related to surface and underground-referred water bodies). The two key objectives are:

– No deterioration of status (or potential) for surface and ground waters;
– Achievement of good status (or potential) by 2021 or 2027, for water bodies currently failing to achieve this status.

The application of the WFD requires a precautionary risk approach for the Waterbodies assessed, which involves consideration of the level of risk for deterioration or prevention of future good status, taking into account the uncertainty of potential impacts and the information available at design stage.

All projects co-financed by EU funds need to comply with EU legislation. This includes, inter alia, meeting the requirements of the WFD. Compliance with Article 4.7 is therefore a compulsory prerequisite for a project proposal to be eligible for co-financing by the EU.

4   WFD IMPACT ON UNDERGROUND PROJECTS

By their nature, tunnels attract water from aquifers, interfering with natural water systems and modifying underground flows, both during construction and operation. Hence, alterations to the level of groundwater are particularly relevant for achieving a good groundwater quantitative status (as defined in WFD Annex V 2.1.2).

In order to comply with WFD Article 4.7, JASPERS developed a four-step approach, with the aim to support improving the quality of investment supported by EU Funds:

1. *Understand the context*: Is there a potential causal mechanism for an effect on ecological or chemical status? If not, keep record for audit purposes but no further assessment required. If yes, proceed to step 2;
2. *Determine scope*: Consider whether effects are temporary or not significant for the water body. WFD assessment is required only for elements that could be affected;
3. *Investigations*: Carry on data collection, evaluation; consider mitigation measures. Is there a residual effect on WFD status? If yes, proceed to step 4;
4. Apply Article 4.7 tests (as set out in the Common Implementation Strategy – CIS Guidance No 20).

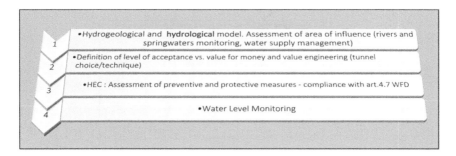

Figure 3.   Flow-chart for the application of the HEC procedure in assessing WFD art. 4.7 criteria.

More in detail, the CIS in its Draft Guidance Note No 36, identifies criteria to be met which *"are particular relevant for failure to achieve good groundwater quantitative status"* and *"in making operational the programmes of measures specified in the river basin management plan for groundwater"* (article 4. 1. B.i,ii,iii from the Directive) and including protected areas (article 4.1.c. from the Directive):

– Available groundwater resource is not exceeded by the long term annual average rate of abstraction;
– No significant diminution of surface water chemistry and/or ecology resulting from anthropogenic water level alteration or change in flow conditions that would lead to failure of relevant Article 4 objectives for any associated surface water bodies;
– No significant damage to groundwater dependent terrestrial ecosystems resulting from an anthropogenic water level alteration;
– No saline or other intrusions resulting from anthropogenically induced sustained changes in flow direction.

In compliance with art.4.7 WFD tests, the HEC methodology described in next section can be used at step 2 and 3 for assessing the criteria described above and whether effects are temporary or permanent, to define the Area of Influence and at step 3 to evaluate and consider the most appropriate mitigation measures (Fig.3).

The proposed approach has been used for projects in Poland (for flood protection), Slovakia and Slovenia (for underground mobility projects) and in Latvia (for port development including dredging), and will be further elaborated for wider application in the near future.

5   HEC METHODOLOGY

In the last decade, a procedure has been developed in Italy, the Hydrogeological Excavation Code (HEC), consisting of a specific set of documents addressing the mitigation of hydrogeological risks, to be developed by the Designer and endorsed by the Client and the National and Local public institutions.

The HEC, as defined by Coli & Tanzini (2016), organizes all the studies on the water inflows in a tunnel into a procedure incorporated in the design process. It analyses the hydrogeological risk in tunneling and the environmental hydrogeological balance, and defines drainage/waterproofing and stabilization systems in the short and long term.

The HEC procedure foresees a series of preliminary studies and consider various excavation modes and procedures (the list is not exhaustive):

– Basins scale hydrogeological balance;
– Permeability definition for each RMZ by in situ tests and geostructural approaches (Coli et al. 2018);
– Prediction of the expected water inflow, and type thereof, for homogeneous tunnel stretches;

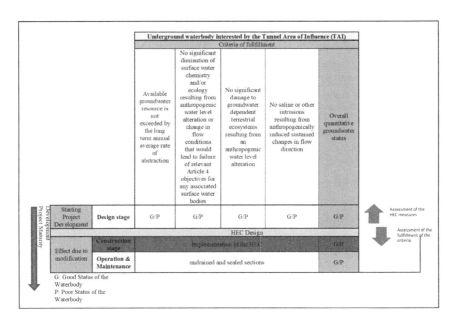

Figure 4. Procedural steps of the WFD/HEC in ensuring water flow control in tunnelling safeguarding environment and safety.

- Analysis of the impact of the tunnel drainage into the water natural resources;
- Definition of type and mode for the prevention of the water drainage operated by the tunnel (grouting, controlled drainage, partial or full sealing of the tunnel from water, waterproofing procedure), to be incorporated in the design as current works, in order to mitigate, reduce or avoid water drainage;
- Water level monitoring before, during and after tunneling for controlling the impacts on ecosystems (streams, rivers, creeks, spring waters, underground water);
- Mitigation measures for the residual impacts on the natural water resources and reuse of the water drained.

The HEC can help identify, predict and control water inflows, allowing for improved cost and time control (value for money safeguards), environmental impacts mitigation and/or protection as well as facilitation of public consent (environmental safeguards). Useful preventive measures should systematically be included in the project from the early stages to avoid issues such as cost overruns and interruptions, negative public consent, claims, and/or economic and environment irreversible damages.

The HEC has been developed for underground project with forecasted water inflows from a few liters to more than 1,000 liters per second under difficult circumstances. In practice, it allows the definition of measures to protect the natural water resources, to address tunnel construction issues aiming at uninterrupted implementation, with a better understanding of risks, thereby increasing public consent (Fig.4).

The HEC compile together all specific studies into a comprehensive document, including reporting on the safeguard procedures to be put in place during tunneling, in order to protect the local water natural resources and the technical execution of the works in safe conditions.

6 HEC APPLICATION

In the recent years, HEC has been introduced and included into the project documentation. The standard applied methodology is described hereunder. The starting point is the identification of the affected classified waterbodies and the assessment of the Area of Influence;

Coli & Tanzini (2016) report a case history and a second one can be downloaded at MinAmb (2015). HEC logical path is described in Fig.5 and the output in Fig.6.

The figure above is an example of WFD art. 4.7 assessment of criteria using the HEC methodology, it is key to report the hydrogeological units and structures which might affect the interested water bodies and if needed, based on the HEC define the environmental safeguards.

These output indicators allow i) a better understanding of the causal effect of drilling a tunnel and ii) precautionary approach on value for money, environmental safeguards and public consent.

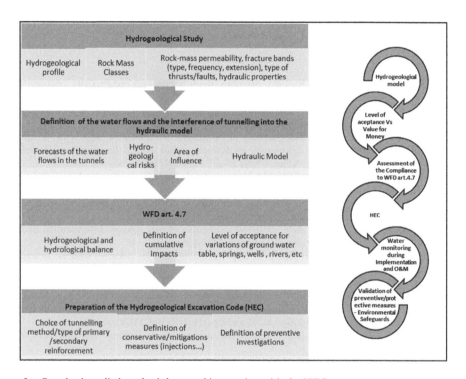

Figure 5.   Standard applied methodology and interaction with the HEC.

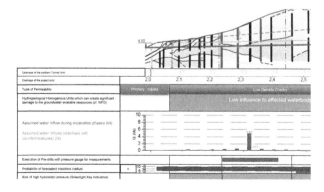

Figure 6.   Example of outputs indicators by applying the HEC procedure referring to compliance with WFD art. 4.7. (Standard hydrogeological profile methodology).

7 CONCLUSIONS

The HEC can be adopted during the entire life cycle of major underground projects, from the design stage, to implementation. It allows the forecast of water inflows and the negative impacts on environment. It also helps ensuring timely and safety execution of the works and ensures project compliance with the WFD.

The WFD regulation, considers the practicability of mitigation measures and efficiency of the remediation costs. The HEC can also play a significant role in this respect, by ensuring the compliance with the directive already at the stage of Costs Benefit Analysis (compulsory for EU funded projects) and the estimation of mitigation/remediation costs, the value engineering/value for money approach and contributing to ensure eligibility for EU Funding.

REFERENCES

Barton, N.R., Bandis, S., Bakhtar, K. 1985. Strength, deformation and conductivity coupling of rock joints. *Int. J. Rock Mech.*, 22: 121–140.
Coli, M., Pinzani, A. 2014. Tunnelling and hydrogeological issues: a short review of the current state of the art. *Rock Mechanics and Rock Engineering*, 47/3: 839–851.
Coli, M., Tanzini, M. 2016. Water safeguarding in tunnelling: for execution and environment. *Proc. GEOSAFE2016, 1st International Symposium on Reducing Risks in Site Investigation, Modelling and Construction for Rock Engineering*, Xi'an, China 25–27, May 2016: 274–287.
Goodman, R.E., Moye, D.G., Van Schalkwyk, A., Javandel, I. 1965. Ground water inflow during tunnel driving. *Eng. Geol.*, 2: 39–56.
Heuer, R. 1995. Approche quantitative, théorique et empirique, del Ronald E. Heuer, sur le venues d'eau en tunnel. *Rapid Exc. and Tunn. Conf.*, San Francisco, Cal., June 18–21.
ISRM 1978. *Suggested Methods for the quantitative description of rock masses*. International Society for Rock Mechanics: Rock Characterization Testing & Monitoring.
Kiraly, L. 1969. Anisotropie et hétérogénéité de la perméabilité dans les calcaires fissures. *Eclogae Geol. Helv.*, 622: 613–619.
Lomize, G.M. 1951. *Water flow through jointed rock*. Gosenergoizdat, Moscow.
MinAmb 2015. https://www.google.it/url?sa=t&rct=j&q=&esrc=s&source=web&cd=1&ved=2ahUKEwj-tJrw8PveAhUyhaYKHUs4AaIQFjAAegQICRAC&url=http%3A%2F%2Fwww.va.minambiente.it%2FFile%2FDocumento%2F162131&usg=AOvVaw3Kp-xHqNRgBpCMEVkuSmQU
Perrochet, P. 2005. Confined flow into a tunnel during progressive drilling: An analytical solution. *Ground Water*, 43(6): 943–946.
Rutqvist, J., Bäckström, A., Chijimatsu, M., Feng, X.T., Pan, P.Z., Hudson, J., Jing, L., Kobayashi, A., Koyama T., Lee, H.S., Huang, X.H., Rinne, M., Shen, B. 2009a. *A multiple-code simulation study of the long-term EDZ evolution of geological nuclear waste repositories*. Special Issue: THE DECOVALEX-THMC PROJECT (Safety assessment of nuclear waste repositories). Environ Geol., 57: 1313–1324.
Rutqvist, J., Barr, D., Birkholzer, J.T., Fujisaki, K., Kolditz, O., Liu, Q.S., Fujita, T., Wang, W., Zhang, C.Y. 2009b. *A comparative simulation study of coupled THM processes and their effect on fractured rock permeability around nuclear waste repositories. A multiple-code simulation study of the long-term EDZ evolution of geological nuclear waste repositories*. Special Issue: THE DECOVALEX-THMC PROJECT (Safety assessment of nuclear waste repositories). Environ Geol., 57: 1347–1360.
Snow, D.T. 1969. Anisotropic permeability of fractured media. *Water Resources Res.*, 5.
Wei, Z.Q., Egger, P., Descoeudres, F. 1992. *Cyclic hydromechanical normal behavior of rock joints*. Proc. ISRM Fractured and Jointed Rock Masses, 2, Lake Tahoe: 389–398.

*Tunnels and Underground Cities: Engineering and Innovation meet Archaeology,
Architecture and Art, Volume 8: Public Communication and Awareness/Risk Management,
Contracts and Financial Aspects – Peila, Viggiani & Celestino (Eds)
© 2020 Taylor & Francis Group, London, ISBN 978-0-367-46873-6*

Bart Silicon Valley Phase II – integrated cost & schedule life-cycle comparative risk analysis of single-bore versus twin-bore tunneling

K. Davey
Santa Clara Valley Transportation Authority

A. Moergeli
MOERGELI CONSULTING LLC

J.J. Brady, S.A. Saki, R.J.F. Goodfellow & A. Del Amo
ALDEA SERVICES LLC

ABSTRACT: An independent integrated cost and schedule life-cycle risk analysis has been used to compare the viability of two competing tunneling options even at different levels of design maturity. This paper describes the process used to provide the Santa Clara Valley Transportation Authority (VTA) with a comprehensive decision-making basis using comparative risk profiles for two tunneling alternatives; a single large diameter tunnel versus two smaller twin tunnels to extend San Francisco's Bay Area Rapid Transit (BART) service into downtown San Jose. VTA also conducted a fact-finding mission along with BART to Barcelona's Metro Line 9 (L9) subway to allow first-hand evaluation of a single-bore system. Construction and system risks were qualitatively assessed and quantified in terms of cost and time to compare the cost and duration of the two options, including the differences in Operations & Maintenance (O&M) costs for the first thirty years of operation.

1 INTRODUCTION

The Santa Clara Valley Transportation Authority (VTA, www.vta.org) is based in San Jose, California. It is an independent special district that provides multi-modal transit services. The Santa Clara Valley Transportation Authority is responsible for the design and implementation of highways and transit projects including the BART Silicon Valley ("BSV") Program. The BSV Phase II Extension project is a 6-mile extension which starts from the Phase I's endpoint, Berryessa Station, as shown in Figure 1. It passes through Downtown San Jose to a new station in Santa Clara. This phase consists of 4.8 miles of running tunnels through San Jose. It includes four stations. Alum Rock, Downtown San Jose and Diridon are underground stations, and Santa Clara is at grade. Phase II has two intermediate ventilation structures and East and West tunnel portals.

VTA re-started the planning efforts for BSV Phase II in 2014 with an update to the project environmental studies. An initial twin bore design effort had been advances to the ~65% complete in 2008 when it was tabled due to funding constraints. In the interim, the continued ongoing community and public concerns about disruption during construction drew VTA's attention towards a single bore (SB) large diameter tunnel as an alternative to the traditional twin bore (TB) option. The advances made by the tunneling industry with respect to developments in larger diameter, soft ground mechanized tunneling in urban settings, encouraged VTA to initiate a feasibility study of a SB alternative. Project alignment, station configurations, emergency egress and ventilation tasks

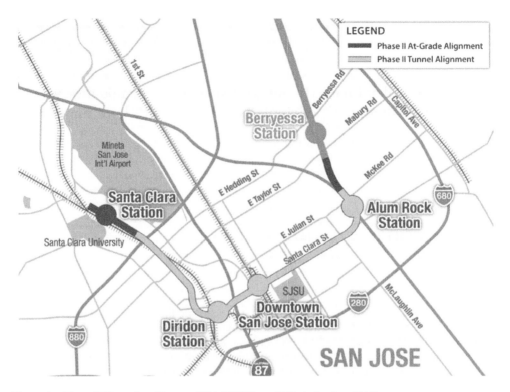

Figure 1.   Phase II Extension (Source: VTA BSV Phase II Tech Studies, 2017).

were studied in the SB feasibility study which was completed in early 2016. The SB feasibility study concluded that a single bore option is technically feasible for the prevailing ground conditions and did not exhibit any fatal flaws (VTA BSV Phase II Tech Studies, 2017). Subsequent technical studies further concluded that single bore might be a viable alternative to the twin bore configuration.

VTA selected Aldea Services, LLC (Aldea) to conduct an independent risk assessment to better inform the decision-making process between the SB and TB tunneling options. The alternative configurations under consideration are a TB tunnel system and a deeper SB tunnel system. A risk assessment comparison was part of VTA's selection process to determine the preferred tunneling alternative. The assessment analyzed, described and compared the qualitative and quantitative risks associated with the two tunneling alternatives. The assessment was carried out within a risk management framework that is intended to proceed throughout design and construction in accordance with the Guidelines for Improved Risk Management on Tunnel and Underground Construction Projects in the United States of America (O'Carroll and Goodfellow, 2015).

2   TUNNEL ALTERNATIVES

The two options are the twin bore (TB) option which constructs two single track 20-foot outer diameter subway tunnels, similar to other tunnels in the BART system, and the single bore (SB) option which constructs a single 45-foot external diameter subway tunnel that is designed to carry two tracks within the same tunnel, using a dividing wall between the trackways. Both alternatives are shown in Figure 2.

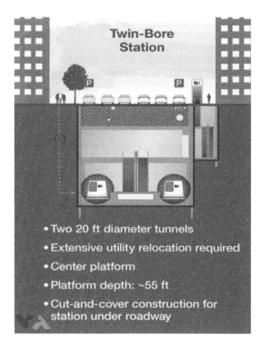
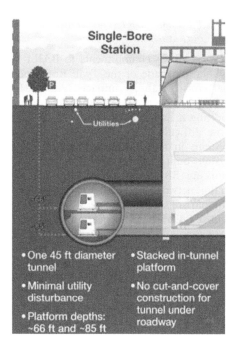

Figure 2. Twin Bore (TB) and Single Bore (SB) tunnel alternatives (Source: VTA Board Workshop and General Public Final Review, Sept 22, 2017).

2.1 Twin bore alternative

The TB design consists of two circular tunnels constructed by two TBMs to interconnect the open-cut stations, mid-tunnel vent structures and portals. The tunnels will be connected to each other by cross passages at regular intervals along the alignment. The project has three proposed underground stations in the 65% Preliminary Engineering Phase; Alum Rock, Downtown San Jose station and Diridon/Arena station.

2.2 Single bore alternative

In addition to the feasibility study which found no fatal flaws, several follow-on SB technical studies performed more detailed evaluations of the SB tunnel option and indicated that a minimum internal diameter of 41 feet was desirable to meet the minimum clearances and vehicle envelopes stipulated in the BART Facilities Standards (BFS) through all of the necessary guideway configurations and transitions along the project alignment (VTA BSV Phase II Tech Studies, 2017).

During early inter-agency coordination discussions, BART indicated a preference for side-by-side rather than stacked track configuration in the running track alignment. This arrangement required transitions from side by side running tunnel to the over/under configuration at the stations which controlled the diameter because the maximum open space was needed to facilitate the transitions. The SB Feasibility Study concluded a minimum depth of cover of 65 feet for the SB tunnel. Subsequent technical studies with more detailed evaluations indicated that a shallower minimum cover depth of 50 feet was constructible and appropriate for further evaluation of a SB tunnel as the design progressed (VTA BSV Phase II Tech Studies, 2017).

2.3 *VTA and BART representatives travel to Spain to inspect Barcelona's Metro Line 9 single bore system*

One of VTA's commitments to BART was to fully investigate the life-cycle impacts in any evaluation made of the Twin Bore and Single Bore options; not simply to assess the initial constructability impacts. At the forefront of these considerations for BART, as it would be for any subway system operator, was long-term Operation and Maintenance. Getting a first-hand account of how a Single Bore system actually works in practice and how difficult it was to service and maintain was no easy task since there were no SB systems in North America. To do this properly would mean travelling to the first such system built in the world; Barcelona's Line-9 (L9) in Spain. In addition to be the longest working SB system, Barcelona met the other essential criteria; they also operated a major network of older non-SB lines (like BART) so they had the ability to make valid comparisons between operating both types of systems. The ability to allow both BART and VTA staff to inspect the L9 first-hand, meet their counterparts in Barcelona, ask questions and engage in free-ranging discussion, led VTA officials to make the effort to bring all these people and experiences together. This culminated in a trip to Barcelona by VTA and BART staff for this purpose in July of 2017.

BART's Observations: BART's review paid particular attention to Line 9's technology requirements, how the system was staffed, and security aspects of line. BART observed that L9 required less service labor due to more automation (for example, trains are driverless). BART learned that L9 has roving staff covering multiple stations for station operations and maintenance.

VTA's Observations: VTA meet with L9 officials and found out that the same challenges of working in a dense urban environment with deep stations (very deep in Barcelona's case) that were not amenable to cut & cover excavation, were the same issues that first led Barcelona to investigating SB for L9. These same officials confirmed that when these conditions combined, significant cost efficiencies were able to be achieved by utilizing an SB configuration.

VTA was impressed with some new features they had not encountered before, particularly the use of Platform Automatic Screen Doors as well as the use of numerous High-Speed Elevators (great for deep stations). VTA made notes of these features (including their own idea of optimizing the timing of elevators for the arrival/departure of trains) and is currently exploring the benefits of incorporating these improvements into the BSV Phase II design.

VTA learned that Barcelona created an 'open' system for L9 O&M Training. They provided everyone a chance to learn O&M for the SB L9 system, but it was completely voluntary. Reviews of this practice were favorable throughout Barcelona's organization. VTA observed a lot of young people, both staff and engineers, that when interviewed said they loved the challenges of working on something new.

VTA investigated the performance of L9's Emergency Exits. During one of their site visits to a deep station, they went through one of the Emergency Exits and travelled to the surface. VTA noted that L9 has emergency ramps (not staircases) and that these deposited into a defined rescue area on the surface (as opposed to the street). These features are also under study for inclusion in the final design of the stations.

VTA concluded that the L9 system functions efficiently, even with deeper stations than those planned for San Jose.

The trip was an important milestone in the selection process. It allowed critics to ask "unfiltered" questions directly to the people involved with running a SB system and receive direct answers to those questions. It allowed both BART and VTA to both tangibly vet as well as look "under the hood" of a system that did not exist in North America as well as get a first-hand account of the realities of maintaining such a system. While it would be incorrect to say that views were changed by the trip, it was critical in removing the abstractive veil surrounding the SB option and sharpened the focus of all parties on the task of evaluating the physical/operational issues that differed between the options.

2.4 *Risk assessment process*

An integrated cost and schedule analysis seeks to identify all risks and uncertainties that might significantly affect the predicted project cost and schedule. Specific methods are used to quantify what each of those impacts might be by using estimates of minimum, most likely and maximum values of cost and schedule. A numerical simulation model aggregates these impacts to risk-based cost and risk-loaded schedule results that are probabilistic distributions (instead of deterministic single value only estimates).

2.5 *Qualitative analysis*

The comparative risk assessment process for cost and schedule first reviewed and evaluated comparative base costs for both alternatives and normalized the costs of both options to a common date (December 31, 2016 in this case). Next, a workshop process was used, including stakeholders (such as BART, VTA, the City of San Jose etc.) and nationally recognized subject matter experts to identify risks and uncertainties that might significantly affect the predicted project cost and schedule for both options.

Risk assessment workshops were used to:

- Identify significant potential events and conditions (both risks/threats and opportunities) that could affect project cost and schedule.
- Assess risk impacts and likelihoods.
- Develop an integrated initial cost and schedule risk register and then identify and discuss mitigation measures for significant risk components and estimate the potential risk reduction from each mitigation measure together with the residual risk after mitigation as shown in Figure 5.
- Identify, discuss and quantify potential opportunities and ways to exploit them.

127 Total Risks, including 64 specific and 63 generic risks, were identified for the TB option, while 121 Total Risks, including 74 specific and 47 generic risks, were identified for the SB option. A "generic" risk was a risk that came from Aldea's generic tunnel "seed" register that was determined to be applicable to the option. The "specific" risks were unique risks identified during risk workshops for the two options.

Risk Control/Mitigation				Residual Risk											Controlling		
Strategy	(Combinations of) Measures/Controls: Apply a Safety System (e.g. by "STOP") to achieve System Safety STOP: S = Strategy / System / Substitution T = Technology O = Organization P = Personnel/Staff	System Safety	Realization	Likelihood	Time Impact	Cost Impact	Reputation Impact	H&S Impact	Environment Impact	Legal Impact	Functionality (Quality) Impact	Comments	Risk Rating	Risk Level	Action Owner	Milestone Target Date	Status
Mitigate	Combination of 1. Minimize use of material subject to tariffs/duties on non-US goods 2. Ask for waiver well ahead in time.	OP	Yes	3		3						By Design-Bid-Build (DBB) per VTA's & Aldea's decision	9	Threat Medium	PM	Part of Preliminary Engineering after system decision	Planning
Mitigate	Carry out supplementary staged site investigations.	OP	Yes	2	3	2						By Design-Bid-Build (DBB) per VTA's & Aldea's decision	6	Threat Medium	PM	Part of Preliminary Engineering after system decision	Planning
Accept	Educate VTA in time.	OP	Yes	3		2						By Design-Bid-Build (DBB) per VTA's & Aldea's decision	6	Threat Medium	PM	Part of Preliminary Engineering after system decision	Planning
Mitigate	Combination of - Reaching out to TBM manufacturer(s) to work out this issue during design of TBM. - Investigate TBM manufacturers on use of real-time monitoring of main bearing. - Reduce required hyperbaric interventions to absolute minimum (emergencies only). - Plan for saturation diving interventions - Plan for safe heavens for planned cutterhead maintenance/tool change. - Monitor cutters in real-time. - Plan for contingencies to handle combination of worn-cut/blocked cutters. - Plan for contingencies to handle obstructions. - Back loading cutters. - Specific devices for cutter handling. - Proper training.	ST OP	Yes	2	1			3				By Design-Bid-Build (DBB) per VTA's & Aldea's decision	6	Threat Medium	PM	Part of Preliminary Engineering after system decision	Planning

Figure 3. Excerpt from risk register after controls (mitigation) implemented.

During the workshops, a numerical ranking method was used to quantify the range of each of those impacts using estimates of minimum, most likely and maximum values for each alternative for cost and schedule threats/opportunities.

2.6 Quantitative analysis

2.6.1 Probabilistic risk analysis approach

A probabilistic approach was used to quantify risks. Conventional construction estimates are presented in terms of a single number. This form of estimating is termed "deterministic" cost estimating. A more reliable way of establishing budget costs is by use of probabilistic forms of estimating that can consider uncertainties and give a range of possible outcomes. These uncertainties can be in the form of pure quantity or material uncertainty or in the form of identified and unidentified risks. When these are combined, a full probabilistic cost and schedule distribution can be developed. The advantage over standard deterministic methods is that it delivers more reliable contextual information because the result is a probabilistic distribution with a range for the risk potential (incl. best case and worst case). The analysis facilitates decision-making in line with the respective project stage. Since actual empirical data for risk analyses is often not available, the exact probability of occurrence can be difficult to estimate. However, use of probabilistic methods allows risks and costs to be depicted for each project phase with individual probability density distributions: larger distributions for larger uncertainties, narrower distributions for smaller uncertainties. Using this approach, reality can be modeled more accurately than with a single deterministic value.

2.6.2 Risk modelling process

Two numerical simulation models were developed using RIAAT software (for more information, please see http://riaat.riskcon.at/) to aggregate these impacts to obtain risk-based cost and schedule estimates for each of the two project options. RIAAT performs numerical simulations to aggregate the contribution of each source of cost and schedule uncertainty to the overall project cost and schedule estimate. Cost impacts of schedule delays including potential changes to the critical path schedule are incorporated in the calculations. The result is an integrated cost and schedule model for the project that includes risk impacts together with the quantified uncertainties in these predictions.

The models aggregated the simulation results of Base + Uncertainty + Risk Costs. In the models, "Risk" includes both Identified Risk and Unidentified (or Known Unknown) Risk. The models also include cost elements to calculate the estimated cost impacts of schedule delays resulting from both Owner-caused and Contractor-caused delay risks. The results are presented in terms of probabilistic distribution ranges rather than single value estimates.

Escalation costs have not been included in the comparative analysis because their calculation is typically a financing calculation reserved for evaluating the time dependent cost of the entire project and the comparative analysis was not based on analyzing the entire project just the subsurface portions.

2.6.3 Quantitative assessment models using RIAAT

The quantitative alternative comparison between the subsurface portions of the TB and SB options was performed using the RIAAT software to analyze 100,000 project cost simulations and 10,000 project schedule simulations for each option.

VTA determined that using the 80th percentile of potential cost distributions would be appropriate for comparative purposes. The P80 level is the result found at the 80th percentile of outcomes, ranked from lowest to highest (i.e., in 100,000 simulations, P80 is the cost result of the 80,000 highest costing project simulation). The relative conservatism of comparing P80 outcomes had a beneficial effect of weeding out any tendency toward "optimism bias" during the process in that participants were never confused that the purpose of this task was not a

Value Engineering (VE) exercise. The comparisons drawn are based on equally less-than-favorable outcomes and that has the benefit of examining overall Risk as a major part of the comparison.

2.6.4 *Comparison of results*

The simulations analyzed the comparative Base Costs which were subject to variable uncertainty in future prices and quantities based upon the level of each option's design maturity or level of design completion (approximately 65% for TB versus 20% for SB). Risks that differentially affected the cost or duration of either option were rated to derive a probability of occurrence and range of possible consequences (should the risk be triggered) and loaded into the models. Risks used in the model underwent a "Basic Mitigation" assessment to filter out that portion of the original unmitigated risk that would be removed or reduced after acknowledging a reasonable minimal level of over-sight and diligence on the Owner's part (equivalent for both options). This is not the Aldea Team's usual practice, nor was it anticipated at the onset. However, we realized that a basic level of mitigation was needed for comparative scenarios because otherwise all the risk uncertainty affected by design maturity level becomes effectively double-counted. In addition to specifically identified risks assessed during the workshops, the model also includes future Market Risk and Unidentified Risk which were based on assessments of project development factors, most notably design maturity. Additionally, a Real Estate Savings Opportunity and a Business Interruption Risk based on assess-ments of the differences in local community impacts expected by each option were evalu-ated and modeled. Finally, there is Schedule Risk which is calculated by RIAAT based upon Owner-caused delays; both Pre-Award and Post-Award of the Heavy Civil (Tunnel & Shafts) Contracts.

Another difference in the evaluation between the two options is that TB was evaluated as a traditional Design-Bid-Build Contract due to the level of design progress (65% Design) while SB (20% Design) was evaluated as a Design-Build Contract to investigate the advantages in potential schedule savings that pursuing this type of contract delivery method might provide. The figures below present the full range of results for the simulated Base + Uncertainty Con-struction Cost (Figure 4), the Construction Program Risk Cost (Figure 5), and Heavy Civil Construction Completion Dates (Figure 6) for both options.

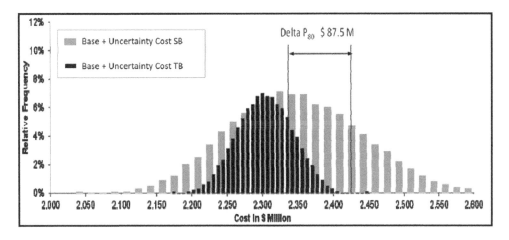

Figure 4.   P0 through P100 comparison SB - TB (construction base + uncertainty cost).

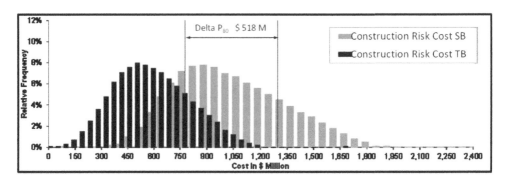

Figure 5. P0 through P100 comparison SB - TB (construction program risk cost).

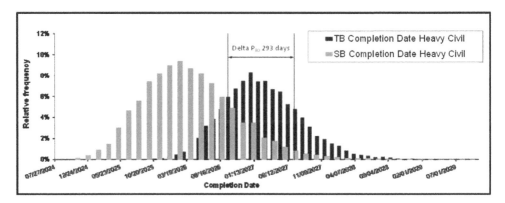

Figure 6. P0 through P100 comparison SB - TB completion dates heavy civil construction.

A summary of the P80 results is presented in Tables 1 and 2.

Twin Bore vs. Single Bore Snapshot Comparison P80 Results Compared	Twin Bore	Single Bore		Delta
COST				
Lower Base Cost		√		-3.5%
Lower Base + Uncertainty Cost	√			-3.6%
Lower Potential Risk Cost	√			-39.9%
First to Revenue Service		√		-8.2%
Lower O&M Cost (1st 30 Years - No Escalation)	√			-2.8%
SCHEDULE				Delta
First to Start of Construction		√	-540	calendar days
Shortest Heavy Civil Construction Duration (Tunnels & Shafts)	√		-247	calendar days
First to Heavy Civil Completion (Ready for Trackwork)		√	-293	calendar days

Table 1. Comparison of twin bore and single bore options.

BSV Phase II Tunneling Alternatives - Comparative Analysis - Independent Assessment			Twin Bore	vs. TB	Single Bore	vs. TB	Delta
Results Summary DRAFT Final (Based on RIAAT_BSVII_Mitigated_V25_F01_R2, August 10, 2017)			(TB)	100%	(SB)	100%	SB - TB
Base Cost	Based on Designers' Estimates	Det.	$ 2,143,407,000	100%	$ 2,071,065,000	96.6%	$ (72,342,000)
	Based on Designers' Estimates + Uncertainties	P_{80}	$ 2,336,793,000	100%	$ 2,424,327,000	103.7%	$ 87,534,000
Total Potential Risk Cost	Based on Risk Workshops & Aldea's Analysis incl. Unknown	P_{80}	$ 779,234,200	100%	$ 1,296,005,000	166.3%	$ 516,770,800
O&M Costs (1st 30 Years)	($2016, i.e., No Escalation) Based on Aldea's Analysis	P_{80}	$ 1,758,099,000	100%	$ 1,807,957,000	102.8%	$ 49,858,000
Heavy Civil Construction Completion Date		P_{80}	07-09-2027		09-19-2026		-293 d
Heavy Civil Construction Start Date		P_{80}	05-20-2021		11-27-2019		-540 d
Heavy Construction Duration (P_{80} Heavy Construction End - P_{80} Heavy Construction Start)		P_{80}	2241		2488		247 d

Note: Risk-based aggregated costs do not equal the sum of the sub-components because probabilistic metrics like P80's are not additive.

Table 2. P80 comparison summary spreadsheet.

3 CONCLUSION

VTA understood that any decision for selecting a tunneling method from competing options at two different levels of design maturity would need to be uncertainty-based to ensure that overall project risk was not overlooked due to lack of design completeness or inadequately evaluated due to lack of independent vetting. This analysis provided the context to the decision makers and can be set alongside non-cost factors such as the desire for geometric consistency within the BART system, or the desire to not disrupt downtown San Jose for extended periods during construction. All identified uncertainties were quantified wherever possible. All cost and schedule cost impacts have been probabilistically aggregated to Comparative Total Cost of Ownership at a Value at Risk 80% (P80) level thereby, providing VTA a risk-based, objective and thorough comparison to support their decision-making process to determine the most advantageous solution. Contracted on March 9, the IRA/CA report was delivered on October 13, 2017 as per VTA's requirements and directions. This validated approach sets the stage for future risk-based project evaluations.

4 PROJECT STATUS AND OUTLOOK

- VTA's Board accepted the single bore solution on April 5, 2018.
- BART Board of Directors approved the single bore tunnel plan on April 26, 2018.
- VTA received additional funding of $730 million from California's Senate Bill 1, the newly passed gas tax, on April 26, 2018.
- VTA received a Record of Decision by Federal Transit Authority (FTA) on June 4, 2018.
- As per September 1, 2018 (date of submittal of this paper), the selection of a General Engineering Contractor (GEC) is ongoing.
- Our WTC presentation will deliver an updated insight into on one of North America's currently biggest infrastructure projects heading for construction.

REFERENCES

BART Silicon Valley PHASE II, Tunneling Methodology Independent Risk Assessment, Pre-Proposal Conference Presentation, November 29, 2016

O'Carroll, J., and Goodfellow, B., 2015. Guidelines for Improved Risk Management on Tunnel and Underground Construction Projects in the United States of America, UCA of SME, Denver

Saki et al., 2018. NAT 2018 BSV Phase II Independent Risk Assessment (https://www.moergeli.com/en/risk-management-iv-rm/61-life-cycle-comparative-analysis-bsvii-i)

NAT2018 BSVII Presentation (https://www.moergeli.com/en/presentations-ii/62-life-cycle-comparative-analysis-bsvii-ii)

VTA's BART Silicon Valley - Phase II Single Bore Tunnel Technical Study, February 2, 2017

VTA Board Workshop and General Public Final Review, Sept 22, 2017

VTA website (www.vta.org) & BSVII project information (http://www.vta.org/bart/)

Tunnels and Underground Cities: Engineering and Innovation meet Archaeology,
Architecture and Art, Volume 8: Public Communication and Awareness/Risk Management,
Contracts and Financial Aspects – Peila, Viggiani & Celestino (Eds)
© 2020 Taylor & Francis Group, London, ISBN 978-0-367-46873-6

Risk management process applied on design and construction of deep stations and TBM tunnel: Analysis and actions to mitigate potential impacts on buildings during the Line 6 underground metro system works, Naples

A. Di Luccio, M. Mele & G. Molisso
Ansaldo STS|A HITACHI GROUP COMPANY, Naples, Italy

ABSTRACT: Risk management includes all the processes concerning the risks' identification, analysis, planning, monitoring and control of a project, with the objectives to decrease the probability and impact of negative events. Risk management is an iterative process to be implemented throughout the project's life, involving all the relevant team members and the main stakeholders to consolidate a set of risks impacting on cost, scope, time and quality.

This paper describes the procedure applied in the project of line 6 of Naples' underground, where the TBM tunnel was implemented in a high-risk urban area and focuses on the geotechnical monitoring system realized to mitigate the risks and execute the works in safety.

The authors describe the action plan implemented when the event of risk occurred, the stakeholders involved, the decisional flow, the execution phase, the monitoring and controlling stage and the overcoming of the risk, considering the related economic project constraints.

1 THE CONTEXT: THE DESIGN, THE TUNNEL'S CONSTRUCTION AND THE MONITORING SYSTEM

1.1 *The project*

Line 6 of Naples underground represents an important element of the public transport railway network outlined by the Municipal Transport Plan (MTP) for the Metropolitan Area of Naples. After its completion, Line 6 will be 11 km long with 12 stations and 5 interchanges.

The operational frequency will be 4.5 minutes in the section 'Campegna'- 'Municipio' and 9 minutes in the section 'Porta del Parco'-'Campegna', with a transportation capability of over 50 million passengers per year.

Line 6 is a typical railway system that acts as an underground light rail system. The Automatic Train Control System is implemented in a Control Central Post that manages operations and maintenance and to which the safety, surveillance and front office services refer. The fail-safe microprocessor signalling system is configured according to the highest standards used in the most recent undergrounds in Italy and abroad, and it is sized in order to ensure an approximate frequency of 3 minutes.

The train traffic is managed automatically with respect to the pre-arranged operation schedule. Any divergences from the agreed times are corrected by using appropriate adjustment strategies, which are transmitted to automatic on-board driving devices in order to ensure the utmost respect of time schedules.

The state-of-the-art driving units have a length of around 39 m, with large interconnecting spaces between the drive cabs and the bodywork at the two sides. They are constructed by 4 bodies distributed over 5 bogies, of which 3 are motor bogies, intended to improve the dynamic behaviour and control noise and vibrations, giving passengers the highest comfort.

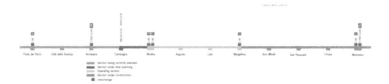

Figure 1. Line 6 functional scheme.

The transport capability of each unit is 300 passengers, and trains will circulate in double composition.

Line 6 contributes to increase the degrees of interconnection of the entire system and configures itself as a segment capable of defining new network elements. It can be construed as a sub-system that can be divided into four parts with different implementation statuses:

– the first section, 'Mostra'- 'Mergellina', already in operation, has a length of 2.2 km with four stations and two interchanges: 'Mostra'– 'Campi Flegrei' interchanges with Line 2 and Line 7, and with Line 2 in 'Mergellina'.
– the second section, currently under construction, 'Mergellina'- 'Municipio', has a length of 3,3 km with four stations, of which the terminal, 'Municipio', is an important interchange with Line 1, funicular F2 and the maritime station, through which the city is connected to the islands;
– the third section, currently under final planning, is 'Mostra'- 'Campagna'. It has a length of 1,1 km with a station - 'Campegna' - which covers over 83,000 square meters of surface with its warehouse workshops. It will be equipped with state-of-the art maintenance systems, in addition to a control center in order to receive technical and administrative professionals;
– the last section, currently under final planning, is 'Campegna'- 'Porta del Parco'. It has a length of 3,9 km, with three stations and 2 interchanges: in 'Acciaieria' with funicular F8, and in 'Porta del Parco' with Line 8.

'Mergellina'- 'Municipio' section
This section is 3,3 km long with 4 stations. It will extend the public operation to 'Piazza Municipio', (Cavuoto et al. 2017) and (Cavuoto et al. 2018) which will be connected to 'Campi Flegrei' interchange through 'Riviera di Chiaia'.

The line gallery was built on a hole basis with a TBM - EPB shield, which ensures a safe, expedient excavation, containment of landslides and distortions of the ground, avoiding damages to the existing infrastructure. The first part of the section, which measures 1,6 km from the front part of the shield to 'Piazza Vittoria', is made with layers of melted materials.

This part extends from the buildings located in Piedigrotta to Largo Torretta, and subsequently covers Riviera di Chiaia.

The second part of the section is 1,4 km long and extends from Vittoria Square to Town Hall Square. It is built in tuff and it goes under important buildings until it reaches Town Hall Square.

In a project with such important technical works, with delicate manufacturing operations, geotechnical monitoring plays an irreplaceable role in ensuring safety during the preparation of the work without any significant interferences with the infrastructure and the urban areas.

Therefore, an advanced data acquisition system has been developed. It controls the vertical and horizontal movements of the points located on buildings, on land, and underground, as well as movements between the opposite sides of damaged buildings and/or manufactured products, and forces action over structural elements, wind bracing of excavations and anchorages in between places located in a certain depth and levels of the groundwater layer.

Figure 2.   The alignment of the tunnel 'Mergellina'-'Municipio'.

1.2   *The tunnel construction: alignment, consolidations, TBM and criticalities*

Line 6 of Naples' underground was built about 30 years ago in the 'Augusto'- 'Mergellina' section; for the subsequent section from 'Mergellina' to 'Municipio' was used a mechanized excavation with TBM Hydroshield able to exert a counter-pressure on the excavation face.

The completion of the tunnel, up to 'Municipio', was realized from 2008 to 2011 with an EPB TBM for a length of about 3300 m.

The examination of the geological profile shows that the depth at which the tufaceous substratum is found increases progressively proceeding towards East, from 'Mergellina' station (about 30 m from ground level) to Vittoria Square (about 60 m from ground level), emerging closer to the surface (about 8 m from ground level) in correspondence of Chiaia street up to the 'Municipio' station, as show (Russo et al. 2015).

In the blanket of loose pyroclastic deposits, on top of the tuff substrate, three parts can be distinguished. The lower part, formed by pyroclastic deposits (pumice, pozzolan of yellow tuff and lapilli), has a thickness that can sometimes reach 20 m.

The intermediate part is made up of pyroclastites generally in a clearly stratified and highly varied granulometry (pozzolans, pumice, lapilli, …) that correspond to the products of the most recent Flegrean eruptive period. The thickness can reach 16–18 m.

The upper part is formed by pyroclastic deposits reworked by alluvial phenomena, whose thickness can exceed 28 m with alternating layers of sand, pumice, and cinerite but with a substantial prevalence of sandy deposits of beach and backwaters.

The presence of pyroclastic products, both loose and lithoid, granulometrically and lithologically different from each other and characterized by very variable geometric ratios, both horizontally and vertically, entails the existence of a complex water circulation, characterized by the presence of several overlapping layers interconnected. The piezometric altitude in the area in question is variable from 1 to 1.5 m above the sea level, therefore most of the line develops below the free surface of the groundwater table.

The alignment intercepts the loose pyroclastic deposits in the route from Piedigrotta street, where the shield is currently located, up to Bausan street for a length of about 1200 m, with variable roofing from 11 m to 19 m with respect to the tunnel cap. They are characterized by a medium-high thickening layer with high permeability. Within this section the lithoid tuff emerges at the altitude of the tunnel between Arco Mirelli and San Pasquale stations for a development of about 300 m. From Bausan street towards the end of the section ('Municipio' station) the tunnel is grafted into the formation of the lithoid tuff with variable roofing from 13 m to 44 m with respect to the tunnel cap. In the terminal part, near 'Municipio' station, the tunnel, once again, appears partially in pyroclastic material with 17 m of roofing. The Neapolitan Yellow Tuff is affected by a series of fractures, essentially sub-vertical, with a non-systematic pattern and spacing that can vary from a few decimeters to many meters. It is equipped, by cracking and porosity, with a high permeability comparable to that of the loose materials above.

1.2.1   *Choice of the methods excavation and support of the excavation face*

In general, the choice of the most suitable excavation method requires the knowledge of a large number of parameters that include the general characteristics of the project. geological-geotechnical characteristics and the machine specifications.

It is therefore important to know the geo-mechanical characteristics of the ground to cross and the interference with the groundwater level. In this sense, the determination of the maximum containment pressure on the front, which depends on the water head, the characteristics of the soil crossed and the roofs, represent a basic element for identifying the machine to be used.

For the completion of the tunnel of Line 6 of Naples' Metro, it was decided to adopt a shield of the EPB-S type (balanced earth pressure), which guarantees the containment of the resentments at the boundary of the tunnel and, consequently, the settlements and distortions at the ground level to avoid repercussions on existing works in a strongly anthropized environment.

The use of the EPB-S constitutes the most suitable system for the excavation of tunnels in soft soils and with the presence of ground water up to pressure values of some bars.

The machine that was used for the double-track tunnel excavated in a blind hole in the construction of Line 6 of the Naples subway, is the **TBM Wirth EPB-S mod. TB816H/GS**

1.2.2 *Final cladding in prefabricated ashlars - Geometric Characteristics and Type*

The cladding of tunnels excavated with a mechanized system consists of a ring of 9 prefabricated segments: 8 segments of average length equal to 2737 mm and the keystone, with a length of 1825 mm.

For the longitudinal dimension, the average value used was1700 mm. Since the ring's extrados diameter is 7850 mm, a ring thickness of 142,5 mm was obtained (157,5 mm with new cutters).

This space, of around 10–15 cm of thickness, is obtained both by the thickness of the shield mantle and by the assembly tolerances of the covering.

When the excavation takes place in rock or in cohesive soils with a self-releasing time of at least a few hours, this annular space is filled with injection mortars, pumped through appropriate holes left in the segments.

This, however, is not possible in loose soils: in this case any extra-excavation would immediately with a high loss of volume and consequent surface subsidence and injury of the existing buildings.

The shield is therefore equipped with a device capable of filling the annular space between the excavation and the covering at the same time the machine moves, leaving the space between the extrados, the segment and the excavation.

It is necessary to pump the cement mixtures through six or eight nozzles placed along the circumference of the shield tail so as to fill the annular space between the excavation and coating immediately behind the shield with suitable pressures depending on the mass and the covering present. The pressurized cement mixture thus counteracts the tendency of the soil to settle around the prefabricated cladding and thus prevents sagging.

The injection takes place under pressure, with values depending on the covering ground, so as to verify the complete clogging of the annular space between the ring extrados and the excavation profile.

This process assumes a primary role in the containment of the volume lost at the excavation, and in the control of subsidence on the surface.

1.2.3 *Criticality related to interference with other works*

The main problems of interference with other works are related to existing buildings those to be realized.

The interference of the line with the existing buildings is the object of the risk analysis developed during the design phase in which the effects of the deformations induced on the buildings by the mechanized excavation of Line 6 of the Naples Metro have been evaluated.

It was therefore necessary to develop specific procedures for assessing the risks of damage connected with the construction of the tunnel.

Together with the prediction of the magnitude of the settlements, the criteria to define corrective actions to be implemented if the extent of the expected settlements would be considered unacceptable were also defined.

The outcomes of the risk analysis in question show that the work performed by the line tunnel does not determine the existence of particular risk situations in the conditions hypothesized regarding the volume lost related to the excavations (Vp = 0.5%). The alignment variation allowed the elimination of the most significant potential risk situations, located in the area of Satriano palace.

Other analysis were also carried out, simulating inauspicious volume lost equal to twice as much as expected; the increase in the degree of risk for the structures was occasional, without however determining conditions that could affect the statics of the buildings. For the above reasons, no specific and preventive works have been found to protect buildings that interferes with the alignment of the line.

1.3 *The monitoring system: measurement, supervision and process*

The development of an urban transportation system in a densely populated metropolitan area involves a wide range of stakeholders, whose needs must be translated into requirements in order to respect and protect the goods and services.

The realization of a great underground work therefore requires the highest level of attention in all the phases of design, construction and testing, with the aim of protecting the essential factors of the technical performance and safety.

The installation of an innovative geotechnical monitoring system is essential, but even more important is its inclusion in a structured process that can help manage the data collected. The nature of the information is such that it is a real-time control allowing a post-processed analysis, whose flow must be addressed with selective criteria designers, civil work contractors, supervisors, guarantors, with the ultimate goal to execute the infrastructure with absolute security.

1.3.1 *Scope of geotechnical monitoring*

The scope of the monitoring system for underground infrastructure and civil works is basically to verify the correspondence among the real feedback of the soil and the structures versus the preliminary and conceptual design evaluation, in consideration also of the influence of the works on the surrounding superficial areas; such verification is conducted during the construction and is finalized therefore to assess the trends of the various measured parameters in relation to the construction phases, materials chosen and the geometry involved.

The monitoring plan follows the entire life cycle of the works; when completed, the same measured parameters during the construction will be controlled and analyzed, recording for a long time, all the variations and ensuring the civil work integrity even during the transportation system operation.

The proposed methodology permits to observe and evaluate the parameters that concern the soil perturbations and the deformations that the civil work can suffer; these parameters, related to the interaction and coexistence of the new structures with the ground, are mainly:

- external soil deformation at the surface in very densely urbanized area;
- soil deformity, both surrounding the void and through the various ground layers between the surface and the void;
- stress-strain structure variations, in relation to the different constructive and geometric components;
- variations of the water level, piezometric surface and soil permeability;
- All types of civil works, their context and their structural components are subjected to a rigorous control and monitoring in tunnels;
- Shafts and stations and on the topographic surface;
- Verification of the stress-strain state of the structure and of the interactions with the soil mass surrounding the excavation;
- Verification of the convergence of void on the tunnel coating;
- Monitoring of the topographic surface changes that can occur during the excavation;

- Verification and monitoring of changes in the piezometric surface and underlying water table in the excavated soil, due to the possible reduction, in hydrodynamic regime, of the capacity to shear strength of some types of ground;
- Verification of the stability of the buildings' foundation ground interested by the presence of the excavations;
- Control of the static of buildings adjacent to the excavation areas, and verification of the possible differential subsidence;
- Monitoring of the variations in opening and closing of lesions in the buildings, existing or determined by the civil works.

1.3.2 *Instrumentation*
The adopted geotechnical instruments are:

- Strain-gauge bars for metal and concrete, installed on the structural elements or inside the concrete, for tension measures of structures;
- Inclinometers with fixed probes, installed external of tunnel and on the topographic surface, in order to measure any horizontal movement of the ground;
- Extensor/inclinometers, placed close to the inclinometers in order to measure any vertical movement within the ground;
- Casagrande and open pipe piezometers, installed close to the extensometers and/or inclinometers in order to measure any variation of the underlying water table;
- Topographic benchmarks on the surface, connected to several external strongholds from the basin of detectably subsidence and consequently considered as fixed, in order to verify any topographic surface subsiding. In addition, a bound for each piezometer, an extensometer and an inclinometer were installed so that all measures of the instruments would be related to the absolute quota;
- Strain-gauges for metal inside the pre-casted tunnel rings for radial stress measures; these tools are important to create a database finalized to post-operam interpretation of collected data with the possibility to develop and improve the design for future works;
- Topographic control systems, used for the historical and artistic buildings along the way; the system is developed starting from specific fixed benchmarks, located in areas outside the influence of the work on the topographic surface, from which points of measurement identified by benchmarks installed on elevated structures are targeted automatically.

All data and measures are collected in a data acquisition centralized system, whose main function will be: i. processing the measured parameters; ii. check all the variations; iii. Compare the variations to the behaviors assumed during the planning stage, both in middle and long term.

1.3.3 *The monitoring process*
Within the decision-making process that governs the construction of a work so complex, a fundamental role is played by the monitoring system that Ansaldo STS has integrated, ensuring the coordination and supervision during the implementation of planned works.

We can assume the monitoring system as the set of instruments, software, applications and people who are responsible for acquiring, organizing, processing and managing data related to the parameters put under control, with the aim of providing operational managers - customers, designers, work directors, civil work structural test commissioners, health and safety coordinators, contractors - all information needed to implement the best solutions.

The actors involved in the procedure of the monitoring system coordinated by Ansaldo STS - those directly involved in construction work and those with an institutional role - are supported by the Department of Hydraulic, Environmental and Geotechnics Engineering, University of Naples, which is been entrusted with the development and validation of mechanisms for the collection, analysis and interpretation of monitoring data. Moreover, the Scientific Committee, proposed by the customer, plays the role of guarantor.

For the acquisition of the measures relating to such a large number of parameters, distributed over a vast territory, different subsystems are necessary for data collection, each with specific characteristics and specialized for static or dynamic parameters; these subsystems, in addition to acquiring the original data - necessary for the control of upstream measures - must be able to pre-process the data through appropriate software, so as to provide a central data processing system of already filtered information.

The pre-processing phase is essential for a correct application of knowledge-based models and to deliver real-time measurements necessary in the daily construction activities.

All processes related to monitoring, the measurements collected directly on-field by specialized operators, their timing and the different stages of works completion, the geological-geotechnical characteristics and those structural of the civil works, are stored and recorded in the central data processing database.

In order to make readable and interpretable all the obtained data, it has been necessary, for each of the monitored areas, to identify one or more control volumes and set up, for each of them, appropriate behavior models developed in advance and implemented in numerical form using specific software.

The control volumes include structures in construction (tunnels, shafts, etc.), existing (buildings, etc.) and the soil between them; this step allows the global interpretation using comparative analysis of all measurements with the numerical models.

2   QUALITATIVE ANALYSIS, MITIGATION ACTIONS, POTENTIAL IMPACT AND PROBABILITY

Project risk is an uncertain event or condition that, if occurs, has a positive or negative effect on the project's objectives. Project objectives include scope, schedule, cost and quality. We define 'Risks' all the uncertain negative events and 'Opportunities' all the positive ones.

Project Risk Management includes all the processes concerned with conducting risk management planning, identification, analysis, responses, monitoring and control on a project.

The objectives of Project Risk Management are to increase the probability and impact of positive events and decrease the probability and impact of negative events in the project.

The risk identification requires the involvement of all the relevant team members and subject experts to consolidate a manageable set of risks directly impacting on cost, scope, time and quality, and is based on: i. detailed analysis of the project scenario and budget assumptions;

ii. clear ownership of risks by appointing Risk Owners; iii. identification of risks performed as early as possible in the project lifecycle to allow more efficient and less costly key decisions; iv. risks have to be described through a complete risk statement, giving evidence of the causes triggering risks and possible consequences; v. risk identification is iterative and repeated throughout the project's life cycle.

After the risk identification, a qualitative analysis is performed to evaluate the importance of each risk and rank them, prioritize individual risks for further attention, evaluate the overall risk level of the project and determine appropriate responses.

The Risk Owner estimates a level of Impact and Probability of occurrence.

The Risk Factor is the result of the mathematical multiplication between impact and probability and is used as indicator to rank the identified risks. This analysis is conducted before mitigation.

The Risk Owner selects a suitable strategy for each individual risk, based on its characteristics and assessed priority, ensuring that the strategy is achievable, affordable, cost effective and appropriate. The possible response strategies to address risks are:

– Avoid = taking the actions required to ensure either the risk cannot occur or can have no effect on the project;
– Transfer = transference to a third party that is better positioned to address a particular risk;

- Mitigate = this is the most widely applicable and used. The approach is to identify actions that will decrease the probability and/or the impact of a Risk;
- Accept = when other strategies are not applicable, acceptance entails taking no actions unless the risk occurs.

In case of mitigation, the Risk Owner carries out the analysis of causes triggering each single risk aimed at decreasing/eliminating relevant impact and/or likelihood; establishes plans that are practical and that can be monitored during their implementation; includes in the budget the cost for the mitigation action to be implemented;

The use of a single strategy that addresses several risks by the identification of a root cause should be considered whenever possible.

Response strategies have to be cost effective, this means that the cost of the action has to be lower than the benefit measured on the single risk through the quantitative analysis.

risk ID	Risk category	Risk sub-category	Risk	Risk description	Trigger event	Trigger event date
01	FIELD ACTIVITIES	Construction costs overrun	TBM assembly at site delayed due to shaft area not available or ready for storage and launching	TBM assembly delay due to lack of component storage area	Local authority authorization	t0 + 3 months
02	PLANNING	Not sustainable schedule	Rings delivery delay	In time materials not being delivery due to transport from remote production location	One month before end of TBM assembly	t0 + 5 months
03	TECHNICAL	Unforeseen development	Excavation works delay due to unexpected geological conditions	Slowdown of excavation activity due to different soil then expected productivity	Hot TBM testing completed	t0 + 6 months
04	FIELD ACTIVITIES	Construction costs overrun	Delay due to machinery breakdown and lack of spares	Several types of breakdown can affect excavation e.g. breakdown of cutters, main bear ring, etc.	Cold TBM testing completed	t0 + 12 months
05	TECHNICAL	Interface management	Additional consolidation measures	On field monitoring data acquisition requires consolidation measures	Cold TBM testing completed	t0 + 6 months
06	FIELD ACTIVITIES	Construction costs overrun	Sealing for loss of watertightness during excavation	Not adequate impermeable rings assembly could imply water inflows	Cold TBM testing completed	t0 + 6 months
07	COST EVALUATION	External costs estimation	Building damages	Repair works required for building damages not recognized before starting TBM works	Cold TBM testing completed	t0 + 6 months
08	FIELD ACTIVITIES	Construction costs overrun	Bad bentonite mix	The bentonite property does not satisfy the specification.	Cold TBM testing completed	t0 + 6 months
09	CLIENT / COUNTRY	Compliance to std/regulations	External stakeholders press campaign against metro works	Local residents and dealers could complain for traffic congestion, noise and air pollution during work execution.	Work sites enclosure.	t0 + 3 months
10	FIELD ACTIVITIES	Construction costs overrun	Ground settlement	Potential ground settlement during tunnel deep excavation could affect near building negatively	Cold TBM testing completed	t0 + 6 months
11	CONTRACTUAL	Backcharges	Delay penalties	Penalties for delay can be applied by client	Local authority authorization	t0 + 24 months
12	FIELD ACTIVITIES	Construction costs overrun	TBM damage and work delay due to unexpected pre-existing interfering structures	Interference during the excavation activity with pre-existing auxiliary and provisional equipment not detectable in the design phases	Cold TBM testing completed	t0 + 6 months
13	FIELD ACTIVITIES	Construction costs overrun	Discrepancy between state of facts building structures and as reported by the drawings	Design changes during construction phase due to state of facts building foundations or other structures in discrepancy with as reported by the drawings	Cold TBM testing completed	t0 + 6 months
14	FIELD ACTIVITIES	Construction costs overrun	Additional TBM front end consolidation	Due to brownfield tunnel previously excavated with front restrainted using previous demobilizated TBM head, more longer and extended consolidation works can be necessary to avoid building subsidence	Cold TBM testing completed	t0 + 6 months
15	PLANNING	Not sustainable schedule	Excavation works delay due to unexpected archeological findings	Change of design and excavation techniques with slowdown of construction activity	Cold TBM testing completed	t0 + 6 months

Figure 3. Project's Risk Assessment Report – step 1.

risk ID	Risk	Quantitative Analisys			Risk owner	Response strategy	Mitigation action description
		Impact	Probability	Risk factor			
01	TBM assembly at site delayed due to shaft area not available or ready for storage and launching	1	1	1	PM, CM	Avoid	Detailed project schedule.
02	Rings delivery delay	3	1	3	SCP	Mitigate	Accurate rings supplier production and logistic plan syncronized with project schedule
03	Excavation works delay due to unexpected geological conditions	2	2	4	PE	Mitigate	Preliminary geotechnical design and survey analysis.
04	Delay due to machinery breakdown and lack of spares	2	2	4	CM	Transfer	Cold and hot TBM test. MTBF component acquisiton and RAM analysis. Spare part list and warehouse stock definition.
05	Additional consolidation measures	1	3	3	PE	Mitigate	Geotechnical monitoring system to observe and evaluate soil perturbations and cw deformations.
06	Sealing for loss of watertightness during excavation	2	2	4	CM	Mitigate	Great accurancy in shutter, rings and tunnel construction as per requirements.
07	Building damages	3	2	6	PM, CM	Transfer	Geotechnical monitoring system to observe and evaluate soil perturbations and cw deformations. Preventive compensation grouting.
08	Bad bentonite mix	1	1	1	CM	Mitigate	Preliminary quality control.
09	External stakeholders press campaign against metro works	1	3	3	EXT REL	Mitigate	Public information on long term metro system advantages.
10	Ground settlement	1	1	1	CM	Mitigate	Geotechnical monitoring system to observe and evaluate soil perturbations and cw deformations.
11	Delay penalties	2	2	4	PM	Transfer	Back to back subcontract with TBM tunnel supplier.
12	TBM damage and work delay due to unexpected pre-existing interfering structures	2	2	4	CM	Accept	
13	Discrepancy between state of facts building structures and as reported by the drawings	3	2	6	PE, CM	Avoid	Detailed design potential interference evaluation with on field accurate prospection
14	Additional TBM front end consolidation	2	3	6	PE, CM	Mitigate	Concrete and chemical injections to consolidate the surrounding excavation.
15	Excavation works delay due to unexpected archeological findings	3	2	6	PE, CM	Accept	

Figure 4. Project's Risk Assessment Report – step 2.

The Quantitative Risk Analysis provides a numerical estimation of the overall effect of risk on the objectives of the project and it is the process used to apportion a specific amount of contingency to a single risk.

Quantification of the economic impact and probability of risk are reported after the implementation of mitigation strategies. The Expected Risk Impact is the result of the mathematical multiplication between Impact and Probability.

Risks whose probability of occurrence is higher than 70% are prudentially considered related to a certain cost and therefore added to the costs budget (100% of the estimated impact).

3 LESSON LEARNED: TIME, COSTS, SAFETY AND BUILDINGS PROTECTION

The construction phase of the TBM required long time compared to what was initially scheduled, since it was necessary to deepen assessments on the reference context, the urban context in which the project was to be inserted and the state of the soil, of the pre-existent structures and of the buildings located in the influence area of the excavation.

The calibration of the various equipment of the excavation machine, the back-up and all the auxiliary systems at the service of the machine, required modifications to the entrance shaft, represented by a technical building obtained inside the Mergellina station, whose dimensions was increased in width to allow the hydroshield and the back-up train to be lowered, in addition to the use of the components most suited the type of excavation and the grounds to be removed. The impact on the budget for the implementation of these changes was about 1% of the total value of the investment necessary for the construction of the tunnel, with a total increase on the schedule of around 2 months.

It is, however, clear that not implementing the changes introduced, in the best case, would have forced a disruption of the final approved design; the adoption of unexplored technical and construction solutions, with potentially increase in time and costs, much higher than what actually occurred. The implemented risk strategy can be classified in "Avoid risk" type.

The deep analysis carried out during the executive design phase, when the manufacturer company was identified, allowed the analysis of some fundamental operational situations, such as the conditions of the front from which the excavation would be started in correspondence of the interruption of about 20 years before, with the abandonment of the old TBM machine about 200 meters after the Mergellina station. The investigations in the field and the consequent in-depth design investigations, recommended the implementation of actions to mitigate the risk of generating dangerous effects on the buildings within the area of the excavation. In particular, an additional consolidation was carried out in jet-grouting of the excavation front and columns of micro piles to defend the buildings foundations.

Only after the completion of these operations, the dismantling phase of the old TBM started and the subsequent start of the new machine.

The impact on the budget for the implementation of these changes was about 2% of the total value of the investment necessary for the construction of the tunnel, with a total increase on the schedule of 6 months.

It should be noted that not implementing the changes described above, would have made the most delicate execution phase of the entire work significantly more risky, with potentially dangerous effects on surrounding buildings, in terms of induced subsidence, in a particularly urbanized context. The implemented risk strategy can be classified in "Mitigate risk" type.

During the tunnel excavation with TBM, far more than 1 km from the start of the works, after passing San Pasquale station, a decisive moment occurred for the continuation of the project.

The deep analysis carried out on the actual certification of the foundations of some buildings located near the excavation, returned parameters close to the risk threshold if the coordinates of the route envisaged by the final project in that section of the tunnel were confirmed.

It was therefore necessary a very quick decision on the need to change the "level" of the track, then proceeding to an altimetric variation of the track down of about 1,5 meters, to increase the distance between the extrados of the tunnel and the foundations of the buildings concerned. The operation, although it did not involve burdensome from the economic point of view, presented some criticalities from the technical and constructive point of view. The analysis carried out in a particularly short time, integrating considerations both on the civil design and on the electromechanical design, allowed to adopt for the implementation of the modification which had no impact on the time and cost of the work, and certainly increased the level of security.

The implemented risk strategy can be classified on "Avoid risk" type.

During the course of the execution of the tunnel, a very delicate theme required a considerable effort for the constructors team to avoid extremely important economic damage that could, in the worst case, have compromised the realization of the work itself. In fact, the findings at the beginning of the excavation of the stations shafts have immediately returned the need to continue the excavation in-depth in "archaeological" modality, sometimes without the use of the mechanical facilities, according to the requirements of the competent Organism.

This event occurred right from the first station encountered by the TBM along its route, the Arco Mirelli station, and then re-proposed also for the San Pasquale and Chiaia stations. The final approved design included the "empty" passage of the TBM inside the volume of these stations, which would have had to be presented to the appointment with the TBM having already completed the excavation inside the perimeter structures (bulkheads).

The heavy delays due to the archaeological findings, on the other hand, meant that, at the arrival of the TBM near the three stations, the excavation had not been completed, so the design had to be modified, proceeding with a "full" passage of the TBM, without stopping the machine near the station but continuing normally the excavation, and then proceeding with the demolition of the tunnel built in the station volume.

The impact on the budget for the implementation of these changes was 10% of the total value of the investment necessary for the construction of the tunnel. However, this additional investment made it possible to save at least 12 months of waiting due to the archaeological excavation, with related costs of "downtime" associated with these times.

The implemented risk strategy can be classified in "Accept risk" type.

4 CONCLUSIONS

In conclusion, the "Lesson learned" emerged from the execution of the work give important suggestions for the cases in question, focusing on:

– deepen the interface aspects, the state of pre-existences, any specific situations of the work to be faced that deviate from a "traditional" excavation right from the final design phase. The experience conducted suggests the importance of involving the constructor company from the beginning, because he is the only stakeholder that knows the actual state of affairs detected in the field, the operating procedures that he intends to implement according to the existing constraints, and the solutions that allow to optimize both time and cost.
– ensure the coordination, in terms of time, between the construction of the tunnel and stations, to effectively allow the sustainability of project hypotheses that, otherwise, remain only theoretical and are destined to be disavowed with inevitable increases in costs and time execution of the work.

REFERENCES

Cavuoto, F., Marotta, P., Massarotti, N., Mauro, A. & Normino, G. Modelling Artificial Ground Freezing: The Case Study of Two Tunnels of the Metro in Napoli (Italy). 3rd Thermal and Fluid Engineering Conference (TFEC) TFEC-2018-22011.

Cavuoto, F., Marotta, P., Massarotti, N., Mauro, A. & Normino, G. Artificial Ground Freezing: Heat and Mass Transfer Phenomena. CHT-17-288. ICHMT 7th International Symposium on Advances in Computational Heat Transfer, in Napoli, Italy.

Russo, G., Corbo, A., Cavuoto, F. & Autuori, S. Artificial Ground Freezing to excavate a tunnel in sandy soil. Measurements and back analysis. Tunn. Undergr. Sp. Technol. (50): 226–238.

Tunnels and Underground Cities: Engineering and Innovation meet Archaeology,
Architecture and Art, Volume 8: Public Communication and Awareness/Risk Management,
Contracts and Financial Aspects – Peila, Viggiani & Celestino (Eds)
© 2020 Taylor & Francis Group, London, ISBN 978-0-367-46873-6

Transposition into national law of the new EU public procurement directives and the impact on subsurface construction contracts – the Portuguese case

G. Diniz-Vieira
CML Lisbon General Drainage Plan, Lisboa, Portugal

R. Pistone
COBA, Lisboa, Portugal

F. Melâneo
Ferconsult, Lisboa, Portugal

C. Baião
TPF Planege, Lisboa, Portugal

ABSTRACT: Last year the Portuguese Public Procurement Code was reviewed and republished by Decree-Law 111-B/2017, which transposed the European Directives 2014/23/EU, 2014/24/EU, 2014/25/EU and 2014/55/EU into Portuguese legislation. The reviewed code came into force in Portugal on January 1st, 2018. This article focuses on the most relevant changes related to Subsurface Contracts for Complex Geotechnical Works. A brief summary of the changes that impact directly the design, tender, execution and risk management of subsurface works (especially for tunnels) is presented. The article also promotes the best international contractual practices recommended by the International Tunnelling and Underground Space Association (ITA). In a second step, the potential appliction of these practices in Portugal will be highlighted and limitations and opportunities will be discussed.

1 INTRODUCTION

It is well known that all Member States of the European Commission were obliged to transpose the latest Public Procurement Directives by 18 April 2016. Despite some delay, by the end of 2017 almost all EU countries have made that transposition, Portugal included.

It is also known that subsurface construction works (in particular tunnels) have some unique characteristics that demand special contractual rules for their successful completion, especially when dealing with complex geotechnical conditions.

But in most cases the public procurement rules of EU countries does not respond to these special contractual needs (Portugal included), and its rigidity makes it very difficult to manage public works properly.

Since 2008, the use of unit price contracts is forbidden in Public Works, which makes it even more difficult to have a flexible contractual tool to deal with unforeseen conditions and uncertainty, which are always present in tunnelling projects.

However, the reform of the Portuguese Public Procurement Code (PPC), which was motivated by the transposition of the new EU Directives, introduced some adjustments to the rules of modification of contracts that has brought a window of opportunity that has to be harnessed, in order to improve the contractual practices of this kind of works.

This law came into force in Portugal on January 1st, 2018. The changes to the existing law are significant and extensive, however in this article we only focus on the most relevant ones related to Subsurface Contracts for Complex Geotechnical Works, in particular tunnels.

The design by scenarios that is advocated by the Portuguese Commission on Tunnelling and Underground Space (CPT) is a compromise solution that makes it possible to take full advantage of current legislation and to move towards the implementation of best contractual practices for this kind of subsurface works in a complex geotechnical environment, as recommended by ITA and, more recently, by FIDIC, in the new Emerald Book.

2 BRIEF BACKGROUND ON THE NEW EU PUBLIC PROCUREMENT DIRECTIVES AND SUBSURFACE PROJECTS

2.1 The new EU Public Procurement Directives

On 26 February 2014 three new public procurement Directives were adopted by the European Parliament and the Council of the European Union: Directive 2014/24/EU on public procurement (known as the 'classic directive'), Directive 2014/25/EU on procurement by entities operating in the water, energy, transport and postal services sectors (the 'sector specific directive'), and Directive 2014/23/EU on the award of concession contracts. Those directives had to be transposed by EU countries into national law by 18 April 2016.

EU member states have stepped up their efforts to implement those new procurement directives, and despite some delay, by the end of 2017 almost all countries (with just four exceptions) have made that transposition, Portugal included.

The 2014 EU public procurement reform pursued several objectives, in order to achieve EU strategic policy goals while ensuring the most efficient use of public funds: i) make public spending more efficient; ii) clarify basic notions and concepts to ensure legal certainty; iii) make it easier for SMEs to participate in public contracts; iv) promote integrity and equal treatment; v) enable contracting authorities to make better use of procurement in support of innovation and common societal and environment goals; and vi) incorporate relevant case-law of the Court of Justice of the European Union.

The new Directives have introduced a number of changes and new rules in order to simplify public procurement procedures and make them more flexible.

For instance, a key aspect of the new directives concerns the article 72 of the 2014/24/EU directive (modification of contracts during their term) that clarifies in which cases contracts and framework agreements can be modified without going through a new procurement procedure, and in which cases a substantial change leads to a retendering of the contract. One example: for additional works (services or supplies which were not included in the original contract) the directive provides considerably more flexibility for contracting authorities in comparison to the previous situation of Directive 2004/18/EC, allowing for a negotiated procedure without prior publication of a contract notice under some conditions. Here, we highlight the first item:

a) *"where the modifications, irrespective of their monetary value, have been provided for in the initial procurement documents in clear, precise and unequivocal review clauses, which may include price revision clauses, or options. Such clauses shall state the scope and nature of possible modifications or options as well as the conditions under which they may be used. They shall not provide for modifications or options that would alter the overall nature of the contract or the framework agreement;"*.

2.2 Particularities of subsurface projects and underground works

Underground works depends on the behavior of the ground materials, and geological, hydrogeological and geotechnical ground characteristics that cannot be precisely known in the design phase. The ground characteristics have a strong influence on the means and methods

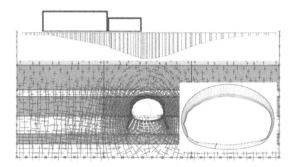

Figure 1.   Design of a tunnel in Lisbon – Red Line extension of Lisbon Metro, 2008.

required for successful implementation of subsurface projects, especially in difficult and complex geotechnical conditions.

In addition, the difficulty in predicting ground behavior and foreseeable conditions, due to several reasons like the nature of the subsurface soil conditions, heterogeneity, previous conditions of over consolidation, depth and extent of the work, etc, leads to inherent uncertainty of underground works which results in unique risks regarding construction practicability, time and cost.

Even a comprehensive subsoil investigation before the design and construction phases does not eliminate all surprises during performance of the underground construction although, in many cases, it will minimize.

For instance, tunnel design is more than a determination of the structural adequacy of component parts. It must meet practical construction requirements to secure economy and safety in a range of heterogeneous ground conditions. When an underground structure is built a redistribution of the natural initial stresses takes place. Due to this redistribution, the underground structure will be subjected to a certain level of stresses and hence loads that influences adjacent structures.

Recognizing these unique characteristics of subsurface projects, especially for those with longer distance and more complex soil-structure interaction like tunnels, that demand special contractual provisions and management for their successful completion, some countries with great tradition in this type of complex geotechnical works (for example, Switzerland with the Suisse Tunnel Code, and others) promote the best contractual practices with the adequacy of the construction methods initially planned to the real ground conditions and behavior found at tunnel face. For instance, by using the observational method (as set out in Chapter 2.7 of the Eurocode 7: Geotechnical Design) in which the design is reviewed during construction.

Therefore, it's recommended the use of flexible legislative solutions adapted to this reality, which follow the best construction practices, and allow a better management of the geotechnical risk allocation.

3   THE PORTUGUESE CASE: TRANSPOSITION AND REFORM OF THE PORTUGUESE PUBLIC PROCUREMENT CODE (PPC)

In Portugal, this need for transposition has been used to proceed with a more thorough reform of the Portuguese Public Procurement Code, and took three years to complete: 2017 was the year that the Portuguese Public Procurement Code was reviewed and republished by Decreto-Lei n.º 111-B/2017. Besides the three new EU procurement directives, the Portuguese legislator also transposed into the same code the 2014/55/EU Directive on electronic invoicing in public procurement. This law came into force in Portugal on January 1st 2018.

During the transposition period some public and private entities and associations in Portugal had the opportunity to contribute and express their opinion on the reform and improvement of public procurement legislation in Portugal.

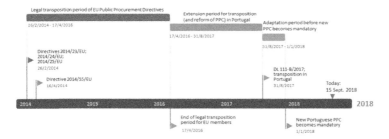

Figure 2. Timescale for transposition of the new EU Public Procurement Directives in Portugal.

Within the Portuguese Commission on Tunnelling and Underground Space (CPT) a specific working group was created; *GT2 - Engineering and Legislation: Contractual Practices* -, which has been working in this topic for the last 3 years and has elaborated some contributions that have been submitted to the legislator. These contributions aimed to facilitate the national legislature to access to the best contractual practices in underground construction, taking into account the rules of the new European Directives and the international experience in other countries and organizations like ITA. The most important aspects are resumed below:

i) The new directives reinforce the possibility of creating more flexible contractual arrangements, based on the rules on contract's modification during their term. They are more open to adaptation to real ground conditions encountered during excavations in the case of complex geotechnical works, if those modifications are clearly provided for in the initial procurement documents [article 72(1)(a)]. This opens the possibility of considering different design scenarios for ground conditions and characteristics with different probability of occurrence.

ii) GT2 of CPT proposed a new type of modifications to the contract during their term, called Special Modified Works (SMW) that can occur due to modifications of the construction method on site as a consequence of the information obtained during excavations, for a scenario different than the baseline, and that can result in additional or less works (considering the baseline construction plan which was based on the most probable scenario for the expected ground conditions).

But more flexibility during construction requires accurate preparation and some conditions must be fulfilled:

iii) The modifications are performed within a Special Technical Assistance (STA) provided to the owner of the works, according to the current Portuguese Legislation (Regulation 701-H) *"(...) to make the analysis of the real geological conditions ... and the assessment of the results provided by monitoring of the work ... to adapt the project to the real underground conditions.";*

iv) In the tender phase, the design should be revised by a qualified and experienced entity, distinct from its author. The revision should follow the provisions of article 43, paragraph 2 of the Portuguese Public Procurement Code and cover all of the design phases, from initial design phase until final design phase;

v) The tender documents must contain all available geological and geotechnical information, including soil investigation and classification, which is the owner's responsibility, compiled in a geological and geotechnical information report;

vi) The baseline ground conditions, that sets out the contractual limits of the most probable conditions believed to be encountered during construction, thus providing clear distinctions in the contract documents between expected and unexpected underground conditions, sustains the design baseline scenario (scenario A), for which the excavation and construction should be designed;

vii) The design must also contain two other possible design scenarios, although with less probability of occurrence, and anticipate the construction methods and the appropriate technical solutions to respond to i) a situation with worst geotechnical conditions (scenario B) and ii) another situation with more favourable geotechnical conditions (scenario C) than the baseline scenario;

viii) The tender documents must contain "Differing Site Conditions" clauses which allocates to the owner the risk if real conditions turn out to be significantly different from expected conditions, and provides a qualitative and quantitative procedure (both financial and technical) by which the contractor can apply for and obtain an equitable adjustment for significantly unforeseen site conditions.

ix) It is advisable to implement a Dispute Resolution Board to facilitate conflict resolution during construction, with three members (as proposed, one may be appointed by the project owner, the other by the contractor and a third one appointed by both parties), technically able to analyse and act quickly as decision maker in situations of objective modifications to the contract due to changes in baseline geotechnical conditions, with respect to the national legal framework.

x) Each contract should regulate the monitoring process of the works during excavation, with periodic measures of settlements and convergences, in order to obtain real-time information about the ground behaviour and, eventually, the influence of induced subsidence on adjacent buildings (in cases of shallow tunnels in urban areas).

xi) The proposals submitted to the tender should include all the necessary works for the baseline scenario design (scenario A) and for the other two less likely scenarios (B and C). All the works will be accounted for and valued with different weights, taking into account the probability of occurrence associated with the design scenarios, as defined by the owner in the tender documents, that should use the criteria of the most economically advantageous proposal to award the contract (for guidance, owner could use a criteria for approximately 70% weighted value for the baseline scenario A and 15% weighted value for the other scenarios B and C);

xii) It is advisable the inclusion of Project Management techniques, in particular those related to formal risk management procedures, covering all the parties involved and all stages of the design (from feasibility studies to final design). It's a systematic approach used as a tool to manage and mitigate identified risks, anticipate scenarios and support the decision making.

The figure below resumes the arguments and the proposed solution:

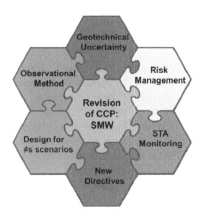

Figure 3.   Puzzle – from the GT2 contribution to the revision of the Portguese PPC, 2016.

The changes to the Portuguese law are significant and extensive. To give an idea of the extension of the revision, more than 50% of the almost 500 articles of the code were adjusted, eliminated or added.

Some of the contribution were considered, some weren't. In the next chapter we will cover the most important contributions for subsurface projects.

Of course, there is no consensual position between all stakeholders, about the benefits and disadvantages of the new Portuguese Public Procurement legislation: it is far too comprehensive (the public entities can buy a pen or build a giant tunnel with the same Code) some say, or it is very specific and rigid say others. One thing is certain, though: it is with this law, this contractual tool, that we will have to manage the Portuguese public works in the next few years, and we'll have to take the best of it (despite trying to continue to change and improve the law).

4 WINDOW OF OPPORTUNITY: CHANGES AND IMPACT FOR SUBSURFACE CONSTRUCTION CONTRACTS (ESPECIALLY FOR TUNNELS)

Taking into account the importance of this revision and the impact on Portuguese subsurface public works with strong geotechnical complexity, a brief summary of the changes that impacts directly the design, tender, execution and risk management of subsurface works (especially for tunnels) is presented.

1. Objective modification of contracts, which can now be carried out *"on the basis of the conditions set out therein ..."* [Articles 312 to 315];

    The novelty of the new wording of Article 312 appears at the beginning of the article: *"The contract may be modified on the basis of the conditions set out therein and further ..."*

    This new justification for the existence of an objective modification of the contract during the construction period transposes into national legislation the assumptions referred to in Article 72 (1) (a) of Directive 2014/24/EU. These recognize the importance of contractual flexibility in order to achieve the objectives of a public contract by accepting the possibility of modifying the contract provided that they have been duly indicated in the procedure.

    This novelty opens the possibility of adopting a design methodology based on the prediction of a scenario of reference (more probable) and two other alternative scenarios (less probable) for construction period, capable of giving a more comprehensive and adaptable response to the situations encountered in accordance with the state of art rules and proper risk management.

    This was one of the measures that Working Group N.2 of the Portuguese Commission on Tunnelling and Underground Space proposed to the government, in the context of the contributions submitted in 2016, in order to respond to predictably uncertain situations occurring in this type of geotechnical complex works. Such measures require a justified adaptation of the construction method due to substantially different geotechnical conditions from those foreseen in the design baseline scenario.

    A very relevant aspect in this type of works that has to be considered is the risk sharing between the parties (as referred in Article 314 (3)), which must be very well defined in the tender documents. For example, another proposal submitted by the WG2 of CPT relates to the allocation of the hydrogeological and geotechnical risk to the owner, while the risk of performance for the expected geotechnical conditions must be assigned to the contractor.
2. New regime of "additional works" and "errors and omissions" on the contract formation stage - introduction of "complementary work" [Articles 61 and 370 to 378];

    With the introduction of a new type of work, called "complementary work", the legislator drops the concept of "additional work" (typically associated with works not foreseen in the contract due to unforeseen circumstances) and "errors and omissions" (typically associated with works: *i)* resulting from situations that are not consistent with reality but could and should have been foreseen, or *ii)* resulting from situations associated with natural constraints with special characteristics of unpredictability such as complex geotechnical works) and broadens the spectrum of circumstances associated with these new complementary work:
    (i) *"(...) where they result from unforeseen circumstances (...)",* their cumulative value does not exceed 10% of the contractual price;
    (ii) *"(...) where (...) they result from unforeseeable circumstances or which a diligent contracting authority could not have anticipated (...)",* their aggregate value does not exceed 40% of the contractual price.

These changes do not respond to the circumstances of predictably uncertain situations, where the owner (or his consultant) cannot accurately predict the actual site conditions (despite having been diligent in the preparation of the tender), but can predict that there may be different situations than those indicated in the project.

3. Alternative dispute resolution through the use of arbitration [Article 476]

The new Article 476 of the Portuguese PPC authorizes the use of arbitration or other means of alternative dispute resolution arising from procedures or contracts. But such a response may not be the most adequate for an expedited resolution of conflicts during the execution of subsurface projects and underground works.

As proposed, for underground construction with high uncertainty potential, there should be implemented a Conflict Resolution Commission with three members (technical experts, one appointed by the Owner of Work, the other by the Constructor and a third named by these two members) capable of technically analyse the situations and to act quickly as a decision maker in situations of objective modifications to the contract due to changes in the initially predicted geotechnical conditions, always respecting the applicable legal regime.

This measure was presented to the government in the 2016 contributions indicated above.

Many other amendments have been introduced in the Code but, as regards to tunnels and complex subsurface works, the most important is to answer the following question:

When uncertainty is certain, what's the right thing to do?

In Portugal, with the recent reform of the Public Procurement Code, the solution is to carry out design for several scenarios, that should take into account the most probable conditions and also other unfavorable conditions (or more favorable) that might occur during construction. Therefore, the design should be done for different scenarios and design options should be updated to be adapted to the real ground conditions.

To resume, some of the adjustments in the law might have opened a "window of opportunity" in Portugal (and perhaps in other EU countries?) so that we have a legislative solution that:

i) meets the objectives of the new European Directives on Public Procurement;
ii) is more prone to the best construction practices of subsurface projects (taking into account ITA's recommendations) for high complexity underground works with uncertainty; and thus
iii) allow better risk management (in particular, the geotechnical risk) and
iv) will lead to *best value for money management* and increased safety of these complex geotechnical public works;

and we have to take it!

5 APPLICATION OF ITA'S CONTRACTUAL FRAMEWORK CHECKLIST FOR SUBSURFACE CONSTRUCTION CONTRACTS IN PORTUGAL: LIMITATIONS AND OPPORTUNITIES

The International Tunnelling and Underground Space Association (ITA) has already prepared Guidelines and Recommendations on Contractual Aspects of Conventional Tunnelling (WG19) and Contractual Framework Checklist for Subsurface Construction Contracts (WG3), that resumes the state of art for contractual purposes on subsurface projects (the revised version of Contractual Framework Checklist will be released for the WTC2019 in Naples by WG3 of ITA).

Also, ITA is working together with FIDIC (TG10) in order to propose a new FIDIC Form of Contract for Tunnelling and Underground Works – The Emerald Book. The motivation for this task group is very well resumed by FIDIC: "*There can be no doubt that subsurface construction projects require specialist contractual frameworks. Within these contractual frameworks, it is of paramount importance to manage the risks specific to underground projects such as uncertainties regarding the geological, geotechnical and structural performance of the subsurface space*".

Both documents will be released soon, and they highlight the importance of promoting equitable risk allocation and the effective dealing with unforeseen conditions in complex subsurface projects.

In this chapter we highlight just two of the most important recommendations, knowing that this subject has already been addressed in the contributions proposed by the CPT in 2016:

1. Allocation of risk:
   The ground and groundwater related risks should be assigned to the Owner, as the party who will most benefit from the completed project and as the party that can best control these risks (he is the only one that can decide to move to another location). The performance related risk arising from expected ground conditions should be assigned to the Contractor.
2. Provisions of a flexible mechanism for remuneration according to ground conditions, foreseen and unforeseen:
   A unit price contract payment system for items that are affected by ground and groundwater conditions should be used. The unit price structure should be organized to facilitate the distinction between fixed costs, time-related costs, value-related costs and quality-related costs.

The difficulty is in applying these principles and recommendations in Portugal, taking into account the new revised Code: the use of unit price contract is prohibited and risk management is neither mandatory nor widespread as it should.

But we will keep trying to move towards the best contractual practices for underground works, specially for tunnel projects.

6 CONCLUSIONS

This article appears in a period of transition to the "new" national legislation on public procurement, in Portugal and also in the other EU countries, due to the necessary transposition of the public procurement Directives. It will require an adaptation of all the stakeholders and, hopefully, an improvement of the contractual practices of this type of complex underground geotechnical works, especially the tunnels.

For these subsurface projects with undergound works we advocate, on the one hand, greater contractual flexibility during the construction phase, but also, on the other hand, a greater rigor in the design and preparation phase of the tender.

There are several recommendations that should be followed, but we highlight the project for several scenarios, that makes it possible to take full advantage of current legislation in Portugal and to move towards the implementation of best contractual practices for this kind of subsurface works: if we can´t use unit price contracts, we can use a "scenario" price contract!

The contributions from GT2 of CPT are summarized in the figure below:

Figure 4.   Several measures from GT2 contribution to the revision of the Portuguese PPC, 2016.

In conclusion, with this article we would like to i) contribute to achieve a better understanding of the legislative changes due to the new EU Directives, and their impact into the Portuguese Public Procurement Code that have great influence on the execution of complex geotechnical works; *ii)* as well as to spread the best international practices in this type of works, mainly in terms of management and risk sharing and alternative conflict resolution procedures.

ACKNOWLEDGEMENTS

The authors would like to thank all those who contributed their experience and dedication to the proposals presented here, especially to the other members of Working Group N.2 of CPT.

Acknowledgment is due to the Portuguese Commission on Tunnelling and Underground Space, in the person of its President Professor João Bilé Serra, for the motivation and commitment that he has always transmitted since the creation of this group within the CPT, and to Ing. Matthias Neunschwander for all the scientific support and orientation for international best practices.

REFERENCES

CCP 2017. Portuguese Public Procurement Code (CCP), reviewed, approved and republished by Decree-Law 111-B/2017, from 31 August 2017.

Diniz-Vieira, G. & Diniz-Vieira, J. 2014. Additional Works to the Contract for Complex Geotechnical Works, in particular Tunnels – Portuguese . *Geotecnia nas Infraestruturas, Proc. 14º Congresso Nacional de Geotecnia, Covilhã, 6-9 April 2014* (CD_ROM).

Diniz-Vieira, G. 2015. Problem to Solve: Specificity of Complex Underground Works and lack of flexibility in Portuguese Legislation. *Proc. of Seminário sobre Obras Subterrâneas Complexas, Riscos Contratuais e CCP: Como Conviver?, Ordem dos Engenheiros, Lisbon, 2 December 2015*.

Diniz-Vieira, G. 2017. Execution of Tunnels and Complex Geotechnical Works - What changes with the revised (Portuguese) CCP?, *Construção Magazine about Tunnels ans Other Geotechnical Works, n.81, October*: 40–47.

Grasso, P., Mahtab, M. A., Kalamaras, G., Einstein, H. H. (2002). On the development of a risk management plan for tunneling. Proceedings of world tunnel congress, Sydney.

GT2-CPT 2015. Initial Contributions to the Revision of the Portuguese Public Procurement Code in the light of the new European Procurement Directives. Working Group n.º2 (GT2) from the Portuguese Commission on Tunnelling and Underground Space (CPT), May 2015, https://geogdv.files.wordpress.com/2015/10/cpt_gt2_englaw_doc_rev_ccp_20150530.pdf

GT2-CPT 2016a. Second Contribution to the Revision of the Portuguese Public Procurement Code in the light of the new European Procurement Directives 2014/23/EU, 2014/24/EU and 2014/25/EU. Working Group n.º2 (GT2) from the Portuguese Commission on Tunnelling and Underground Space (CPT), March 2016, https://geogdv.files.wordpress.com/2016/04/cpt_gt2_englaw_contribu tos_20160331-_f.pdf

GT2-CPT 2016b. Third Contribution to the Revision of the CCP in the scope of the Public Consultation to the Preliminary Draft of its Review. Working Group n.º2 from CPT, September 2016, https://geogdv.files.wordpress.com/2016/09/contributos-da-cpt_consulta-publica-revisao-do-ccp_set2016.pdf

FIDIC 2011. "New Red Book" Condições Contratuais para Trabalhos de Construção; "New Yellow Book", Condições Contratuais para Instalações e Concepção-Construção - Portuguese version.

FIDIC 2015. Motivation for Proposing a new FIDIC Form of Contract for Tunnelling and Underground Works. *Task Group 10, April 2015*.

ITA-AITES 2011. Contractual Framework Checklist for Subsurface Construction Contracts. *Report Nº06 of ITA WG Nº3, April*. www.ita-aites.org, accessed in 20/03/2016.

ITA-AITES 2013. Guidelines on Contractual Aspects of Conventional Tunneling. *Report Nº13 of ITA WG Nº19, May*. www.ita-aites.org, accessed in 20/03/2016.

Marulanda, A. 2013. Exploring the applicability of the Swiss Tunnel Code principles in other jurisdictions. *Proccedings of WTC2013, Geneva*.

NFF 2012. Contracts in Norwegian Tunneling. *Publication N.º 21 Norwegian Tunneling Society*. http://tunnel.no/wp-content/uploads/2014/01/Publication_21.pdf, accessed in 20/03/2016.

http://ec.europa.eu/growth/single-market/public-procurement/rules-implementation_en, accessed in 10/09/2018.

http://ec.europa.eu/regional_policy/sources/docgener/guides/public_procurement/2018/guidance_public_procurement_2018_en.pdf, "Public Procurement Guidance for Practitioners", accessed in 10/09/2018.

http://fidic.org/content/new-standard-tunnelling-contracts-tunnelling-journal-april-may-2017, accessed in 15/09/2018.

Tunnels and Underground Cities: Engineering and Innovation meet Archaeology,
Architecture and Art, Volume 8: Public Communication and Awareness/Risk Management,
Contracts and Financial Aspects – Peila, Viggiani & Celestino (Eds)
© 2020 Taylor & Francis Group, London, ISBN 978-0-367-46873-6

The Geotechnical Baseline Report in the new FIDIC Emerald Book – suggested developments

G. Ericson
iC group, Lund, Sweden

ABSTRACT: The Emerald Book is a new FIDIC form of Contract for Tunnelling and Underground Construction. One of the key features of the Emerald Book is the remeasurement of time and quantities for excavation, support and lining, depending on the encountered geotechnical conditions compared to the contractually predicted – "foreseen" – ground conditions. The Geotechnical Baseline Report (GBR) will play a critical role as the only contractual definition of "Foreseen Ground Conditions" valid for the chosen design and construction methodology of excavation and. The GBR thereby defines the border line between ground related risks retained by the Employer and the performance related risks for which the Contractor shall be responsible.

To match these requirements, the GBR will have to be developed and elaborated further than what is traditional. The paper will discuss ad give examples of the suggested development of a GBR needed for being part of an Emerald Book Contract.

1 INTRODUCTION

As is well known to the Tunnelling Industry, there is an ever-growing demand for utilizing underground space for infrastructure. The difficulty in predicting underground behaviour and physical conditions poses unique challenges regarding construction practicability, time and cost. Thus, allocation of underground risks among the stakeholders becomes critical in underground construction. To address these unique risks the International Tunnelling and Underground Space Association (ITA) and the International Federation of Consulting Engineers (FIDIC) joined forces to draft the new FIDIC Form of Contracts for Underground Works (the "Emerald Book"). To accomplish this, the two organizations setup a joint task group (TG10).

The Emerald Book has been modelled on the 2017 FIDIC Yellow Book (Conditions of Contract for Plant & Design Build) but with significant innovations tailored to the specifics of underground construction. Consistent with FIDIC's philosophy of achieving a fair allocation of risks among the parties, the Emerald Book has been drafted with a view to promoting a balanced risk allocation that is specifically adapted to the risks inherent and unique for underground works.

While a balanced risk allocation reduces the overall cost of the project and the risk of disputes, the uncertainty and risk inherent to underground works mean that projects comprising such works remain, to an even greater degree than other construction projects, prone to claims and disputes.

A key element for an improved contractual practice in underground construction is a clear definition of "Foreseen Physical Underground Conditions" in the contract which then forms the firm basis for the tenders and which is also the central tool for adjustment of construction time and cost from the assumptions made in the bid to the physical conditions actually encountered during construction. In the FIDIC Emerald Book, the Geotechnical Baseline Report (GBR) takes this role. However, to achieve these goals, it has been found that some developments of a GBR under the Emerald Book are required compared to the way GBR's have been written and used so far.

This contribution aims at presenting the historic development of the GBR and to describe the further development needed, as identified during the drafting of the Emerald Book and the associated Guidance Notes for Preparation of Tender Documents.

2 BACKGROUND

The use of Geotechnical Baseline Reports (GBRs) has gradually increased in the Tunneling Industry in recent years. Initially, following the Guidelines introduced by ASCE 1977 and the second edition 2007, the GBRs were mainly adopted in the USA. The Joint Code of Practice for Risk Management of Tunnel Works produced in the UK by the Association of British Insurers and the British Tunneling Society 2003, the Code of Practice for Risk Management of Tunnel Works by the International Tunneling Insurance Group 2006 and the Guidelines for Tunneling Risk Management: ITA Working Group No. 2, have subsequently promoted the use of GBRs also in Europe and elsewhere.

The main objective of a GBR is to define the contractual allocation of ground related risks between the Employer and the Contractor.

Important feedback of experience has, for example, been reported in;

- Recent development in the use of Geotechnical Baseline Reports (Essex & Klein 2000)
- Crossrail's experience of Geotechnical Baseline Reports (Davis 2017)

The number of significant projects having used GBRs has increased in recent years and numerous lessons have thereby been learnt, in particular that;

Contractual allocation of risks determines fundamental matters such as;

- The Contractor's Bid Contingency
- Contractual behaviour and attitudes
- One sided contracts shifting all risk to Contractor lead to dversarial relationships, speculative bids and a false sense of security
- Effective contracts promoting fair and equitable risk-sharing mechanisms lead to partnering approach by the involved parties which is benificial for all stakeholders in a project
- Risk should be allocted to the party in best position to manage it

In Lump Sum Contracts all geotechnical risk is often shifted to the contractor. He then recovers through claims arguing differing site conditions which in turn leads towards litigation, loss of time and more costly projects (and even uncompleted projects)

The main conclusions from this situation, which have been important drivers for the development of the FIDIC Emerald Book are that Contracts should;

- Accept that the Ground belongs to the Employer
- Accept that the Employer has responsibility to pay reasonable costs required to handle ground conditions encountered during construction
- Include a clear contractual definition of foreseen physical ground conditions as the basis for Tenders
- Encourage cost-reimbursable contracts
- Include differing ground conditions clauses
- Include contractual procedures for handling unforeseen conditions
- Disclose all available data

3 FIDIC EMERALD BOOK SPECIFICATION OF THE GEOTECHNICAL BASELINE REPORT

The following is a summary of how the Geotechnical Baseline Report is specified in the Emerald Book Guidance Notes for Preparation of Tender Documents.

The Geotechnical Baseline Report (GBR)is intended to form the basis for establishing the ground related risks in the design and construction for the execution of Underground Works, and as such is critical in providing a balanced allocation of risk between the Parties.

The GBR shall be the single source contractual document that defines what sub-surface physical conditions are to be assumed to be encountered in the execution of the Works through referenced baseline statements for the contractual allocation of the foreseeable ground related risks of physical conditions of the ground between the Parties.

The GBR thereby allocates risks between the Employer and the Contractor for specifically defined physical conditions and related elements of work as set out in the Baseline Schedule.

Consequently, the GBR is the only contractual definition of the foreseeable physical conditions for Underground Works and shall be considered as the basis for the preparation of the Tender and execution of the Works. The procedure for the management of these allocations shall be described in the Contract Risk Management Plan.

The design concept selected by the Employer and the interpretations stated in the GBR collectively represent the Employer's preferred risk allocation for the physical conditions of the ground. This shall apply to the Employer's reference design and any alternative design and method of construction submitted in the Contractor's Proposal by a tenderer.

The GBR baseline statements may also be based on previous experience or exploration data from other sources of relevant information on the physical conditions of the ground. This implies that the GBR might deviate from the factual geological data contained in the Geotechnical Data Report or (GDR).

The GBR also serves to convey and highlight the key project constraints and requirements to enable the tenderer appreciate the key project issues.

The GBR shall include parameters that state the physical characteristics of the ground and ground water conditions, as well as the most likely ground behavior to be encountered during the various Excavation and Lining stages in an adequate format. Focus shall be on the behavior of the ground and ground water caused by or having impact on the method of construction for the Excavation and Lining Works rather than on pure scientific data.

Physical and behavioral baseline statements in the GBR shall be described using quantitative terms (with limits where appropriate e.g. water inflow, speed of deformation) to the maximum extent possible. As far as possible the selected baseline parameters shall have the ability to be confirmed by the physical conditions encountered quantitively in the field, to reduce ambiguity in the scope of work, avoid delays and potential for disputes, and improve time and cost certainty.

The GBR shall also, where possible, present summaries of relevant, local construction experience encountered physical conditions similar to those anticipated to be encountered for the proposed Works.

Each risk of foreseeable ground related physical conditions shall, with the ambition to achieve a balanced risk, be allocated in the Contract Risk Register to the Party that is best positioned to control it, which leads to more effective risk control. Balanced and equitable allocation of ground related risks, by experience, leads to lower cost of the Works and more competitive Tenders. The Employer should therefore avoid establishing an overly conservative GBR, as this would render it ineffective. Instead, the Employer is advised to provide sufficient rationale in the GBR for how the baseline statements have been set to give the tenderers confidence in the fairness of the Baseline Schedule as the basis for the Tender.

The location of the ground for the construction of the Underground Works is selected and made available by the Employer. Consequently, the Employer's documents i.e. the GBR and the Employer's reference design need to be compatible with the other Employer's Requirements as these constitute the basis for the assumptions in the Tenders. The Contractor's Proposal of design methodology, the detailed means and measures including methods of construction for Excavation and Lining and the associated production rates for any given set of circumstances, are then selected by the tenderer and submitted in the completed Baseline Schedule consistent with the Employer's reference design and/or alternative design (if any).

If the encountered physical conditions of the ground vary within the limits stated in the GBR, there will be an influence on time and cost for the Excavation and Lining Works.

This difference in the time and cost for Excavation Works is due to how the selected means and measures (necessary to excavate and ensure the long-term stability of the surrounds of the space created by the Excavation process) depend (compromised or constrained) on the physical nature of the ground encountered (including everything contained in the ground, like e.g. water, gas, natural or man-made obstacles etc.).

The contractual time and cost for Excavation, (including all necessary support) shall be adjusted accordingly based on the variations between the physical conditions of the ground encountered and those stated in the GBR.

Any physical sub-surface conditions that are outside the limits stated in the GBR shall be considered Unforeseeable physical conditions, and the "differing ground conditions clause "of the Emerald Book General Conditions of Contract will be applicable.

The Contractor, when making any claim based for Unforeseeable sub-surface physical conditions, should be able to demonstrate:

i) that the Contractor in his Tender relied on the physical conditions defined in the GBR and
ii) the impact of any changes in scope, time, risk allocation or cost was caused by the difference of the encountered sub-surface physical conditions against the limits stated in the GBR. Purely numerical differences in the physical conditions encountered do not provide sufficient basis for any compensation to be adjusted.

Variations within the baseline limits stated in the GBR, for example regarding a given percentage distribution of excavation and support classes, will be re-measured according to a procedure defined in the Emerald Book.

The physical conditions stated in the GBR shall be described, and later monitored, measured and recorded, in terms of the means and measures required for the Excavation and ensuring the stability of the space created by the Excavation, and the effect on the surrounds including adjacent property affected by the Excavation. The data should generally be allocated to the different variations and combinations of homogeneous and heterogeneous zones along the length of the different drives.

Baseline statement conditions shall be described in terms of the:

i) anticipated methods of construction
ii) logistics determining access to the working faces and ensuring suitable working conditions

The parameters contained in the GBR shall for each type of excavation and support and each type of construction methodology focus on ground behavior and/or ground response rather than geologically oriented parameters. For example, rather than to establish permeability parameters for the ground, seepage estimates and grouting requirements should be stated to the maximum extent possible, baseline statements should best be stated using quantitative terms that can be measured and verified on Site during construction. For example, maximum allowed convergence or settlement for different support classes at specific locations should be included.

The GBR will need to use a contractual ground classification system that properly reflects the effort (time and cost) of excavating and supporting the cavity in the expected ground conditions. For this purpose, the definition of a ground classification system, together with the associated quantitative criteria for the application on Site, is convenient.

The ground classes should be first established in accordance with:

i) the type of Excavation Works (e.g. open cut, shafts, portals, tunnels, other openings, caverns with top headings and benches, etc.);
ii) the associated anticipated methods of construction, for each type of excavation works, e.g. tunneling or mining techniques, drill and blast or mechanical excavation; full face or partial excavation including pre-treatment (if any), and
iii) the method of support (where stability, ground movement, squeezing etc. should be considered).

The ground classification system should then consider:

i) the behavior of the ground when excavated,
ii) the support measures required for stabilization of the surrounds, and
iii) conditions generated by unique geological features, such as fault (active or passive) or shear zones.

The description of the contractual ground classification system for Excavation Works should include:

i) The quality and structure of the soil and/or rock in relationship to the excavation process, influence of water on the excavation process, methods of Excavation, average types and quantities of support, different ground classes, installation sequence, expected/allowable deformations, and ancillary methods of construction (i.e. grouting, soil improvement, freezing etc., if any);
ii) A percentage distribution of ground classification classes based on the foreseeable physical conditions of the ground and represented spatially in information associated to the Excavation profile and location within specific Sections of the Works.;
iii) Drawings showing the ground classes for different working conditions and methods of construction in soils, rock and/or mixed ground for each drive.

The nature of the foreseeable physical conditions of the ground and the measures to control them, will have significant impact on how to define the contractual ground classification system. The use of geomechanical classification systems as a contractual ground classification system may have to be integrated with the selected design and construction methodology in order to serve the purpose of the GBR.

The contractual ground classification system should typically include:

- Profile type
- Ground behavior;
- Ground behavior for TBM intervention (where relevant);
- Geological hazard scenarios;
- Stand up time, if applicable;
- Ground excavation and support sequence;
- Support type;
- Grouting type and sequence;
- Expected deformations for the ground type and the means to control the deformations;
- Expected ranges of ground mass parameters that could affect the productivity and cost of the methods of construction for each type of Excavation and potential situation during the drive (e.g. include abrasivity, bit & cutter wear, drillability, groutability and cuttability, risk of slurry loss during D-wall construction)
- Geotechnical "Hold Points" with attention and alarm values related to reference values established in the Employer's reference design and/or in the Contract Risk Management Plan;
- Required monitoring.

The GBR should:

- avoid the inclusion of design parameters
- present the general description of the geology and hydrogeology of the Works and in the GDR;
- include a thorough discussion on anticipated ground water levels, seasonal and/or tidal (if any) variations, the baseline physical conditions, including items such as inflows, estimated pumping volumes and rates, anticipated ground water chemistry and temperature;

The GBR should provide referenced baseline statements on:

i) acceptable construction impacts on adjacent property and facilities;
ii) parameters such as maximum/differential settlement, rate and tendency of deformation etc., on affected property;

iii) and other known natural geotechnical, hydrogeological, hydrological and/or man-made
    sources of potential difficulty or hazard that could impact the construction process such as:
    • Natural Hazards:
      Boulders, cavities and other obstructions, high or low top of bedrock, mixed face
      physical conditions, occurrence of hard strata, geological contact zones, highly stressed
      or permeable physical conditions, gas, coal seams or oil deposits, high temperatures and
      hot water inflows, intrusions, isolated aquifers, seismic conditions etc.;

    • Man-made Hazards:
      Other obstructions such as: identified or unidentified deep foundations and/or aban-
      doned piles, exploratory shafts, pits, wells, boreholes, buried utilities, buried debris,
      unexploded ordnance, engineered and/or reclaimed ground, waste tips, contaminated
      ground and ground water within the impact zone etc.;
      as well as
    • Items of Value and/or Interest according to governing Laws:
      Any anticipated items of value or interest described under Sub-Clause "Archaeological
      and Geological Findings".

All conditions that are not explicitly included in the scope of the GBR are considered as
Unforeseeable physical conditions.

It should be noted that the Geotechnical Data Report (GDR) is one of the "any other documents forming part of the Contract", and as such has the lowermost priority of contract
documents.

The GDR should be issued to the tenderers as part of the Invitation to Tender only to
enable tenderers to make their own interpretations and assessment of the risks associated with
the referenced baseline statements in the GBR.

4 CONCLUSION

As can be seen above, a GBR under the Emerald Book will be a comprehensive and highly
qualified document, setting considerably increased requirements on how it is written compared to what is normal today.

It typically means that the writer(s) need to;

• Understand and be fully aligned with the Employer's preferred risk level as it is illustrated
  by the Emloyer's Requirements, the Employer's Reference Design (or, as may be the case,
  the Contractor's Alternative Design) and the suggested methods of construction
• Understand in what way each specific construction method chosen for the project and for
  each type of excavation and support within the project may be compromised by unforeseen
  conditions and provide baselines for such conditions
• Understand the general geological/geotechnical setting of the site, the variations within
  the site
• Understand the limitations of the factual site investigation data presented in the GDR
  and what may have to be considered by baselines outside what is described by the ground
  investigation data
• Understand that the baseline statements are the only definition of what is foreseen and
  conditions outside the baseline limits, as well as any conditions not covered by the GBR
  baselines are by definition unforeseen physical conditions
• Understand the link between the GBR and the compensation mechanisms of the
  Contract, i.e. both the mechanisms for re-measurement within the baseline limits set in
  the GBR and the mechanism for compensation for conditions outside the GBR baseline
  limits, i.e.unforeseen conditions

This broad and in-depth scope suggests that the writing of a GBR needs an integrated team-
work - and therefore that it will probably not be possible for a single individual to cover all
necessary aspects.

Major projects, split into many contracts, will further require several GBR's. In that case there is in addition a need for consistency in approach and contents between the different GBR's. Thus, it is suggested that the responsibility for producing the GBR should be given to a limited core team which then provides the various involved parts of the project organization with a document framework where they can see the overall context and then fill in their relevant details. The core team shall also be responsible for organizing necessary reviews and revisions, in close liaison with the concerned parts of the project organization.

Acknowledgement

This contribution is based upon the work of the FIDIC Task Group 10 "New Form of Contract for Tunneling and Underground Works". The author wishes to thank FIDIC, the ITA and his colleagues Hannes Ertl (D2 Consultants, Linz, Austria), James Maclure (Independent Consultant, Durham, United Kingdom), Andres Marulanda (Ingetec, Bogotà, Columbia), Charles Nairac (White & Case LLP, Paris, France), Matthias Neuenschwander (Neuenschwander Consulting Engineers Ltd, Bellinzona, Switzerland) and Martin Smith (Matrics Consult Ltd., Seoul, Republic of Korea) for their important contributions.

Caveat: at the moment of writing of this article, the FIDIC Emerald Book is still under review. Part of the content may therefore be in contrast with the published Form of Contract. The reader should always consult the published FIDIC Form of Contract for Underground Works.

REFERENCES

ASCE. 2007 Geotechnical Baseline Reports for Construction.

Davis, J. 2017 Crossrail's experience of Geotechnical Baseline Reports. *ICE Publishing.*

Essex, R.J. & Klein, S.J. 2000 Recent development in the use of Geotechnical Baseline Reports. *North American Tunneling*

FIDIC-ITA TG10. 2018. FIDIC Emerald Book unpublished draft and unpublished working papers of FIDIC TG10.

ITIG. 2006 A Code of Practice for Risk Management of Tunnel Works.

Tunnels and Underground Cities: Engineering and Innovation meet Archaeology,
Architecture and Art, Volume 8: Public Communication and Awareness/Risk Management,
Contracts and Financial Aspects – Peila, Viggiani & Celestino (Eds)
© 2020 Taylor & Francis Group, London, ISBN 978-0-367-46873-6

Risk allocation in the FIDIC forms of contract, and the Emerald Book's place in the Rainbow Suite

H. Ertl
D2 Consult International, Linz, Austria

ABSTRACT: The previously available FIDIC contract forms (Red, Yellow, Silver, Gold, Green Books) are different between themselves, in particular with regards to risk allocation between the Employer and the Contractor. However, they have not been drafted with regards to the risks related to sub-surface conditions. In order to prepare a contract form that is suitable for Underground Works a joint task group put in place by FIDIC and the ITA developed the "Emerald Book". The balanced risk allocation, which is one of the core principals of the Emerald Book, will contribute to significantly lowering the project cost, improving cost stability and will allow for easier project implementation. This paper aims to highlight the approach to risk in the different books of the Rainbow Suite and show the changes implemented in the Emerald Book in order to apply the principle of balanced risk allocation to Underground Works.

1 INTRODUCTION

As is well known to all members of the ITA, there is an ever-growing demand for utilizing underground space for infrastructure. The difficulty in predicting underground behavior and conditions poses unique challenges regarding construction practicability, time and cost. Thus, allocation of underground risks among the stakeholders becomes critical in underground construction. To address these unique risks the International Tunnelling and Underground Space Association (ITA) and the International Federation of Consulting Engineers (FIDIC) joined forces to draft the new FIDIC Form of Contracts for Underground Works (the "Emerald Book"). To accomplish this, the two organizations set up a joint task group (TG10). The Emerald Book has been modelled on the 2017 FIDIC Yellow Book (Conditions of Contract for Plant & Design Build) but with significant innovations tailored to underground construction. Consistent with FIDIC's philosophy of achieving a fair allocation of risks among the parties, the Emerald Book has been drafted with a view to promoting a balanced risk allocation that is specifically adapted to the risks inherent and unique to underground works.

2 RISK

2.1 *What is risk*

In general, risk can be defined as possible events during the execution of a project that can lead to loss or damage or a possible gain to the parties involved. These possible events, which can have an impact on the project, are a consequence of uncertainties and unknowns before the execution of the works. These multiple uncertainties are to be dealt with throughout the project implementation from the stage of the feasibility study throughout to the preparation of the tender documents until the completion of the works.

By implementing a risk management system throughout the project development certain risks can be eliminated and dealt with, however there will be risks that have to be handeled

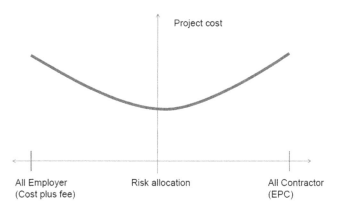

Figure 1. Contract price in relation to risk allocation.

throughout the execution of the works. During the construction, it can only be two parties that will be able to carry the consequences of uncertainties, namely the Client and the Contractor.

2.2 Why risk allocation is important

If risks materialize during the execution of the works these events, become a liability and will result in cost and/or loss of time. In order to have a fair and balanced contract it is important to clearly allocate risks to a party, which will allow the responsible part to account for these liabilities.

An unbalanced risk allocation will lead to escalation of project cost and can even make a project economically unviable.

As it can be seen on the above graph, that not only one-sided risk allocation towards the Contractor will increase the cost of a project, but all risks carried by the Client increase the cost as well.

Improper risk allocation may also result in extended construction times, high number of claims and disputes as well as wastage of resources.

Proper risk identification, management and fair distribution of the responsibility for certain risks will increase the efficiency of project execution and will reduce the disputes significantly which will allow for a smooth project implementation.

2.3 Who shall be responsible for certain risk scenarios

Allocation of risk is a principle that developed over the last decades and throught the time the distribution became more and more precise. In the early days it has been suggested by Abrahamson that risk should be allocated to a party if:

- the risk is within the party's control;
- the party can transfer the risk, for example, through insurance, and it is most economically beneficial to deal with the risk in this fashion;
- the preponderant economic benefit of controlling the risk lies with the party in question;
- to place the risk upon the party in question is in the interests of efficiency, including planning, incentive and innovation; and/or
- if the risk occurs, the loss falls on that party in the first instance, and it is not practicable, or there is no reason under the above principles

Which was later simplified by Bunni to these four principles which are followed in the newer contract forms:

- Which party can best control the risk and/or its associated consequences?

- Which party can best foresee the risk?
- Which party can best bear that risk?
- Which party ultimately most benefits or suffers when the risk eventuates?

2.4 *How risk allocation is handled in FIDIC contract forms*

FIDIC documents if used correctly and not amended too much by the Employer are generally considered as fair and balanced contract forms. Depending on the type of work different contract forms have been developed by FIDIC which are drafted with the mindset of fair risk allocation in regard to the type of work.

Over the evolution of the different contract forms, the allocation of risk has been amended and became more detailed to be up to date with construction developments.

As a principle the main risks are distributed between the parties in the General Conditions of Contract. However as no project is alike these general principles must be adapted to suite the project requirements, risk acceptance by the parties and local regulations. These modifications are handled in the Particular Conditions, which are usually drafted by consultants for the Client. However, there is a tendency that the PCC are misused in a way to assigned risk unilaterally to the Contractor.

2.5 *FIDIC golden principles*

FIDIC has realized the tendency described under 2.4 and has developed following principles which it strongly recommends that the Employer, the Contractor and all drafters of the Special Provisions take all due regard of the five FIDIC Golden Principles. These principles are part of the new set of the rainbow suite as well as the Emerald book:

GP1: The duties, rights, obligations, roles and responsibilities of al l the Contract Participants must be generally as implied in the General Conditions, and appropriate to the requirements of the project .

GP2: The Particular Conditions must be drafted clearly and unambiguously.

GP3: The Particular Conditions must not change the balance of risk/reward allocation provided for in the General Conditions.

GP4: All time periods specified in the Contract for Contract Participants to perform their obligations must be of reasonable duration.

GP5: All formal dispute's must be referred to a Dispute Avoidance/Adjudication Board (or a Dispute Adjudication Board, if applicable) for a provisionally binding decision as a condition precedent to arbitration.

These FIDIC golden principles are described and explained in the publication FIDIC's Golden Principles, and are necessary to ensure that modifications to the General Conditions:

– are limited to those necessary for the particular features of the Site and the project, and necessary to comply with the applicable law;
– do not change the essential fair and balanced character of a FIDIC contract; and - the Contract remains recognizable as a FIDIC contract.

3   RISK ALLOCATION IN DIFFERENT FIDIC CONTRACT FORMS

The risk allocation in the FIDIC rainbow suite was drafted with the mindset of fair distribution of risks and liabilities considering the relevant contractual model. Therefore, it is obvious that risks are allocated differently in design-bid-build contracts (FIDIC Red Book), design-build contracts (FIDIC Yellow Book) and EPC/turnkey contracts (FIDIC Silver Book).

In order to understand the place of the Emerald Book in the rainbow suite the risk relevant to underground constructions and their allocation in the different contract forms are described below. In general, all risks outlined below are related to the uncertainty of precise prediction of ground conditions and the behavior of the ground in regard to the excavations.

3.1 FIDIC Red Book

For the execution of projects using the Red Book, the Employer will engage a designer to develop the design and tender the works based on BOQ rates. The Contractor will execute the works as per the design and instructions by the Employer/Engineer and will therefore be relieved on any of the risks related to changing ground conditions as these are dealt with by new applicable design to be provided by the Employer and by an increase of quantities.

Generally, this contract form would be a good option for underground works as the ground related risks are to be handled by the Employer, however, it does over impose the Employer with risks and does not allow for innovative solutions from the Contractor. The Red Book solution also requires an Employer that is keen to take the responsibility of the design. The industry sees the tendency that the Employers like to hand over the design responsibility to the Contractor as this reduces the problems of suitability of the design for the Contractor's equipment.

3.2 FIDIC Yellow Book

For the execution of projects using the Yellow Book, the design responsibility lies with the Contractor, which allows for innovative solutions from the Contractor and a design that is suitable for the Contractor's equipment.

As per the Yellow Book's contract conditions there is a clause to deal with unforeseen ground conditions, however this is generally a point of disputes between the parties what was foreseeable for an experienced Contractor. As these changed ground conditions usually imply changes to the cost and time for completion these need to be claimed by the contractor, which often leads to long lasting arguments between the parties and are often problematic for the smooth executions of projects.

Claims and in many cases, their slow administration have a negative impact on the cash flow of the Contractor which is additionally hampering the execution of the project.

These claims lead to an "increase" in project cost and "longer" time for completion, which is often problematic for the Employer as he is running over his budget. However, these are no real increases of cost and time as these are a necessity due to the ground conditions encountered.

3.3 FIDIC Silver Book

For the execution of projects using the Silver Book, most of the risks are to be handled by the Contractor. The Silver Book is, as already outlined in the introductory note, NOT suitable for underground constructions. However, several Clients use this contract form to push all risks towards the Contractor, which is not fair and will generally bring problems during construction and/or results in high project costs to the Client.

4   THE EMERALD BOOK

It was one of the first tasks by the task group developing the Emerald Book to decide if the new Emerald book shall be based on the Red or Yellow Book principles. The Task Group decided on the Yellow Book, which was confirmed by the FIDIC Contract Comity and by the availability of the new version published in 2017 it was jointly agreed to use the new Yellow Book as the bases. This "godmother" document was amended in several aspects to ensure that the Emerald Book is suitable to be a fair and well-balanced contract form for underground works.

The fundamental difference between underground works and most other kinds of works lies in the fact that the realization of underground works involves largely the creation of the necessary space within the ground, the behavior or response of which is impossible to know perfectly in advance. Therefore, the ground related risks and related aspects have been intensively looked at.

The two main principles of the Emerald Book are:

- The ground and groundwater related risks are assigned to the Employer, as the party who will most benefit from the completed project and as the party that can best control these risks.
- The performance related risk arising from expected ground conditions are assigned to the Contractor.

The provisions drafted considering following points:

4.1 Design

Based on the Guidance for the Preparation of Tender Documents which is part of the Emerald Book the Employer will provide an Employer's reference design which shall be compatible with the Employer's Requirements and be consistent with the GBR. This is however no detailed design which should be the responsibility of the Contractor (or the contractor's designer)

4.2 Geology and geotechnical behavior

It is recommended that all available geological and geotechnical information are disclosed to all tenderers for information only.

Next to that, a geotechnical contractual baseline is to be included in the tender/contract documents that sets out the contractual limits of the conditions anticipated to be encountered during construction, thus providing clear distinctions in the contract documents between expected and unexpected underground conditions and behavior.

The Contractor shall be entitled to rely on the contents of the Geotechnical Baseline Report, including the anticipated sub-surface conditions as set out in the Baseline Schedule, irrespective of any discrepancies or contradictions that may exist between such conditions and the conditions described in Site Data or other documents made available by the Employer under Sub-Clause 2.5.

4.3 Remeasurement of items that depend on sub-surface conditions

The risk of quantities for the excavation and support of the underground structure lies with the Employer and to honor this principal a remeasurement clause was introduced to remunerate the Contractor for the works that are directly influenced by the underground conditions.

This can be either done by definition of excavation and support classes which are paid as units or even BOQ items for additional support measures.

The concept of remeasurement applies for direct as well as indirect cost items and by applying this method the time related cost which are a matter of disputes in a lot of contracts are dealt within the well defined system.

This remeasurement only applies for activities related to excavation and support and all other works to be performed by the Contractor are to be remunerated considering the lump sum offer.

4.4 Adjustment of Time

Sub clause 13.8.3 outlines the procedure how the ground and production related risks are to be considered during the execution of the project.

The time allowed in the Completion Schedule (as amended as a result of any previous adjustments under this Sub-Clause, or any extension of time granted to the Contractor) for the completion of the Works, Sections and Milestones (if any) comprising the Underground Works, shall be reassessed (reduced or extended) by applying the production rates provided by the Contractor in the Baseline Schedule, to the actual quantity of each item of work necessarily carried out and measured, as recorded, agreed and/or confirmed by the Engineer pursuant to Sub-Clause 3.2.2 [Engineer's Specific Duties and Authority for Underground Excavation and Lining].

Based on this reassessment, and if and to the extent that the Time for Completion of the Works, Section or Milestone is impacted, an adjustment (reduction or extension) shall be calculated for such Time for Completion based only on the logical sequential links provided in the Completion Schedule.

By the adjustment of time rather than a variation, the time for completion is an easy mathematical function of the ground conditions encountered and the production rates offered by the Contractor. So as long as the ground conditions are inside the GBR no claim is required to deal with an extension or shortening of time for completion.

5 CONCLUSIONS

During the development of the Emerald Book the drafters tried to follow the main principals of ground related risks are to be borne (or benefited from) by the Employer and production related risks are to be carried by the Contractor.

In order to achieve this and to prepare a document which will be widely used by the underground industry the Emerald Book was based on the FIDIC Yellow Book, however applying kind of Red Book philosophy (for time and cost) for the excavation and support.

Based on the principles applied the Emerald Book stands in between the Red Book and the Yellow Book in regard to risk allocation.

The balanced risk allocation, which is one of the core principals of the Emerald Book, will contribute to significantly lowering the project cost, improving cost stability and will allow for easier project implementation.

ACKNOWLEDGEMENT

This contribution is based upon the work of the FIDIC Task Group 10 "New Form of Contract for Tunneling and Underground Works". The author wishes to thank FIDIC, the ITA and his colleagues Gösta Ericson (IC Consultants, Lund, Sweden), James Maclure (Independent Consultant, Durham, United Kingdom), Andres Marulanda (Ingetec, Bogotà, Columbia), Charles Nairac (White & Case LLP, Paris, France), Matthias Neuenschwander (Neuenschwander Consulting Engineers Ltd, Bellinzona, Switzerland) and Martin Smith (Matrics Consult Ltd., Seoul, Republic of Korea) for their important contributions.

Caveat: at the moment of writing of this article, the FIDIC Emerald Book is still under review. Part of the content may therefore be in contrast with the published Form of Contract. The reader should always consult the published FIDIC Form of Contract for Underground Works.

REFERENCES

Baker, E. & Bobottom, L. & Lavers, A. 2017 Allocation of Risk in Construction Contracts. *The Guide to Construction Arbitration.* www.whitecase.com

Bunni, N. 2009. The Four Criteria of Risk Allocation in Construction Contracts. *International Construction Law Review*, Vol 20, Part 1, p. 6

Bunni, N. 2001. FIDIC's New Suite of Contracts – Clauses 17 to 19. *International Construction Law Review "ICLR"*, Vol 18, Part 3.

Femeena, M. 2012. Risk allocation in FIDIC forms of contract. https://www.academia.edu/11410276/Risk_Allocation_in_different_FIDIC_Contract_Forms

FIDIC TG 10. 2018. FIDIC Emerald Book unpublished draft and unpublished working papers of FIDIC TG10.

*Tunnels and Underground Cities: Engineering and Innovation meet Archaeology,
Architecture and Art, Volume 8: Public Communication and Awareness/Risk Management,
Contracts and Financial Aspects – Peila, Viggiani & Celestino (Eds)
© 2020 Taylor & Francis Group, London, ISBN 978-0-367-46873-6*

A new approach to fire safety in Dutch tunnels

H.M. Hendrix

Rijkswaterstaat – Dutch Ministry of Infrastructure & Water Management, Utrecht, The Netherlands

ABSTRACT: The fire safety concept in tunnels situated in the Dutch highway network consists of non-spalling classified concrete with passive fire protection in designated areas, concrete with added polypropylene-fibres or a combination of spalling classified concrete with passive fire protection. Active fire protection is not used in the tunnels of Rijkswaterstaat. After several years of fire safety research, Rijkswaterstaat concluded that the existing fire safety concept is no longer valid, because the concrete that was classified as non-spalling does show spalling. This affects the fire safety concept in such a way, that fire testing is obligatory for all new tunnels. Also, the existing tunnels will be tested for fire safety. In this paper, the development of the previous fire safety concept in Dutch tunnels owned by Rijkswaterstaat is explained, the fire testing sequences that have been performed in recent years are described and the current standings are clarified.

1 INTRODUCTION

Until the mid-1970's, Dutch tunnels were not equipped with fire protection of any kind. The occurrence of large fires in tunnels was not considered a threat since trucks with hazardous goods were not allowed to pass through tunnels. Also, concrete spalling was considered not to be an issue. In 1974, the Dutch Ministry of Traffic and Water management started an investigation on the transportation of hazardous goods through tunnels. The amount of such transports had increased since the mid-1960's and in the surrounding municipalities of tunnels, resistance grew as people were aware of the increased risks involved in these transports.

In 1978, an accident between two lorries (one loaded with flowers and one with soda) and three cars happened in the Velsertunnel causing a fire which took the lives of five people. The tunnel was severely damaged by the fire (one emergency cabinet was destroyed and the sound-absorbent ceiling cladding was severely damaged).

In 1981, the Minister of Traffic and Water management decided that all new tunnels as well as all existing tunnels should be equipped with fire protection. Protection of tunnels was considered necessary because of the fire in the Velsertunnel and the additional costs for additional fire protection to allow for the transport of hazardous goods through tunnels were relatively low. After some pilot testing on available fire protection materials, both existing and new tunnels were equipped with either fire protection boards or sprayed fire protection after 1981.

1.1 *Design philosophy*

Until the early 1990's, highway tunnels in the Netherlands were mainly used for river crossings. Up to then, the most widely used tunnelling methods in the Netherlands were the immersed tube tunnelling method and the cut-and-cover method. In both methods, the tunnel consists of reinforced concrete box-shaped tunnel segments.

A typical cross section of a Dutch highway tunnel is seen in Figure 1.

The distribution of bending moments in the rectangular cross-section of immersed tube tunnels allowed for the outer concrete walls to be unprotected because on the inside of the tunnel

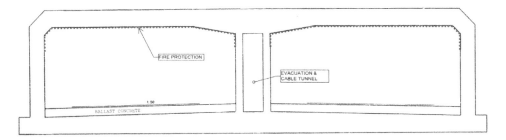

Figure 1. Typical cross-section of a Dutch highway tunnel.

tube no tensile reinforcement is required. Also the concrete can be easily repaired on the inside of the tunnel in case the concrete becomes damaged after a fire. The commonly used concrete mixture was non-spalling classified, which makes this a credible assumption. The ceiling (inside of the tunnel roof) and the upper 1 m of the walls are equipped with passive fire protection to keep the reinforcement temperatures < 250°C and the concrete temperatures in the tensile zone < 380°C – after two hours of RWS-curve fire (see Figure 2). In the corner connecting the walls and the ceiling, a large bending moment is transferred. This causes large compressive stresses. It was decided to limit the concrete temperatures to < 380°C, to limit cracking on the outside of the tunnel, where cracks are impossible to repair.

It was decided to use a passive fire protection system because at the time of implementing the fire protection, only sprinkler systems were common practice as active fire protection systems. During the decision-making process, fire fighters claimed that visibility in the tunnel would be nihilated in the case a sprinkler system is activated, making it nearly impossible to support evacuation or rescue victims in the event of an emergency.

Since the second half of the 1990's, tunnels are no longer built only as river crossings. The concept of land tunnels is becoming more common. Land tunnels for highways are usually built with the cut-and-cover method. This makes sense because land tunnels are usually shallow. A typical feature of land tunnels is, that the surcharge on the tunnel is quite low in comparison to an immersed tube tunnel (for a comparison: 20 kN/m² for surface loading and/or traffic loads in case of a land tunnel, 10–20 m of water pressure (100–200 kN/m²) for immersed tube river crossings).

For land tunnels, the design philosophy regarding fire is different from tunnels under open water. The main difference in philosophy is based on the fact that, should a part of a land tunnel collapse because of a fire, a land tunnel can be repaired using common techniques. For comparison, should an underwater tunnel partly collapse, then it will be flooded. Repair of such type of damage is nearly impossible within an acceptable amount of time.

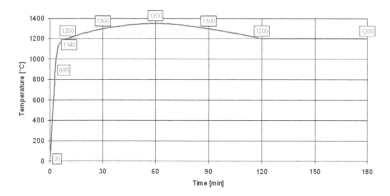

Figure 2. RWS fire curve (time-temperature curve).

Because of this difference in design philosophy, the temperature and time requirements are less stringent for land tunnels in case of a fire (reinforcement temperatures < 250°C, concrete temperatures in the tensile zone < 380°C at the structural cover of the reinforcement). Thanks to these less stringent requirements, other solutions for fire safety have become more common (unprotected non-spalling classified concrete (with an increased amount of concrete cover thickness), concrete with added PP-fibres). These solutions also found their way in new river crossing tunnels.

In 2003, the Western Scheldt tunnel was opened, the first bored highway tunnel in the Netherlands with a length of 6,6 kilometres. The tubbing elements of the bored tunnel were of a high concrete strength (B55). To verify the structural safety of the tunnel in case of a fire, the tubbing elements were fire tested and classified as severely spalling concrete (interface temperature behind fireproofing at the time of spalling approx. 225°C). After testing several options for fireproofing, the tunnel was fitted with passive fire protection (a sprayed thermal insulation layer of 45 mm thickness).

So since the early 2000's, three systems for fireproofing of a tunnel are common in the Netherlands:

1. Non-spalling classified concrete, possibly with insulation by application of fireproofing or increased concrete cover to meet temperature requirements;
2. Concrete with added PP-fibres which can prevent spalling;
3. Spalling classified concrete with an insulation layer of fireproofing material.

Timeline fire safety in Dutch highway tunnels (RWS)

- 1974: RWS assigns an orienting research to TNO regarding the transport of hazardous goods through road tunnels
- August 12th, 1978: Fire in de Velsertunnel.
- 1979: Fire tests carried out by TNO, development of the RWS fire curve
- 1981: Outcome of the TNO research (started in 1974): tunnels should be protected against fire by application of fireproofing.
- 1981: RWS decides to retrofit all important highway tunnels with fireproofing
- 1984: Kickoff of the "Working group Execution Tunnels"
- 1984: 'Decision model for the transport of hazardous goods' is published by TNO.
- 1993: 'Regulations for the transport of hazardous goods through road tunnels' is published by the Ministry of Transport, Public Works and Water Management.
- 2000: Verification fire test non-spalling concrete: ROK concrete
- 2003: Validation of the RWS fire curve in the Runehamar tunnel in Noorwegen during the UPTUN research project.
- 2017: start research program on concrete spalling by RWS

In 2000, a fire test was performed to verify the spalling behaviour of the common concrete mixture used in tunnels. During this test, the concrete showed no spalling (only surface damage of several mm depth). This concrete mixture (so-called ROK concrete) was adopted as an RWS standard non-spalling classified mixture which could be used in tunnel construction and where a further verification of the spalling behaviour was assumed to be unnecessary. The mixture description left room for variations, but dictated the concrete strength to be ≤ C35/45, no fillers to be added and a w/c (water/cement) ratio of 0,5. Actually, the ROK concrete mixture description represented a whole family of concrete mixtures which could be optimized for use in tunnels.

1.2 *Fire testing for spalling*

In 2015, some questions arose regarding the verification of the RWS standard concrete mixture. These questions were supported by the results of fire tests for another research program which had been executed before by RWS. Also, a signal came from the concrete industry itself, since improvements in concrete technology had been ongoing since 2000. This signal was mildly supported in 2008 when spalling occurred at an ROK concrete sample carrier slab. These slabs are used for temperature ingress testing (high temperature insulation performance) by a fire test laboratory. Since this occurrence was considered inconvenient, it was not documented nor researched and the concrete mixture for these sample carrier slabs was changed to a PP-fibre concrete mixture and the problem was solved.

Based on the abovementioned reasons, a test program was started for the verification of the non-spalling classification of the ROK concrete. The results of the first test series were a huge eye-opener.

1.2.1 *Test program*

In the test program, three concrete mixtures were tested for spalling resistance: two gravel aggregate mixtures and a limestone aggregate mixture which was designed as an 'escape' in case the other two mixtures would show spalling.

All three concrete mixtures were to be tested at a sample age of three months and twelve months. The current test protocol requires fire tests to be carried out in double, with the second test acting as a verification of the first one.

Fire tests were carried out on unprotected concrete under RWS curve fire loading conditions. The samples were fitted with thermocouples on six locations and six levels (concrete surface, 35 mm, 70 mm, 100 mm, 150 mm, 200 mm) in depth. The samples were loaded with an even distributed compressive load of 10 MPa parallel to the long side of the samples. Sample dimensions were 5000 mm x 2400 mm x 400 mm (see also Figure 3).

1.2.2 *Results*

All six tests showed severe spalling (Rijkswaterstaat 2018), see also Figure 4. Roughly 22 minutes after the start of the fire test, the cover concrete (70 mm) of the gravel aggregate samples had spalled off, leaving the reinforcement exposed to the fire. The limestone aggregate mixtures showed even more spalling: after about 15–17 minutes, the reinforcement became exposed to the fire.

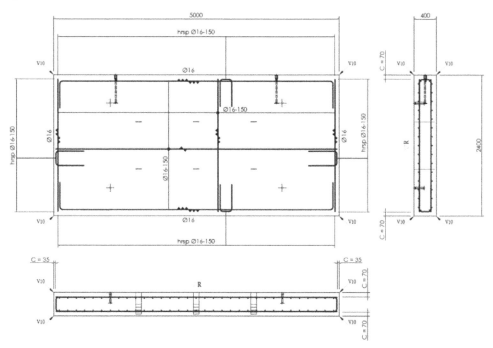

Figure 3.   Sample dimensions for spalling tests.

Figure 4. Concrete samples: left before fire test, right after fire test (Rijkswaterstaat 2018).

2 PROBLEM STATEMENT

Based on the results of the laboratory fire tests, the impact of these results on the tunnels owned by RWS needed to be defined. RWS owns tunnels which are built over a period of 60 years, which makes it unlikely that the concrete of all RWS tunnels is susceptible to spalling. The fire test of 2000 indicates, that the concrete up to and around that time can be classified non-spalling and somewhere between 2000 and 2017, a tipping point is expected.

Since there was a mild signal from the testing laboratory around 2008, the assumption was made to first narrow down the scope of the problem to the tunnels which have been built with ROK concrete since 2008. This reduced the amount of 'suspicious' RWS tunnels to four tunnels. After the verification of the fire resistance of these tunnels, the tunnels of earlier date will be checked as well.

The mild signal of 2008 proved to be hard to confirm since nothing has been recorded about the occurrence of spalling at that time, nor were there any indications at the production facility of the precast sample carrier slabs.

Once the reinforcement becomes exposed to a large fire, it can no longer fulfil its structural function. For the suspicious tunnels, the structural integrity has been studied qualitatively (related to the results of the fire tests) in the case of a fire. The result of this study was, that each of the tunnels would collapse locally in case one structural element would fail due to the fire. A tunnel is usually embedded in soil and depends for its structural stability on strut action of both the roof and the walls. If either one fails, the whole section fails (i.e.: one tunnel segment will collapse).

3 QUESTIONS ARISING

For a proper definition of the problem and how to tackle it the following questions arise:

3.1 *Are the suspicious tunnels safe to use?*

For each of the suspicious tunnels a qualitative risk analysis based on the existing quantitative risk analysis and with conservative assumptions has been made which shows that in the case of an emergency with fire, the tunnel can be left before the tunnel segment collapses. This means, that the suspicious tunnels are safe to use, even in the event of an emergency with fire.

3.2 How to ensure the fire safety in Dutch tunnels again?

In august 2017, two additional fire tests were carried out to test a possible repair method. These tests were successful; a layer of fireproofing boards can provide sufficient insulation to the concrete to prevent spalling.

3.3 What to do with existing tunnels with spalling-classified concrete?

The scope of the problem was reduced to four RWS tunnels which were built between 2008 and 2017 in which the ROK concrete mixture has been applied. With the repair method mentioned in the previous paragraph and after fire testing and verification testing, the tunnels will be retrofitted with fireproofing boards.

3.4 Which requirements are valid for existing tunnels and for new tunnels?

In the Dutch tunnel law, a new tunnel situated under open water should withstand a fire according to the RWS curve for two hours without failing. A land tunnel (not situated under open water) should withstand a fire for one hour without failing. The difference has been made due to economic reasons (if a tunnel segment of a tunnel under open water collapses repair is extremely difficult or even impossible. If a tunnel segment of a land tunnel collapses, the tunnel can be repaired with conventional techniques.).

3.5 Is it still possible to make non-spalling concrete?

Since the start of the fire test program in 2017, none of the fire tests on unprotected concrete without the addition of PP fibres has been successful. Unfortunately, the mechanism behind spalling is not known, nor is the influence of the material properties known. On beforehand it's not possible to predict whether a concrete mixture will spall or not.

Next to the poor predictability, test results are sometimes ambiguous, e.g. when spalling occurs at relatively low temperature on a location while at the same moment the temperature at another locations is much higher (and no spalling is found at that location yet). However, as soon as spalling is initiated at the weakest spot (trapped water, stress concentration) and the fireproofing is damaged, spalling propagates over the whole sample.

3.6 Which design solutions are available for new tunnels?

For new tunnels, some possible solutions to reach the required fire safety level are:

- Concrete with added PP-fibres;
- An insulation layer of fireproofing boards or sprayed fireproofing.

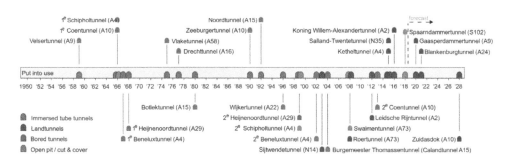

Figure 5.   Timeline highway tunnels in the Netherlands.

Both solutions require laboratory testing of the complete system, i.e.: not only the insulation properties of the fireproofing are tested, but the response of the concrete on the temperature intrusion needs to be tested as well in a full system test.

In these full system tests, the concrete mixture used in the samples must be the exact same concrete mixture as the concrete that is going to be used in the tunnel. This requires making decisions in an early stage of a project about mixture properties, chemical additions to adopt for environmental influences etc.. The mixture requirement also poses a challenge for the procurement of the ingredients for the concrete. The reason for all this is, that the influence of apparently small changes in the ingredients might have an important influence in the behaviour of concrete under fire conditions.

3.7 Which repair solutions are available for existing tunnels?

For existing tunnels, the required fire safety level can only be reached by isolating the concrete from high temperatures. This means, that an insulation layer of fireproofing material needs to be installed in the tunnel.

To be able to define a repair solution for an existing tunnel, in-situ fire tests are required to learn the temperature at which concrete spalling initiates. Usually, one fire test is performed on unprotected concrete to be able to classify the concrete as spalling or non-spalling. Once spalling is shown, further testing is usually done on a dummy fireproofing board.

With the mobile furnace (dimensions 1 m x 1 m) a design test can be carried out with the dummy board. The insulating properties of the dummy board are known and so is the temperature dissipation behind the board into the concrete. With this knowledge, the temperature dissipation into the concrete of an RWS fire curve on a fire proofing board with a certain thickness can be imitated, thus making the test a design test (van der Waart van Gulik 2016). After some design testing, the interface temperature (between the board and the concrete) at which spalling initiates is known for a certain location in a tunnel (note: spalling is influenced by both material properties as well as structural properties). As long as the concrete is protected against this temperature, spalling will not occur.

3.8 Which test methods are valid? Laboratory test or in-situ test?

For new tunnels, laboratory testing of full size concrete samples with their fireproofing system should be carried out. For existing tunnels, only in-situ testing with a mobile furnace is possible. At the moment, research is in progress to check to what extent the outcome of a full size test is comparable with a mobile furnace test to make sure both tests are valid for design and verification purposes.

Figure 6.   In-situ testing: test location prepared with dummy board (left) and mobile furnace in action (right).

3.9 *What about the ramps of tunnels?*

Ramps at the exit and entry of the tunnels are usually open concrete trenches. RWS design regulations require that the concrete ramp structure should withstand a fire according to the hydrocarbon curve. The temperature range of a fire according to the hydrocarbon curve is less high than the temperature range of the RWS fire curve. In a closed tunnel section the heat of the fire can't dissipate to the environment (concrete is a bad conductor for heat), which creates circumstances that can be compared to a furnace. In an open concrete trench, the heat can dissipate easily into the open, leading to lower temperatures than in a closed section.

Usually, the ramps are built with the same concrete mixture as the closed section of a tunnel. Since the concrete for the closed section is tested with the higher temperature demands of the RWS fire curve, it is not necessary to perform fire tests for the hydrocarbon curve unless a significant cost reduction can be obtained by doing so (e.g. when a substantial reduction of insulation thickness can be obtained through testing).

4 CURRENT STANDINGS AND FORESEEN SOLUTIONS

Currently, in situ testing in the first of the four suspicious tunnels has been finished and testing of the second tunnel is in preparation. All ongoing tunnel projects either have finished a test program or their test program is in progress. At this moment, not all test results are released for publication yet, so these data can not be shared.

Testing of existing tunnels will not only cover the four suspicious tunnels, but also some older tunnels will be subjected to a fire test. This will be done to verify the test results of the fire test in 2000.

Next to that, material research is carried out to possibly gain some clues or an indication for the spalling behaviour of concrete. The main aim for this research is, to be able to indicate a tipping point between non-spalling concrete and spalling concrete.

And finally, research is carried out to analyse the impact of smaller fire loads (e.g. a car or a small delivery truck) on spalling classified ROK concrete.

5 CONCLUSIONS

Tunnel construction industry in the Netherlands is facing a new challenge since the implications of the fire testing program regarding spalling of concrete became clear. Building a tunnel in which fire safety is intrinsically taken care of by using a certain concrete mixture is no longer possible without fire testing. Currently, RWS asks for a fire test for each new project and each different concrete mixture that is used in a tunnel project. The results of the fire test are only valid for three years afterwards, to make sure that every tunnel project executes their own test program.

REFERENCES

Rijkswaterstaat, "Website Rijkswaterstaat - information on spalling concrete in tunnels," 2017. [Online]. Available: https://www.rijkswaterstaat.nl/wegen/wegbeheer/tunnels/betonkwaliteit/index.aspx. [Accessed 16 august 2018].
van der Waart van Gulik, T 2016, "Design, assessment and application of passive fire protection in the Port of Miami tunnel according to NFPA 502," Bleiswijk.

*Tunnels and Underground Cities: Engineering and Innovation meet Archaeology,
Architecture and Art, Volume 8: Public Communication and Awareness/Risk Management,
Contracts and Financial Aspects – Peila, Viggiani & Celestino (Eds)
© 2020 Taylor & Francis Group, London, ISBN 978-0-367-46873-6*

Risk engineering surveys: A project's lifecycle risk management tool with multilateral merits and tangible benefits for insurers and all project stakeholders

T. Konstantis
Risk Engineering Consultant, Marsh, London, UK

ABSTRACT: All tunnels and underground works, from concept to operation, carry inherent risks albeit diversified through time. These risks may vary, alternate and transform upon the project's maturity level, affecting and reshaping its overall risk profile. The transition from a high level to a more detailed risk allocation, the timely identification, evaluation and mitigation is of utmost importance to all project stakeholders such as project owners, contractors and insurers. A powerful and proven tool that can contribute in this direction, as integral part of the project's lifecycle risk management and inextricably linked to its insurability is risk engineering surveys. The particulars of a well-structured and professionally executed risk survey, such as planning, structuring and objectives are presented and analyzed. Their focal role and contribution is clarified with explicit references to the tangible benefits gained and concurrent reduction of the total cost of risk for the interest of all major project stakeholders.

1 INTRODUCTION

Underground and tunneling works are by their own nature risk-bearing projects. They comprise of various distinct, concurrent and/or overlapping activities with each one carrying its own risk dynamic and exhibiting its own risk profile.

As the project progresses and matures, these risks transform, alternate and manifest a varying range of appearances and potential. Some of the identified project-specific risks are accepted, some are the subject-matter of a mitigation process whereas others are transferred. Insurance is a key means for transferring risks from the broader construction industry into the insurance market which has the capacity to cover the quantified consequences of these risks in case they are realized. As a general principle adherent to insurance is that it exist in order to insure against loss or damage where the identified hazards/risks exposures, as the outcome of a thorough and profound professional risk assessment/risk management process, cannot be totally eliminated and/or satisfactory mitigated. The allocated capacity can then be determined on the basis special requirements and concerns pertaining to the large underground projects.

The following chapters of this paper focus on the risk management area, and more specifically on the project's lifecycle risk management approach and its manifestation in the form of risk engineering site surveys.

2 RISK MANAGEMENT REMIT

2.1 *General*

Risk Management, in broader terms, is considered to be a multilateral process spanning a diverse range of disciplines and construction areas. Therefore, a clear definition is essential with explicit references in the way it is applied and of its particular context.

In brief and simplistically addressed, risk management should ideally start from the project's development stage (early stage) and follow it through until its final completion and transfer to its owner/operator. Stated in insurance terms, the above mentioned period could be divided into two distinct phases; the Insurance policy pre-placement phase and the Insurance policy post-placement phase. More details regarding each phase are given in the following chapters of this contribution.

2.2 Pre-placement phase

This phase encompasses the development and tendering stage of the project and clearly addresses matters prior to the commencement of the construction works. Essentially, this is a phase under the explicit and clear responsibility of the principal/project owner. He must conduct and implement his own risk management procedures and identify certain hazards and risks to the project, emphasizing on the major ones which host the potential to substantially impact on the project and its development route. Some of the identified risks may then be decided to be transferred through insurance, with the aim being the "risk proofing" of the project over those risks and their consequences on the construction phase (for more details on that refer to ITA WG2. (2006).

Nonetheless, this phase and its particulars, lie outside the scope and purpose of this manuscript.

2.3 Post-placement phase

At this phase, the insurers are getting involved with their involvement being critical and highly beneficial.

The general principle of insurance is to insure against loss or damage where the identified risks and exposures, scrutinized through a professional and profound risk assessment and risk management process, cannot be eliminated or satisfactorily mitigated (at least to an acceptable level).

The risk management approach and responsibilities of the insurers/underwriters form part of this phase and starts directly after the commencement of the physical activities, i.e construction and erection works. The efficiency of their involvement towards the reduction of the potential failures and losses highly depends upon the adopted approach along with their expertise and experience.

This paper focuses on that approach and the manner it is manifested and implemented. The best and most widely-spread and recognized way is the undertaking of physical on-site risk engineering surveys. The following chapters elaborate on the various particulars of these activities, the procedures to be followed, the key factors that are inextricably related to them as well as their tangible key benefits.

3 RISK SITE SURVEYS PARTICULARS

3.1 General

A robust process to maintain control over the project's development and progress is considered a given and an inextricable element in the tunnelling and underground sector. To date, all existing guidelines and procedures used for underground and/or geoengineering works require, at a minimum, the following items to be available and implemented:

  i. Solid design with established and clear checking and approval procedures
  ii. A well-structured monitoring plan with corresponding trigger levels and action plans
iii. Applicable and project-specific technical specifications
iv. Detailed construction works method statements
  v. Experienced contractors awarded the contract (as emerged from the tendering process)

vi. Clear supervision scheme of the overall construction work activities with clearly allocated responsibilities.

Nonetheless, accidents could still occur with their consequences being unknown and practically impossible to be precisely estimated. A detailed and quantitative evaluation and analysis of the publicly available tunnel losses has been carried out by Konstantis, Konstantis & Spyridis (2016). Various reasons could contribute towards this direction; however, some rather prominent route causes could be identified as follows:

i. Variability and uncertainty of the geo-material's conditions and characteristics
ii. Deviation in the predicted behavioral mode of the underground opening compared to the real one
iii. Insufficient process control measures
iv. Human factors and errors

For each underground and tunnel project, a bespoke and appropriate insurance program should be in place prior to any commencement of the construction activities. The terms and conditions of this policy should meticulously reflect the needs and requirements of the project, in a manner that all insurable events are addressed with the provision of sufficient and adequate coverage. In order for the underwriters to be in a position to clearly and properly assess the risks and hazard exposures associated with the project, comprehensive information needs to be provided as relate to the all major and important aspects of the project. These, for instance, could relate to the organizational structure of the project team, the geological information, the construction methods, etc.

Nonetheless, this route, albeit adequate to generally create realistic assessments and insurance coverage assertions, does not have the dynamic and characteristics to follow the project throughout its construction period. The true picture and "real time" observation/monitoring of the project's risk profile and its transformation through time stems only from regular site visits and site surveys on the construction area throughout the entire duration of the construction works.

3.2 *Technical services roadmap*

A risk site survey should, first and foremost, be project-specific, correlated to the project's particularities and cater for all requirements of the insurance policy. As an element of insurance, its scope should be clear, specific, targeted and channeled through the explicit particulars of the insurance program. Its objectives should be kept clear, explicit and free of any trivial generalizations and/or specializations.

The end product of the site surveys should be mutually valuable to both the insurers and the insured; hence a non-negotiable prerequisite is that these tasks are carried out by experienced and knowledgeable risk engineers with expertise pertinent to the technology used and the means and methods of the design & construction of the project being surveyed. On top of that, the same risk engineer should master the way insurance market works and be a connoisseur of the requirements of the specific insurance policy.

The information flow between the insured, the insurer and the risk surveyor can be simplistically illustrated at the following flowchart of Figure 1.

3.2.1 *Project Documentation*
Prior to the commencement/undertaking of any site survey activity, the risk surveyor ought to be fully acquainted with the project and obtain a detailed picture of its current progress status and maturity level. In that context, the risk engineer should be meticulously looking to receive all the necessary latest updated project documentation which will guide and assist him and also facilitate its significant and multilateral task.

However, it has to be acknowledged and accepted that projects tend to produce a diverse range and a very large number of documents, as part of the project management procedures and contractual requirements. In order to avoid onerously and unnecessarily overburdening

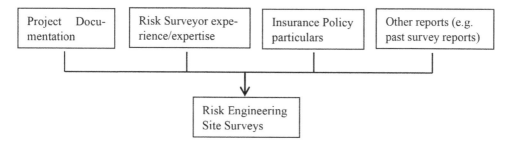

Figure 1. Risk survey information flow.

the risk survey task as well as maximizing the benefits gained from it, a targeted selection and shortlisting of the requested documentation is clearly essential. This selective attitude should be tied in with the requirements of the insurance program as well as the project risks dynamics and exposure limitations.

Briefly, some of the important and critical documents required could be as follows:

i. Monthly progress reports
ii. Risk registers
iii. Project timeline
iv. Construction works method statements

What also needs not to be neglected is the regular flow of specific information between the successive risk site surveys. This is deemed very important and crucial for the sake of clarity, continuity and coherence among these activities.

3.2.2 Surveyor's competence profile and expertise

The primary factor for a successful and constructive risk survey concerns the skills and qualification of the surveyor, with only experienced and professionally competent engineers should be assigned the risk survey tasks. Their specialization and expertise in specific areas is considered an invaluable and important asset.

The undertaking of a survey is a multilateral task which requires the combination of various skillsets and expertise. The survey needs to go in-depth in items that are of material importance to the project and its insurance program, without compromising the technical adequacy and details.

A well-established professional work history in the realms of engineering, integrating both design and construction exposure, is considered a fundamental requirement. A good level of soft skills, such as negotiation and communication capabilities, is also considered to be among the top competences a risk surveyor should master.

In addition, a risk engineer ought to be well versed in contractual items, such as construction contracts, as their requirements could materially affect the risk profile of the project by increasing or reducing the risk potential (e.g risk allocation clauses, unforeseen physical conditions, etc).

The robust and solid technical knowledge, as it relates to design and construction particulars (such as applied methods and technologies, geotechnical evaluations and assertions, etc) is the key parameter for a technically complete and holistically approached site survey. A risk surveyor should be capable of comprehending the technical details of the project, its distinct and unique features, the correlation between design concept and construction methodology and the impact and applicability of the design assumptions to the project's constructability.

Last but not least, the risk surveyor's comfort and proficiency around the particulars of the insurance program could be considered the bonding agent among all previously mentioned capabilities. Proper and realistic interpretation and assessment of the risk profiles of the projects, holistic evaluation of the provided coverage, analysis of the DSU modes, identification of the third party (TPL) risk exposures and liabilities comprise the cornerstone of a suitable and desirable risk survey responsibility.

3.2.3 *Insurance Policy Particulars*

Each project is bound by a dedicated and bespoke insurance policy. The success of the insurance product is of utmost importance to all project stakeholders, leaving aside ambiguities and unnecessary acrimonious disputes.

There are various characteristics and design requirements that are stipulated in the insurance policy and conform to the Insured's risk appetite. Among the numerous other details, the following insurance particulars are considered to be directly interlinked with the risk survey requirements. For more details refer to Konstantis (2018).

One of the most prominent parameters of the insurance policy is the type and extent of the provided coverage. The most common types relate to the physical loss or damage coverage (CAR), the third party liability (TPL), the delay in start-up (DSU) or otherwise called advance loss of profit (ALOP) and the business interruption (BI). Each aforementioned type includes its own distinct characteristics and details, which ought to be elaborated and addressed combined or in a stand-alone basis in the context of the risk survey.

The period of the provided insurance coverage is a key factor which, in the majority of the cases, reflects the actual timeline and construction period of the project. Depending upon the total project duration, along with the escalation of the works and the built-up value of the project, it can be a critical parameter that influences the frequency and content of the risk survey.

A major element of the insurance policy, which unfortunately becomes neglected or underestimated, is the various exclusions stipulated in the insurance policy document. As such, it is imperative to fully comprehend the extent and validity of these exclusions and determine how the policy terms alter and consequently affect the insurability of the project. In various case, a specific activity and/or hazard may be excluded whereas its consequences may fall under the insurable remit.

Last but not least come the stipulated sublimits, particularly those pertaining to the underground and tunnelling activity. Some aspects of the overall project, such as the excavation works, may require the implementation of bespoke terms and conditions followed by a specific insurance limit (i.e tunnel works limit). More details on that topic can be found in Konstantis (2017). Hence, due care should be given to how the construction sequence and order may apply to each category.

3.3 *Key elements – Risk Control Plan*

Like every other professional task, the proper planning and structuring of the survey is the first step towards its success and efficiency. There are specific requirements and criteria that ought to be met and fulfilled for each survey, starting from its early planning until the delivery of the final product, i.e the survey report. In brief, there are four identifiable areas where a risk engineer should focus on:

  i. Planning works and preparatory meeting activities (pre-survey phase)
 ii. Conduct of on-site risk survey (at-survey phase)
iii. Loss control recommendations – risk awareness level (at-survey phase)
 iv. Preparation of the survey report and follow-up activities (post-survey phase)

Each phase is manifested in a diverse manner and features its own specific weight and importance. In brief, an indicative and illustrative flowchart regarding the general Risk Control procedures is presented in the following Figure 2, as excerpted from IMIA WGP 15(01) (2001).

At the pre-survey stage, the actual frame and structure of the site survey is set. This includes preparatory meetings with the insurers (and more particularly with the lead) in order to identify and establish all key topics that ought to be covered and addressed during the survey. No ambiguities are allowed as these could obscure and compromise the efficiency of the survey. Once the agenda is concluded, it is then timely communicated to the project team/insured for his information and action. Due care should be given to the preparation of the information

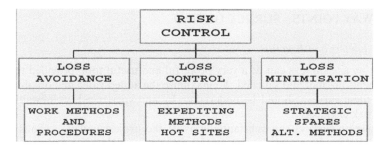

Figure 2. Role and Technical Services of Risk Engineer.

list that is requested to be received prior to the survey, as thoroughly explained in previous sections of this paper.

The actual survey is a twofold endeavor. It comprises office discussions with the key project personnel and a physical on-site tour to the construction sites. All items as per the visit's agenda are thoroughly discussed and all relevant information which is deemed essential for obtaining a complete picture of the corresponding item is obtained. The main target is the detailed and in-depth elaboration of all project-related items, either on a stand-alone basis or as per their interfaces and interlinks.

The site survey is finally concluded in the wrap-up meeting. All findings are highlighted and all concerns identified by the surveyor are discussed and explained. A roadmap of the further actions to be taken along with their assessment is then concluded by all involved parties, which additionally forms the basis of similar future visits.

3.4 Site Survey Frequency

As an insurance-driven element, risk engineering site surveys are dictated and set by the lead insurer (with the consent and agreement of the following insurance markets). At previous sections of this contribution, the essential areas and characteristics of the site surveys have been covered and explained. Nonetheless, another important factor concerns the total required number of site surveys along with their conduct frequency and the time interval between all consecutive visits.

There are various factors that determine their execution frequency, with the main ones being as follows:

i. Project complexity & risk type
ii. Project nature
iii. Project duration and progress
iv. Project value
v. Insurance policy coverage & terms

In general terms, it is acknowledged that a typical frequency of 6 to 12 months for each survey could be considered as common practice and a time interval that allows the maximum capitalization of the gained benefits.

Nonetheless, there are numerous factors, entirely bespoke and project-specific that could alter significantly the required frequency. In accordance with the insurance program and the provided coverage, the survey could be limited only to specific aspects of the project (such as construction works, third party liability, construction plant and equipment, etc). Furthermore, the sequencing of the works, their nature and their corresponding level of "risk attraction" could also elongate or shorten the required time intervals between consecutive surveys.

The timeline of the built-up and/or insurable aggregate project value is of primary import-ance, especially at projects of high construction value and corresponding increased insurance exposure. In cases where the levels of the DSU exposure are significant, a closer and more frequent monitoring of the project may be enforced and requested.

4 TAKE AWAY POINTS – SURVEY OUTPUT

4.1 *Key Parameters and Benefits*

Risk engineering site surveys can be a very powerful and useful tool, provided that some specific requirements are met and satisfied. To maximize the added value and obtain the best possible results, these surveys should be undertaken by an experienced, well-established and of good quality risk engineer.

Risk engineering surveys can substantially assist the project and its various stakeholders multilaterally and multi-parametrically, navigating through the extremely volatile environment. From that process, invaluable and bespoke key benefits can arise, as described below.

Instead of being a "passive" tool that will be perhaps necessitated after an endured loss to the project, a timely executed risk survey can increase the confidence level of insurers by assisting the project and by rendering its risk management as effective as possible. It could also clarify issues that may have been left 'open' and/or 'disputable' during the insurance placement period.

A risk survey can complement and/or verify the initial findings and conclusions during the initial stages of the project (e.g. Benchmarking exercise) and update them (if required) accordingly. When the overall task includes references to confidential and commercially sensitive information, such as the case where DSU/ALOP coverage is provided, the experienced risk surveyor can handle it in a very discrete and professional manner, preventing any undesirable situations of being created.

Since the risk surveys are insurance-related and initiated activities, due care is given in the actual policy wording and the stipulated terms and conditions. A properly and meticulously executed survey can follow the general and particular trends in the tunnelling industry and combine them with the actual developments and provisions in the insurance industry. Allowing for the correct interpretation of the Insured's business and operation model, the risk survey's outcome hosts the dynamic of acting as a "buffer", ameliorating any potential consequences the insurance industry may endure.

One of the key merits of the risk surveys are the timely recognition of what can go wrong and what set of suitable measures may be adopted to minimize the risk. The proper and designated manner for this is through the identification and articulation of targeted and project-specific risk mitigation recommendations. Due to the importance and criticality of this remit, the loss control recommendations particulars are presented in more detail below.

4.2 *Loss Control Recommendations*

The core element of the site surveys relate to the highlight and raise of targeted and project-specific loss control recommendations. These are made solely to maintain and further improve the insurability of the site and, as such, they are frequently the subject-matter of debate between the project team, the risk engineer and the insurers.

Risk recommendations should be viewed and approached with a critical eye and always on the basis of reality and practicability. It goes without questions that it is considered pointless and burdensome for recommendations to be raised that are trivial, impractical and/or onerously expensive to be implemented. In order to serve their purpose and be well received and accepted and acknowledged by the insured, risk recommendations ought to be prioritized with a clear description of the survey 'findings', a targeted proposal for its mitigation, an escalating rating in terms of severity and importance as well as a definite timeframe for implementing any rectification measures.

Loss control recommendations heavily depend upon the surveyor's engineering judgement, his expertise and his previous work experience. However, it would be entirely an onerous stance if these recommendations were perceived as being direct instruction and/or undisputable obligations, as the final decision of the specific action, its timing and extent lies solely with the project team and not the risk engineer. As excerpted from "IMIA WGP 28(03)": *"Survey recommendations are made to maintain and improve the insurability of the site. They are not a safety review, are not exhaustive, and do not purport to identify all hazards, present or future, which may exist or occur on the site. Prior to initiation of implementation, the insured must analyse and develop in detail all required safety, engineering and working procedures necessary for implementation. The insured is solely responsible for any potential hazard at the site and for compliance with all applicable laws and regulations".*

Risk recommendations could be raised and applied to any general and/or specific area and aspect of the project. Nonetheless, an indicative categorization into three generic areas could be identified as follows:

i. Organizational structure remit
   This involves changes to the organization structure and project team staffing. If changes occur during the construction period and/or gaps are revealed at critical position, a significant modification of the project's risk profile could be created.
ii. Project-specific risk exposures
   This involves items on the project documentation and/or construction site practices that could be further improved or should be mitigated as they could pose notable risks at specific areas of the project.
iii. Standards and Codes Compliance
   This involves cases where obvious and significant misalignment/breach with the followed standards, codes and specifications are identified and which go beyond acceptance.

A very critical matter that ought to be at the spotlight is the cost implication of any recommendations raised. Any recognizable potential increase of the project budget as a result of the absolute adherence to the recommendations could be tricky and distort their purpose and effectiveness. Hence, the risk surveyor should be mindful of that and always strive to consider and contemplate other viable alternatives. More details on that can be found on the work carried out by Radevsky (2011). Furthermore, it is widely recognized that a risk cannot be totally eliminated and a residual risk will always be present. In that context, the risk recommendation should be raised on a cost-benefit-type basis, with the ALARP principle forming the risk acceptance criteria. Figure 3 below illustratively presents the ALARP concept and the applicability remits of the risk recommendations and anticipated measures.

5 DISCUSSION AND CONCLUSIONS

The structuring and undertaking of a bespoke risk engineering survey is a difficult and complex task, especially at the early stages of a project. Each project is unique with its own distinct

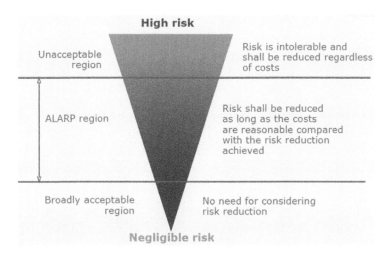

Figure 3. ALARP Principle.

characteristics and the risk survey program should be equally tailored and targeted in order to capture and reflect its needs. From that perspective, the insurers and insureds involved in the project see the substantial benefit of avoiding accidents and losses or minimizing their impact and consequences.

Various aspects and particulars of a risk survey program have been presented and analyzed in this contribution, with the corresponding benefits and merits clearly identified and explained. Nonetheless, these tasks are not free of any problems or concerns about its objectives and process, especially among those who are not familiar with this area. These concerns may include feelings of suspicion and criticism, fear over loss of confidentiality and onerous implications of the raised risk recommendations, worries over the practicability and/or cost implications of these recommendations as well as sentiments of disbelief against any meaningful contribution of the conducted survey to the project.

Notwithstanding the above concerns, the general assertion and sentiment from insurers is clearly in favor of the execution of risk engineering site surveys. Furthermore, it comes with great pleasure to realize that its purpose and objectives become more and more recognized and acknowledged by an increasing number of project stakeholders (such as project owners and contractors) in the context of evaluating the introduced benefits and merits to their project or by preceding ones.

REFERENCES

ITA WG 2. 2006. Guidelines for Tunneling Risk Assessment, *ITA – AITES 2006*, Seoul

IMIA WGP 28 (03). 2003. Risk Management approaches in CAR/EAR projects, *IMIA Conference – Stockholm, 2003*

IMIA WGP 15 (01). 2001. The role of the engineer in the future of engineering insurance, *IMIA Conference*, 2001

Konstantis, T., Konstantis, S., Spyridis, P. 2016. Tunnel Losses: Causes, Impact, Trends and Risk Engineering Management, *World Tunnelling Congress 2016*, San Francisco, California, USA

Konstantis, T. 2017. Tunnel Projects: Risk Exposure, Risk Management and Insurance Coverage – A realistic roadmap, *World Tunnelling Congress 2017*, Bergen, Norway

Konstantis, T. 2018. Underground Projects: Insurance coverage, Insurance market trends and risk cost reduction through proactive risk management, *World Tunnelling Congress 2018*, Dubai, UAE

Radevsky, R. 2011. Risk Engineering for major construction projects, *International Association of Engineering Insurers 44th Annual conference*, Amsterdam, The Netherland

Tunnels and Underground Cities: Engineering and Innovation meet Archaeology,
Architecture and Art, Volume 8: Public Communication and Awareness/Risk Management,
Contracts and Financial Aspects – Peila, Viggiani & Celestino (Eds)
© 2020 Taylor & Francis Group, London, ISBN 978-0-367-46873-6

The role of the Engineer in the Emerald book

J. Maclure
Consultant, Durham, UK

ABSTRACT: It is normal in a design and build contract for the contractor to be solely responsible for the adequacy not only of the design but also the construction operations. However in the FIDIC form of Contract for Underground Works (the "Emerald Book") the Parties agree a Geotechnical Baseline Report (GBR) and the Employer (Capitalized terms used in this paper have the meaning ascribed to them in the Emerald Book) accepts the risk of subsurface conditions that fall outside those agreed in that GBR. Accordingly the Engineer has a specific role to monitor the actual physical conditions encountered and record the corresponding measures taken by the Contractor, in order to assess if and to what extent those measures correspond to the expectations of the GBR and are in accordance with the Contractor's Proposal and the Contract. The Engineer will use these records and the Contractor's quantum measurements to determine adjustments to the Contract Price and times for completion of the Underground Works

1 INTRODUCTION

As is well known to all members of the ITA, there is an ever-growing demand for utilizing underground space for infrastructure. The difficulty in predicting underground behaviour and conditions poses unique challenges regarding construction practicability, time and cost. Thus, allocation of underground risks among the stakeholders becomes critical in underground construction. To address these unique risks the International Tunnelling and Underground Space Association (ITA) and the International Federation of Consulting Engineers (FIDIC) joined forces to draft the new FIDIC Form of Contract for Underground Works (the "Emerald Book"). To accomplish this, the two organizations setup a joint task group (TG10). The Emerald Book has been modelled on the 2017 FIDIC Yellow Book (Conditions of Contract for Plant & Design Build) but with significant innovations tailored to underground construction. Consistent with FIDIC's philosophy of achieving a fair allocation of risks among the parties, the Emerald Book has been drafted with a view to promoting a balanced risk allocation that is specifically adapted to the risks inherent and unique to underground works.

At the time of drafting this paper, the Emerald Book had not yet been published. Therefore there may be some differences in wording between the sub-clauses referenced in this paper and the wording in the published Form of Contract.

Largely as a consequence of the deserved status and reputation of individual engineers, the 18th century saw a separation of design and construction activities. Works contracts – both public and private - were highly individual, drawn up on the advice of an Engineer and reflecting that he was wholly responsible, on behalf of the Employer, for the design, approval of construction and certification for payment of the completed Works. By the end of the 19th century standard forms of traditional Design Bid Build contracts were in common use. Design was usually carried out on behalf of the Employer by a Consulting Engineer. It comes as no surprise that FIDIC, a federation of national Associations of Consulting Engineers, is committed to maintaining the role of the Engineer – indeed it sees this role as critical!

However many iconic structures have come about as a result of Design Build projects where both the design and the construction are the responsibility of the Contractor - and this form of procurement is increasingly perceived by both public and private developers as having several advantages:

(1) A single point of responsibility and liability – no separation of liability between the design and the build.
(2) An enhanced design standard – under a typical design appointment the designer is obliged to do no more than exercise 'reasonable skill and care'. Under a design and construct arrangement, the completed construction must, unless the parties agree otherwise, be reasonably fit for its intended purpose.
(3) The Contractor's experience and specialist knowledge of construction techniques can lead to better problem solving and savings in both time and cost.
(4) And it is not necessary for the Contractor to do all the design - but he must be willing to assume responsibility for the Employer's concept design (if any), and complete the detailed design.

Some have argued, however, that for underground works the risk allocation is unsuitable for Design Build – there are too many unknowns in the ground conditions. The Emerald Book seeks to provide a balanced and equitable allocation of the risks arising from ground conditions in such a way that the advantages of Design Build can be realized.

2   UNDERGROUND WORKS DESIGN RISK ALLOCATION

A fair allocation of risk requires that if underground works in uncertain or difficult ground conditions are likely, the risk of adverse unforeseen physical conditions should be borne by the Employer who owns the project, selected the Site, carried out the pre-tender ground investigation, and is the party that should benefit most if an anticipated risk does not occur.

In the Emerald Book the Geotechnical Baseline Report becomes, at Contract award, the agreed measure of the foreseeable ground conditions, on which the Contractor's Proposal is deemed to have been based. Since the anticipated response of the ground during excavation is dependent on the excavation profile and construction methodology, the initial GBR will have reflected the Employer's reference design and anticipated construction methodology. When completing his Proposal, the Contractor is therefore required to suggest changes to suit his preferred design and construction techniques, and the final 'Contract' GBR must therefore be negotiated and agreed before Contract award. Sub-Clause 4.10.2 states

"The Contractor shall be entitled to rely on the contractually anticipated sub-surface physical conditions as set out in the Geotechnical Baseline Report irrespective of any discrepancies or perceived contradictions that may exist between such physical conditions and the physical conditions described in site data or any other documents including all documents made available by the Employer under Sub-Clause 2.5 [Site Data and Items of Reference] or 4.10.1 [Use of Site Data]." (Unless stated, all quoted extracts are from the FIDIC 2018 Emerald Book)

The Employer's Requirements will include (see Sub-Clause 1.1.40) *"the scope, the preliminary design carried out by or on behalf of the Employer (Employer's reference design), and design and/or other performance, technical and evaluation criteria for the Works"*. The Contractor's design obligation is stated in Sub-Clause 5.1 *"The Contractor shall carry out, and be responsible for, the design of the Works to the extent specified in the Employer's Requirements, and, where applicable, in accordance with the Geotechnical Baseline Report."*

3   THE ENGINEER AND HIS APPOINTMENT

The appointment of the Engineer may have a significant influence not only on the tendered prices, but also on the successful outcome of the project. One of the first questions a bidder asks is "who is the Engineer?", because the reputation of the Engineer in administering the contract, dealing promptly and fairly with issues that may arise, and in issuing instructions and processing payments and variations will have a significant impact on the Contractor's progress and overhead costs.

In keeping with the other FIDIC forms of contract, the Engineer has a double role in the Emerald Book:
(1) On the one hand he is engaged by the Employer, defined as 'Employer's Personnel' (Sub-Clause 1.1.39) and entrusted with the responsibility of administering the Contract.
(2) On the other hand he is required to act 'neutrally' between the Parties when it comes to agreeing or determining any matter or Claim under Sub-Clause 3.7.

The Engineer can be a legal entity (such as a firm of Consulting Engineers) in which case a natural person must be appointed and authorized to represent the Engineer on site and to act on his behalf in carrying out the duties assigned to him under the Contract.

He/she must be 'a professional engineer' – (not just a professional project manager) with 'suitable qualifications', 'experience' and 'competence', and when carrying out duties or exercising authority under the Contract the Engineer must act as a 'skilled professional' (Sub-Clause 3.3.1).

These qualifications reflect the importance of his duties - he must be able to give instructions when necessary, order variations, make fair assessments of payments due, and clarify apparent ambiguities and discrepancies. The scope of the Engineer's duties are very broad – and naturally he will need support from assistants with different specializations such as geologists, mechanical engineers, electrical engineers, planners, and surveyors. If the Engineer is an entity (firm) then a natural person (individual) must be still be appointed to be 'the Engineer'. He/she may appoint and duly authorize an Engineer's Representative (Sub-Clause 3.3) who must be based at the Site for the whole time that the Works are being executed, and must be replaced by a suitable temporary replacement during periods of absence from Site. The authority to make agreements or determinations under Sub-Clause 3.7 cannot be delegated to the Engineer's Representative.

4 THE ENGINEER'S AUTHORITY

Over the years there has been a trend for Employers to restrict the authority of the Engineer by requiring him to get written approval prior to exercising a duty under the Contract. The result has been that responses or decisions were often delayed or withheld, with a resultant breakdown of trust between the Engineer and the Contractor.

In the Emerald Book if the Engineer is required to get the consent of the Employer before exercising a specific authority, this must be clearly stated in the Particular Conditions. However Sub-Clause 3.2 states clearly:

"There shall be no requirement for the Engineer to obtain the Employer's consent before the Engineer exercises his authority under Sub-Clause 3.7 [Agreement or Determination]."

Under the FIDIC 1999 conditions (and if required by the Employer's Requirements), the Engineer could be required to 'approve' Contractor's Documents. However it has been contended that in some jurisdictions such approval carries with it a legal assumption of liability.

The emphasis is now on the Engineer's duty not to 'approve' but to assess if and to what extent a document or submission complies with the Contract, and may be used for the Works. If the submission does not comply, the Engineer must give the Contractor a Notice that the submission fails - to a stated extent - to comply. A 'Notice' under the Contract is now to be identified as such, and must meet certain requirements as set out in Sub-Clause 1.3 [Notices and Other Communications]. In the Emerald book the Engineer does not give "approval' – instead he is required to 'Review' (defined in Sub-Clause 1.1.83) a Contractor's document or submission, or give a 'Notice of No-objection' (Sub-Clause 1.1.67), or to give his 'consent' (defined in Sub-Clause 1.2(g)).

5 THE ENGINEER'S RESPONSE AND TIME LIMITS

Contractors have often complained in the past that time-limits applied against them but not against the Employer or Engineer. This has changed in the 2017 FIDIC conditions, and time-limits now apply to the Engineer's response to Contractor's requests or Notices. Table 1 sets out some of these time-limits.

A failure of the Engineer to respond within the stated time period leads to a deemed outcome which may be not what the Engineer wanted or intended! In general a lack of response leads to a deemed outcome in favour of the Contractor, the notable exception being that a failure by the Engineer to make a determination of a Claim under Sub-Clause 3.7.3 will mean that the Claim is deemed to have been rejected. Table 1 also details the default outcome if the Engineer fails to respond within the specified period.

6 THE ENGINEER'S AUTHORITY TO AGREE OR DETERMINE MATTERS – SUB-CLAUSE 3.7

This is probably the single most demanding duty allocated to the Engineer, and one where he/she is required to be '*acting neutrally and deemed not to be acting for the Employer*'. The Guidance for the Preparation of Particular Conditions Part B suggests that the intention is that "*the Engineer treats both Parties even-handedly, in a fair minded and unbiased manner*" (FIDIC Yellow Book 2nd Ed. 2017 Notes on the Preparation of Special Provisions p21).

Many Sub-Clauses include the requirement for the Engineer to agree or determine. Indeed every situation in which a Party claims an entitlement under Sub-Clause 20.1 leads to the Engineer being required to agree or determine the matter or Claim under this Sub-Clause.

Table 1.   Time Limits on Engineer's Responses and Default outcome.

Sub-Cl.	Description	Response period	Default outcome (If Engineer fails to respond within period)
1.16	Submission of Contract Risk Management Plan for Review	14-days	Deemed Notice of No-objection
3.2.2	Contractors Notice of inaccuracies in Engineer's records	14-days	Deemed confirmed as corrected by Contractor.
3.5	Contractor's Notice – objecting to (or considering instruction is a Variation)	7-days	Deemed revocation of instruction
3.7.3	Consultation to reach agreement	42-days 1st period	Engineer required to proceed to determination
3.7.3	Engineer's fair determination	42-days 2nd period	Claim? – deemed rejection Matter to be agreed or determined?– deemed to be a Dispute
4.3	Contractor's Representative	28-days	Deemed consent
4.4	Sub-contractor – prior consent?	14-days	Deemed consent
4.9.1	Quality Management System	28-days	Deemed Notice of No-objection
4.25	Milestone Completion Certificate	28-days	Deemed Milestone Completion Certificate
5.2.2	Review of Contractor's Documents	Period Ne 21-d	Deemed Notice of No-objection
6.12	Key Personnel	14-days	Deemed consent
7.5	Review proposal – Remediation of Defects	14-days	Deemed Notice of No-objection
8.3	Review of initial programme	21-days	Deemed Notice of No-objection
	Revised programme	14-days	Deemed Notice of No-objection
9.1	Tests on Completion – ready to carry out tests	21-days	Deemed Notice of No-objection
9.1	Review – test certificates	14-days	Deemed Notice of No-objection
10.1	Taking Over Works & Sections	28-days	Deemed Taking Over Cert
10.3	Interference with Tests on Completion	14-days	Immediate issue of Taking Over Certificate
20.2.2	Engineer's initial response to Notice of Claim	14-days	Notice of Claim deemed valid
20.2.4	Fully detailed Claim	14-days	Notice of Claim deemed valid even without statement of contractual/legal basis.

The procedure starts with the Engineer attempting to get the Parties to resolve the matter by agreement. Negotiation and mediation skills are needed first! Only if no agreement is achieved must he move on to make a fair determination. A step-by-step procedure is required:

Step 1- consultation with the Parties to encourage them to reach agreement – The Engineer must record the consultation and give Notice of the agreement or record the Parties failure to reach agreement, within the time limit for consultation (42-days unless otherwise proposed and agreed). Sub-Clause 3.7.1 [Consultation to reach agreement] *"The Engineer shall consult with both Parties jointly and/or separately, and shall encourage discussion between the Parties in an endeavour to reach agreement"*. If agreement is reached the Engineer has to give a Notice to both Parties with a record of the agreement which the Parties must sign.

Step 2- making a fair determination (in the absence of agreement) within the time limit for determination (42-days from the Notice of step 1, unless otherwise proposed and agreed). Sub-Clause 3.7.2 [Engineer's Determination] says *"The Engineer shall make a fair determination of the matter or Claim, in accordance with the Contract, taking due regard of all relevant circumstances."* The consequences of the Engineer's failure to meet the time limits is stated in Sub-Clause 3.7.3 [Time limits] *"If the Engineer does not give the Notice of agreement or determination within the relevant time limit*:

(i) in the case of a Claim, the Engineer shall be deemed to have given a determination rejecting the Claim; or

(ii) in the case of a matter to be agreed or determined, the matter shall be deemed to be a Dispute which may be referred by either Party to the DAAB for its decision under Sub-Clause 21.4." Dispute which may be referred by either Party to the DAAB for its decision under Sub-Clause 21.4."

Step 3a- giving effect to the agreement or determination. Sub-Clause 3.7.4 [Effect of the agreement or determination] says *"Each agreement or determination shall be binding on both Parties (and shall be complied with by the Engineer) unless and until corrected under this Sub-Clause or, in the case of a determination, it is revised under Clause 21 [Disputes and Arbitration]. "*

Step 3b- if either Party is dissatisfied with the determination, then within 28-days after receiving the determination, the dissatisfied Party must give a Notice of Dissatisfaction ("NOD") under Sub-Clause 3.7.5 [Dissatisfaction with the Engineer's determination].

Step 4- a NOD allows either Party to then refer the matter to the DAAB for decision under Sub-Clause 21.4[Obtaining DAAB's Decision].

7   THE ENGINEER'S SPECIFIC DUTY AND AUTHORITY FOR EXCAVATION AND LINING

The Contractor will have prepared his Design Build proposal on the basis of the GBR and the Employer's Requirements, entered production rates against the activities set out in the Baseline Schedule, and completed the Schedule of Rates and Prices, based on his own design and construction methodology. The Accepted Contract Amount is deemed to cover all his obligations under the Contract.

However under the Emerald Book form of contract, the Employer expects not only to pay for increases in actual quantities of Excavation and Lining Works (within the boundaries of the anticipated conditions described in the GBR), but also that if the actual conditions are better than anticipated he will be entitled to benefit from reduced quantities and a shorter Time for Completion. If the physical conditions fall outside the GBR, the Contractor will be entitled to claim under Sub-Clause 4,12 for Unforeseeable Physical Conditions

Therefore, the Employer entrusts to the Engineer the duty to monitor the Underground Works and to record the actual conditions encountered and measures taken by the Contractor to assess to what extent those measures are compliant with the Contractor's Proposal and the Contract.

Accordingly Sub-Clause 3.2.2 [The Engineers Specific Duty and Authority for Excavation and Lining] provides *"The Engineer shall monitor and record progress of the execution of the Excavation and Lining, for compliance with the Contractor's obligations under Sub-Clause 4.24 [Excavation and Lining]."*

The Engineer is not specifically required to monitor and record all the Underground Works, only the "Excavation and Lining" Works, and both are defined terms in the Emerald Book. "Excavation" is defined in Sub-Clause 1.1.44 as *"all work undertaken to excavate, support and secure the space for the Underground Works, including but not limited to exploratory investigations, preliminary mitigation measures, ground treatment, excavation, ground support measures, seepage treatment, temporary and ancillary works (in each case if any)* and "Lining" is defined in Sub-Clause 1.1.61 as *"the permanent lining works of the Excavation, whether constructed at the same time or at a later stage, including waterproofing, contact grouting and backfill (if any)"*.

These two underground work activities are considered to be those which are directly influenced by the ground conditions and whose design and planned construction methodology is dependent on the baselines agreed in the GBR. The Contractor's obligations are more fully described in Sub-Clause 4.24 [Excavation and Lining]

"The Contractor shall take whatever measures are stated in the Employer's Requirements, Particular Conditions, Geotechnical Baseline Report and/or are agreed in a method statement, and/or are necessary for the safety, stability, timely progress and/or execution of the Works. The Contractor shall submit on a daily basis his interpretations of the data from the sub-surface and surface monitoring programmes, if any.

The Contractor shall endeavour to reach agreement with the Engineer that such measures are (or, if the measures have already been taken, were) necessary for the execution of the Excavation and Lining in accordance with the Contract. The agreement or lack thereof shall be recorded as per Sub-Clause 3.2.2 [Engineer's Specific Duties and Authority for Excavation and Lining]"

A copy of the records will be provided by the Engineer to the Contractor within an agreed time and the *"Contractor shall examine the records and be deemed to have accepted them unless within 7 days from receipt of the records he gives a Notice to the Engineer of the respect in which the records are asserted as being inaccurate."*

After receiving the Contractor's Notice the Engineer has 14-days to review, vary or confirm the records. Failure to respond within the 14-day time period will mean that the records are confirmed as corrected by the Contractor. If the Engineer disagrees with the Contractor's Notice and wishes to maintain the records without change, he must follow the procedure set out in Sub-Clause 3.7 [Agreement or Determination] as described in Paragraph 6 above, with the first time limit for consultation commencing on the date the Engineer receives the Contractor's Notice.

8   THE ENGINEER AND UNFORESEEABLE PHYSICAL CONDITIONS SUB-CLAUSE 4.12

If conditions are encountered which are outside the possible conditions foreseen in the GBR, the provisions of Sub-Clause 4.12 [Unforeseeable Physical Conditions] come into play. The Contractor will be required to give a Notice setting out the reasons why he considers the physical conditions to be Unforeseeable, the Engineer must then inspect and investigate, give instructions as necessary, and agree or determine whether and to what extent the physical conditions encountered were Unforeseeable, and the extent of the Contractor's entitlement to Payment and or Extension of Time, all in accordance with Sub-Clause 4.12. Sub-Clause 4.12.1 [Contractor's Notice] provides

"After discovery of such physical conditions, the Contractor shall give a Notice to the Engineer, which shall:

(a) *be given as soon as practicable and in good time to give the Engineer opportunity to inspect and investigate the physical conditions promptly and before they are disturbed;*

(b) *describe the physical conditions, so that they can be inspected and/or investigated promptly by the Engineer;*

(c) *set out the reasons why the Contractor considers the physical conditions to be Unforeseeable; and*

(d) describe the manner in which and the extent to which the physical conditions will have an adverse effect on the progress and/or increase the Cost of the execution of the Works.

The FIDIC Yellow Book at Sub-Clause 4.12.2 required the Engineer to inspect and investigate *"within 7 days"*. However if such conditions are not anticipated by the GBR, the measures required and the Contractor's entitlement to time and payment will depend on the Engineer's acknowledgement of the unforeseeable physical condition, and the Contractor's operation may be halted waiting for the Engineer's subsequent instruction. The Guidance for the preparation of Particular Conditions Part B – Special Provisions, *"Each time period stated in the General Conditions is what FIDIC believes is reasonable, realistic and achievable in the context of the obligation to which it refers, and reflects the appropriate balance between the interests of the Party required to perform the obligation, and the interests of the other Party whose rights are dependent on the performance of that obligation."* Accordingly in the Emerald Book at Sub-Clause 4.12.2 (Engineer's inspection and investigation), safety is the first priority of both parties, but thereafter the Engineer should make his inspection "as soon as possible" – the first paragraph now states:

"The Engineer shall inspect and investigate the physical conditions (if safe to do so) as soon as possible or as agreed with the Contractor, after receiving the Contractor's Notice."

9 AGREEING OR DETERMINING MEASUREMENTS - SUB-CLAUSE 13.8.1

Since the Parties have agreed prior to the award of the Contract that the Geotechnical Baseline Report represents the agreed measure of the foreseeable ground conditions, upon which the Contractor is deemed to have based his proposal, then the Contractor is entitled to payment for the actual quantities of 'foreseeable' Excavation and Lining Works carried out. Accordingly the actual Excavation and Lining Works carried out in conditions conforming to those described in the agreed baselines in the GBR are subject to measurement.

In the FIDIC Red Book (FIDIC 2017 2nd Edition Conditions of Contract for Construction for Building and Civil Engineering Works designed by the Employer) the responsibility for measurement is allocated to the Engineer and measurements are made of the net theoretical quantity of the Permanent Works shown on the drawings or other records prepared by the Engineer. In the Emerald Book the responsibility for the design is with the Contractor, so responsibility for the measurement of Underground Works is given to the Contractor under Sub-Clause 13.8.1 [Responsibility for Measurement] which states *"Unless agreed otherwise, the Contractor shall be responsible for the measurement and shall submit relevant measurements with full supporting records to the Engineer at the intervals stated in the Contract Data (if not stated then at monthly intervals)."*

The Contractor will provide full supporting details with the measurement, and the Engineer is then tasked with reviewing, agreeing or varying by determination the Contractor's measurement. Sub-Clause 13.8.1 goes on to provide *"On receipt of the measurement records, the Engineer shall proceed under Sub-Clause 3.7 [Agreement or Determination] to agree or determine the measurement (and, for the purpose of Sub-Clause 3.7.3 [Time limit], the date when the Engineer receives the measurement records shall be the date of commencement of the time limit under Sub-Clause 3.7.3 [Time limit]."*

10 RISK MANAGEMENT

The Emerald book promotes good practice by the identification of risks, their allocation between the parties to the Contract and the management and control of risks as recommended by the ITIG Code of Practice for Risk Management of Tunnel Works, through the use of a Project Risk Management Plan in accordance with the details set out in the Employer's Requirements (ITIG 2nd Ed May 2012 A Code of Practice for Risk Management of Tunnel Works, clause 9.3 p.17).

Sub-Clause 1.16 [Contract Risk Management Plan] provides for the Contractor to complete a Contract Risk Register and a Risk Management Plan and provide it to the Engineer for his Review. The Engineer is required to respond within 14-days with a Notice of No-Objection or a Notice stating that the Risk Register or Risk Management Plan fails to comply with the Contract with reasons.

The Contract Risk Management Plan will lead to Risk Management meetings and/or to Advance Warnings in accordance with Sub-Clause 8.4 [Advance Warning] requiring either Party or the Engineer to advise the others of any known or probable future events or circumstances that may adversely affect the work, or cause a delay or increase in the Contract Price.

11 THE ENGINEER AND CLAIMS

The claims procedure in Clause 20 now applies equally to Contractor's claims, and to Employer's claims. So the 28-day time bar applies equally to a late Notice of Claim from either the Contractor or the Employer under Sub-Clause 20.2.1 [Notice of Claim] which provides

"The claiming Party shall give a Notice to the Engineer, describing the event or circumstance giving rise to the cost, loss, delay or extension of DNP for which the Claim is made as soon as practicable, and no later than 28 days after the claiming Party became aware, or should have become aware, of the event or circumstance (the "Notice of Claim" in these Conditions).

If the claiming Party fails to give a Notice of Claim within this period of 28 days, the claiming Party shall not be entitled to any additional payment, the Contract Price shall not be reduced (in the case of the Employer as the claiming Party), the Time for Completion (in the case of the Contractor as the claiming Party) or the DNP (in the case of the Employer as the claiming Party) shall not be extended, and the other Party shall be discharged from any liability in connection with the event or circumstance giving rise to the Claim."

Upon receipt of a Notice of Claim, the Engineer is required to give the claiming Party a Notice within 14-days if he/she considers that the Notice of Claim was not given in the required 28-days. However if the claiming Party disagree with this ruling it may provide, with its fully detailed claim, reasons that in its opinion justify the late submission. If the Engineer does not give such an initial response in 14-days the Notice of Claim is deemed to be valid.

The Claiming Party is required to keep contemporary records, and allow the Engineer to inspect them when required

Within 84-days (or such other period that may be proposed by the claiming Party and agreed by the Engineer) of the claiming Party becoming aware of the event or circumstance, the claiming Party is required (Sub-Clause 20.2.4) to submit a fully detailed claim, or risk receiving a Notice from the Engineer that the Notice of Claim is considered to have lapsed and no longer be valid. The fully detailed claim must also contain a statement of the contractual and/or legal basis of the Claim. If the claiming Party fails to provide this statement within this time period the Engineer should, within 14-days of the expiry of the 84 days, give a Notice stating that the Claim Noticehas lapsed and is no longer valid. Again, if the Engineer does not give such a Notice within 14-days the Notice of Claim is deemed to be valid, and once again if the Claim Notice has been declared lapsed, but the claiming Party disagrees, it must give a Notice to this effect, with details, to the Engineer, who must then review the circumstances. So the claiming Party faces two time-bar provisions – the first relating to the date of the Notice of Claim, and the second relating to the submission of a statement of the contractual and/or legal basis, but in both cases the claiming Party can submit justification of the circumstances to the Engineer to review, agree or determine.

On receipt of the fully detailed Claim the Engineer will proceed under Sub-Clause 3.7 [Agreement or Determination] to agree or determine the matter of the claim.

12 CONCLUSIONS

The role of the Engineer in the Emerald Book is extraordinarily challenging. Not only does he/she have the normal duties expected of the Engineer in a Design Build setting – such as to

Review Contractor's Documents, to clarify and resolve ambiguities, to witness tests, monitor progress, and give consent or No-objection to the Contractor's operations, but also, in view of the agreed risk allocation represented by the GBR, the Engineer has the added duty to monitor and record the measures taken by the Contractor in response to the actual conditions encountered. This duty will be crucial to the determination of foreseeable physical conditions. In addition, since the Excavation and Lining Works are to be measured by the Contractor for the purposes of payment and adjustment of Time for Completion, the Engineer will receive the measurement records made by the Contractor, agree or determine the measurement and adjust the Time for Completion and Contract Price accordingly.

The key qualification requirements of the Engineer will undoubtedly be competence and experience. In the Emerald Book (following FIDIC 2017 2nd Editions) the focus ends up on the Engineer's duty in connection with agreeing or determining in which he/she also requires communication, negotiation and mediation skills. The Engineer, a natural person, may not delegate that authority – even to the Engineer's Representative!

ACKNOWLEDGEMENT

This contribution is based upon the work of the FIDIC Task Group 10 "New Form of Contract for Tunneling and Underground Works". The author wishes to thank FIDIC, the ITA and his colleagues Gösta Ericson (IC Consultants, Lund, Sweden), Hannes Ertl (D2 Consultants, Linz, Austria), Andres Marulanda (Ingetec, Bogotà, Columbia), Charles Nairac (White & Case LLP, Paris, France), Matthias Neuenschwander (Neuenschwander Consulting Engineers Ltd., Bellinzona, Switzerland) and Martin Smith (Matrics Consult Ltd., Seoul, Republic of Korea) for their important contributions.

Caveat: at the moment of writing of this article, the FIDIC Emerald Book is still under review. Part of the content may therefore be in contrast with the published Form of Contract. The reader should always consult the published FIDIC Form of Contract for Underground Works.

Contractual time for completion adjustment in the FIDIC Emerald Book

A. Marulanda
Ingetec, Bogotá, Colombia

M. Neuenschwander
Neuenschwander Consulting Engineers Ltd, Bellinzona, Switzerland

ABSTRACT: The Emerald Book, the new FIDIC Standard Form of Contract developed for Underground Works, was prepared following a balanced risk allocation approach, whereas the Employer retains the ground related risk and the Contractor is responsible for the performance related risk for given ground conditions. The contract uses a Geotechnical Baseline Report (GBR) to set out the sub-surface conditions anticipated under the contractually agreed Excavation and Lining design and construction methods and states the allocation of the risks contemplated for sub-surface conditions between the Parties.

If the encountered ground conditions differ from those set forth in the GBR, the contract contemplates that the Contractual Time for Completion should be adjusted, increased if more unfavorable conditions are encountered than foreseen in the GBR, or shortened if more favorable conditions are encountered, of and to the extent that the critical path is affected.

This paper describes the philosophy behind this contractual mechanism and the time adjustment procedures contemplated in the Emerald Book.

1 INTRODUCTION

It has long been recognized that underground construction projects exhibit unique characteristics that demand special contractual considerations for their successful completion. Contractual principles and practices effective in other types of construction projects are neither necessarily applicable nor adequate for tunneling projects. Moreover, contractual practices and provisions play a key role in determining the efficacy and efficiency of design and construction methods in underground projects.

Underground construction is highly dependent on the geological and geotechnical ground mass characteristics that have a defining influence on the required means and methods needed for tunneling. Moreover, the difficulty of predicting ground behavior and unforeseen conditions imply an inherit uncertainty for tunneling projects, leading to unique risks.

The difficulty in predicting underground behavior and conditions poses unique challenges regarding construction practicability, time and cost. Thus, allocation of underground risks among the stakeholders becomes critical in underground construction. To address these unique risks the International Tunneling and Underground Space Association (ITA) and the International Federation of Consulting Engineers (FIDIC) joined forces to draft the new FIDIC Form of Contracts for Underground Works (the "Emerald Book"). To accomplish this, the two organizations set up a joint task group (TG-10).

The Emerald Book has been modelled on the 2017 FIDIC Yellow Book (Conditions of Contract for Plant & Design Build), but with significant innovations tailored to underground construction. Consistent with FIDIC's philosophy of achieving a fair allocation of risks among the parties, the Emerald Book has been drafted with a view to promoting a balanced risk allocation that is specifically adapted to the risks inherent and unique to underground works.

The most relevant aspects of the Emerald book will be covered in a series of articles prepared for the WTC 2019 - Naples. This article expands on the discussion of how the proposed mechanism for adjusting the time for completion works in the new Emerald Book depending on the encountered ground conditions.

2 BACKGROUND PRINCIPLES BEHIND THE EMERALD BOOK

To effectively deal with the nature of underground works, the task group identified a series of concepts and issues that were the basis for the modifications of the new standard for of contract for Underground Works. The most relevant include:

- Allocation of risk. The ground and groundwater related risks should be assigned to the owner, as he will be the one benefitting from the completed project and is the party that can best control them. The performance related risk for the expected ground conditions should be assigned to the Contractor.
- Inclusion of a contractual geotechnical baseline. If a differing site condition clause is included in the contract documents of an underground project, it has to be accompanied by a geotechnical contractual baseline that sets the contractual limits of the conditions anticipated to be encountered during construction, thus providing clear distinctions in the contract documents between expected and unexpected subsurface conditions.
- Disclosure of all available geological and geotechnical information. All available information should be shared and transmitted to contractors during the tendering stage. The use of exculpatory language in the geotechnical baseline document should be avoided
- Inclusion of a contractual unforeseeable physical conditions clause. In case actual encountered ground conditions differ from the predicted ones, a differing site conditions clause should be incorporated in the contract documents to allow relief from the unforeseen conditions and allow the contractual flexibility to compensate for them.
- Implementation of a ground classification system for supporting particular conditions that properly reflect the effort for excavation and stabilization. If the ground conditions are contractually classified based on the measures the contractor has to take in order to excavate and support the ground, claims and disputes are reduced.
- Time for completion is largely influenced by ground conditions. For this reason, time adjustment according to ground conditions should be regulated by the contract.
- A flexible mechanism for remuneration according to ground conditions, foreseen and unforeseen should be implemented. Thus, a unit price contract payment system for items that are affected by differing site conditions should be used. The unit price structure should be organized to facilitate the distinction between fixed costs, time-related, value related costs, and quantity-related costs.

Some of these ideas and principles are expanded in the following sections.

3 RISK ALLOCATION IN THE EMERALD BOOK

One of the unique aspects of Underground Works that distinguish them from other types of construction is the lack of knowledge regarding the sub-surface physical conditions and the reaction of the surrounding ground mass to the excavation and support.

A well-accepted principle in risk management considers that a risk should be allocated to the party in the best position to manage it. Thus, in the interest of economy, the related risks should be allocated to the Party that is best prepared to control them:

a) The risk of the sub-surface physical conditions not corresponding to the expectations before tenders are submitted should be allocated to the Employer. The Employer is the Party who had the possibility of assessing those risks during the preparation of tender documents.

b) The risk of the production rates and relevant cost not corresponding to the expectations within a given set of sub-surface conditions should be allocated to the Contractor. The Contractor is the Party who has the experience in the detail design and construction for such conditions, including the avoidance and/or limitation of impact on third parties.

This risk allocation philosophy implies that either Party shall with respect to the risk the Party has been entrusted with, bear the loss resulting from the encountered conditions that are worse than the expected conditions and shall on the other hand enjoy the benefit resulting from the encountered conditions that are better than the expected conditions. This principle is incorporated in the New FIDIC Emerald Book.

4 SETTING THE BASELINE FOR THE CONTRACTUALLY FORESEEN PHYSICAL CONDITIONS

To set the allocation of subsurface risk in the contract documents, a Geotechnical Baseline Report (GBR) is used. This document shall be the single source of allocation of sub-surface risks due to physical conditions. It is very important that the GBR be part of the contract documents, not just a reference. In fact, the GBR and the Baseline Schedule shall have the highest possible priority amongst the Contract documents. A companying paper also prepared by TG-10 for WTC 2019 expands on the Geotechnical Baseline concept and provides more detailed guidance on how it should be prepared and what should it include (Ericson 2019).

A geotechnical baseline is a contractual baseline, not necessarily geotechnical fact (Essex, 1997). Baselines are partly developed from the factual data derived from the subsurface investigations, but they also incorporate interpolations and extrapolations of that data, precedent, and engineering judgment.

The reaction of the surrounding ground to the excavation and support processes depends not only on the geotechnical parameters of the ground mass, but also on the design concept proposed by the Employer and the methods of construction for excavation and support. The GBR shall consequently contain the Employer's proposal for a feasible method of excavation and support differentiated between Excavation and Lining Works for the purposes of measurement.

The anticipated quantities of Excavation and Lining items of work for the Underground Excavation included in the GBR shall be itemized in one or more Baseline Schedules. The tenderer shall then enter the corresponding production rate against each different item of work.

Due to the overriding issue of Time for Completion, the Baseline Schedules, which provides the time link to the Completion Schedule, takes precedence over the GBR.

If the encountered physical conditions of the ground vary within the limits stated in the GBR, there will be an influence on time and cost for the Excavation and Lining Works.

The contractual time and cost for Excavation, (including all necessary support) shall be adjusted accordingly based on the variations of the physical conditions of the ground encountere within the limits stated in the GBR. This adjustment shall be performed according to the amounts and production rates stated in the Baseline Schedules.

Variations within the baseline limits stated in the GBR, for example regarding a given percentage distribution of excavation and support classes, will be re-measured according to a procedure defined in the Emerald Book.

The GBR will need to use a contractual ground classification system that properly reflects the effort (time and cost) of excavating and supporting the cavity in the expected ground conditions. For this purpose, the definition of a ground classification system, together with the associated quantitative criteria for the application on Site, is necessary.

5 COST STRUCTURE TO FACILITATE REMEASUREMENT

Cost items associated with ground conditions, such as the underground excavation, support and lining should be regularly remeasured depending on the actual conditions encountered in comparison to the anticipated ones in the GBR. All other works shall be handled in the lump sum component of the contract price.

A major portion of the costs incurred by Contractors is related to time rather than to the quantities produced (in particular, technical, commercial and administrative overheads, depreciation and maintenance of Contractor's Equipment, leasing costs etc.). The Contract documents shall consequently provide for time-related items which, allow for corresponding adjustments in the Contract Price according to the adjustments in Time for Completion.

Hence, to facilitate remeasurement, the contract cost schedules presented with the tendering documents should differentiate quantity related, times related and value related cost items (Neuenschwander & Marulanda 2019).

6 ADJUSTMENT OF TIME FOR COMPLETION

The evaluation of the time for excavation and support depends on the expected subsurface physical conditions and on the production rates submitted by the Contractor. As the subsurface physical conditions are within the risk sphere of the Employer and the production rates are within the risk sphere of the Contractor, the time available to the Contractor for the Underground Excavation and Lining stages shall be measured and adjusted against the difference between the sub-surface physical conditions expected (as quantified in the Baseline Schedule) and the sub-surface physical conditions actually encountered.

The adjustment mechanism shall be clearly stated in the tender documents. No adjustment shall be provided for any difference between the performance rates stated in the Contractor's Proposal and the performance rates actually achieved for the specified set of sub-surface physical conditions in the Baseline Schedule.

The Milestones that are relevant to the Employer shall be clearly stated in the Completion Schedule, which shall be compatible with and based on the quantities itemized in the Baseline Schedule. Time adjustments and Extensions of Time, if any, shall be calculated and determined by reference only to the Baseline Schedule and the Completion Schedule.

Unlike events leading to an Extension of Time (i.e. Unforeseeable Physical Conditions and/or Variations), time adjustments may also lead to a reduction in the Time for Completion if more favorable conditions than foreseen in the GBR are encountered.

7 ADJUSTMENT OF TIME IN PRACTICE

Understanding the application of the time adjustment mechanism proposed in the Emerald Book is facilitated using a simplified example. Consider a 1000m long tunnel. The GBR prepared by the Employer anticipated two support sections A and B as shown in Figure 1. The Employer has anticipated in the Baseline Schedule (or BS) that both support sections will be 500m long. However, section exhibits A more favorable ground conditions than B. Consistently, the Contractor offered in his tender an advance rate for Section A of 10m/day and 1m/day for section B. The Time for completion for the foreseen bidding conditions would the 550 days (50 days for section A and 500 days for Section B).

During construction, encountered conditions could vary from that foreseen in the Baseline Schedule. Two scenarios are evaluated to clarify the time adjustment mechanism proposed in the Emerald Book for better and worse ground conditions with respect to conditions foreseen in the Baseline Schedule.

7.1 *Encountered Conditions are Better than foreseen in the Baseline Schedule.*

As explained before, the Emerald Book proposes to reduce the Time for Completion if encountered conditions are better than foreseen in the Baseline Schedule. If the Employer is

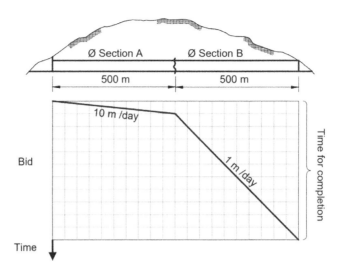

Figure 1.   Base Scenario. Bid with Ground Conditions as per anticipated in the GBR.

bearing the risk of ground conditions, he should also benefit if more favorable conditions are encountered. The more favorable conditions should have allowed the Contractor to advance according to his proposed rate for better ground, thus reducing his foreseen excavation and support time and costs as well. This would the case in a scenario where the encountered length for Section A was 600m and Section B 400m as illustrated in Figure 2. In this case, the Time for Completion would be 460 days.

7.2   *Encountered Conditions are worse than foreseen in the Baseline Schedule*

Conversely, if encountered ground conditions are worse than foreseen in the BS, the Time for Completion should be extended. The Contractor should be compensated in time and cost for encountering more challenging conditions than foreseen in the BS. This would the case if the length of section A was 400m and Section B 600m as shown in Figure 3 (Scenario 2a). In this scenario the Time for Completion would be 660 days.

An alternative scenario (2b) is presented in Figure 4. In this example conditions are worse than anticipated, same section lengths as Scenario 2a, but the Contractor performed with

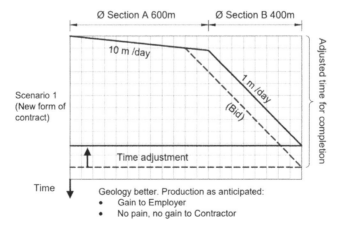

Figure 2.   Scenario 1. Encountered Ground Conditions better than anticipated in the Baseline Schedule.

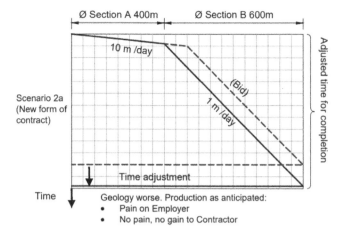

Figure 3.   Scenario 2a. Encountered Ground Conditions worse than anticipated in the BS.

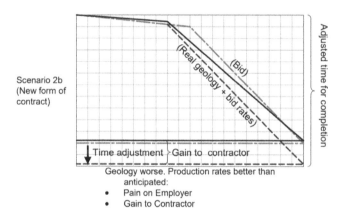

Figure 4.   Scenario 2b. Encountered Ground Conditions worse than anticipated in the BS, but Contractor advances with higher rates than offered.

faster advanced rates than offered. In this case, the Contractor would perceive a gain, while the Employer should recognize the extra cost and time for the more challenging conditions encountered.

8   CONCLUSION

The Emerald Book, the new FIDIC Standard Form of Contract developed for Underground Works, was prepared following a balanced risk allocation approach, whereas the Employer retains the ground related risk and the Contractor is responsible for the performance related risk for given ground conditions. The contract uses a Geotechnical Baseline Report (GBR) to set out the sub-surface conditions anticipated under the contractually agreed Excavation and Lining design and construction methods and states the allocation of the risks contemplated for sub-surface conditions between the Parties.

If the encountered ground conditions vary within those set forth in the GBR, there will be an influence on time and cost for the Excavation and Lining Works. The Emerald Book contemplates that the Contractual Time for Completion should be adjusted, increased if more unfavorable conditions of those foreseen in the GBR or shortened if more favorable conditions are encountered, of and to the extent that the critical path is affected. On the other hand, if the ground conditions differ from those specified in the GBR, they shall be considered "Unforeseeable Physical Conditions", and the relevant sub-clause in the Emerald Book (4.12, "Unforeseeable Physical Conditions" shall apply.

The reader is recommended to read this article in conjunction with the series of other articles prepared by TG-10 on the Emerald Book for WTC 2019 in Naples.

ACKNOWLEDGEMENT

This contribution is based upon the work of the FIDIC Task Group 10 "New Form of Contract for Tunneling and Underground Works". The authors wish to thank FIDIC, the ITA and their colleagues Gösta Ericson (IC Consultants, Lund, Sweden), Hannes Ertl (D2 Consultants, Linz, Austria), James Maclure (Independent Consultant, Durham, United Kingdom), Charles Nairac (White & Case LLP, Paris, France) and Martin Smith (Matrics Consult Ltd., Seoul, Republic of Korea) for their important contributions.

Caveat: at the moment of writing of this article, the FIDIC Emerald Book is still under review. Part of the content may therefore be in contrast with the published Form of Contract. The reader should always consult the published FIDIC Form of Contract for Underground Works.

REFERENCES

Ericson G. 2019. "The Geotechnical Baseline Report in the new FIDIC Emerald Book – Suggested Developments". ITA, WTC 2019. Naples.
Neuenschwander M. and Marulanda A., 2019. "Measuring the Excavation and Lining in the Emerald Book". ITA, WTC 2019. Naples.

Tunnels and Underground Cities: Engineering and Innovation meet Archaeology,
Architecture and Art, Volume 8: Public Communication and Awareness/Risk Management,
Contracts and Financial Aspects – Peila, Viggiani & Celestino (Eds)
© 2020 Taylor & Francis Group, London, ISBN 978-0-367-46873-6

Managing a large-scale engineering project: Standards, methods and technologies

O. Mazza
Spea Engineering SpA, Italy

A. Selleri
Autostrade per l'Italia SpA, Italy

ABSTRACT: The design and approvals process for the Genoa Gronda motorway bypass –
from the public debate in 2009, through definitive design and the environmental impact study
of 2014, to the project implementation plan in 2018 – kept Autostrade and Spea engaged for
ten years and required hundreds of thousands of engineering hours, with hundreds of service
providers carrying out investigations, analyses, and specialist studies. This document describes
how this organisation - in terms of the resources and technologies used - was able to face up
to this colossal engineering challenge, managing the enormous flow of information generated
by a succession of increasingly detailed design studies, whilst at the same time interfacing with
the authorising bodies. The document begins by describing the regulatory framework, identifying the most important actors in the project and their responsibilities, before moving on to
the most important logical sequences that led from the engineering concept design to the final
design. Finally, the main contract-related aspects and the risk-management processes applied
to the development of the project are described.

1   THE GRONDA PROJECT AND THE APPROVAL PROCESS

The Gronda Genoa project consists in a new section of motorway with two lanes for each direction that doubles the section of the current A10 Genova-Savona motorway where it crosses
through the city of Genoa, thereby streamlining the sections of the A7 and A12 motorways that
lie between the Genova Est, Genova Ovest and Bolzaneto junctions. The special morphological
conformation of the terrain that the new infrastructure will occupy has forced the designers to
resort to intensive use of underground structures, so much so that the new road system is almost
entirely underground and crosses through 23 tunnels that cover a total of 54 km and that feature
sections ranging in diameter from 12 m^2 of the emergency shafts to the 200 m^2 of the TBMs
that will excavate the doubling of the A10, all the way to the 500 m^2 of the chambers where the
motorways interconnect. The work plan envisages the excavation of over 11 million cubic
metres of rock bank, of which about 50% poses the risk of containing asbestos fibres.

An innovative design solution was adopted for transporting this material from the tunnel
excavations to the sedimentation beds: a "slurry duct" using a pipeline approximately 9100 m
long overall that will mainly follow the same line as the Polcevera stream.

The route of the Genoa Gronda is the result of a long period of design and of exchange
with the regional and local authorities and with the citizens of Genoa that has lasted about
thirty years, during which many design plans were conceived but never approved. In December 2008, the Municipality of Genoa and Autostrade per l'Italia decided to submit to the
public five design ideas for the upgrading of the motorway hub of Genoa. The Public Debate,
supervised by an independent panel headed by Prof. Luigi Bobbio (University of Turin),
lasted three months (from February to April 2009). This was the first time in Italy that the

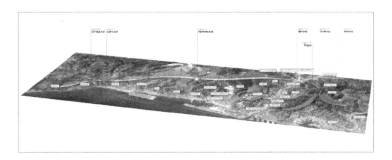

Figure 1.   Three-dimensional view of the route.

public was consulted in the authorization phase of a large-scale infrastructure. The Debate made it possible, albeit with some difficulty, to examine and study further the various routes proposed, to review the traffic volume estimates, to reduce the environmental and, most importantly, the social impacts deriving from the construction of the infrastructure, and to reach an agreement regarding the forms of monitoring and of control of the various design and construction phases of the works. The dialogue with the territory then continued through the creation of a local Observatory, which is still in operation, run by the local authorities and by citizen representatives.

The Public Debate ended with the presentation to the City Council of 2009 of the solution that Autostrade per l'Italia identified as being the "best", after which the Preliminary Design was quickly drafted, shared and approved by all of the players involved (the Ministry of Infrastructure and Transport, the Liguria Region, the Province of Genoa, the City of Genoa, the Port Authorities of Genoa, Anas SpA and Autostrade per l'Italia SpA), thereby resulting in the undersigning of a Memorandum of Understanding in 2010.

Immediately afterwards, the Final Design activities were launched, and its final validation by Anas arrived in 2011, followed by the positive Environmental Impact Assessment by the Ministry for the Environment in 2014, by the authorization of the 'Conferenza di Servizi' by all of the entities involved in 2015, and by the approval of the Ministry of Infrastructure in 2017, which led to the start of the Executive Plan.

2   LARGE-SCALE PROJECTS: A CHALLENGE FROM A TECHNICAL AND ORGANISATIONAL POINT OF VIEW

The Gronda Executive Plan that – as we have seen – was created after a very long approval process, has been listed by the client 'Autostrade per l'Italia' as a maximum criticality project so as not to risk invalidating, right at the last stage of the engineering development phase, the fruit of so many years of study and of work. Moreover, the sheer size and cost of the project places it among the largest currently under way in Europe and fully justifies the very high level of attention placed on it by the client.

The preparation of a Project Implementation Plan for a large-scale infrastructure project – i.e. for an initiative with a budget in excess of 100 million euros, in line with the definition of *'work of significant value'* found in article 104 of the Contract Code for Competitive Bidding – goes hand in hand with the preparation of numerous aspects relating to civil engineering and represents one of the greatest challenges in technical and organisational terms faced by the Project Designer.

Under Italian legislation, which pending the decree promised in art. 23, paragraph 3 of Legislative Decree 50/2016 continues to apply the provisions contained in part II, title II, chapter I and title XI, chapters I and II and the annexes referenced therein by Presidential Decree no. 207 of 5 October 2010, the Project Implementation Plan *'constitutes the embodiment in engineering terms of all the work and, therefore, comprehensively defines the project due to be implemented in each of its architectural, structural and engineering aspects*' (article 33 of Presidential Decree 207/2010) and therefore requires an in-depth technical appraisal to enable the Contractor to proceed without delay, in a way that respects the environment and ensures safety during the completion of the work, irrespective of its complexity or difficulty.

The Project Implementation Plan must also take into account all the deadlines, requirements and observations that have arisen during the preceding approval procedure, often extremely long and torturous, given that – as also set out in article 33 of Presidential Decree 207/2017 – '*[it must be] drafted in full compliance with the final project and the requirements stipulated in the planning permission documentation or during the assessment of its conformity with urban planning requirements or emanating from the "conferenza di servizi" [conference for coordinating and reconciling the various public players involved] or those relating to compliance with environmental standards, if contemplated*'.

Therefore, apart from a sufficient level of competence and technical preparation, the Project Design Team must also possess an extensive logistical and organisational structure with a view to ensuring that they have:

– an articulated and hierarchical corporate structure capable of covering and coordinating, in accordance with legislative requirements and the job specifications provided for therein, the various specialist engineering job profiles that will collaborate on the project;
– a quality certification system complying with ISO9001 incorporating the main processes that will provide guidance during the design development process into predefined procedures;
– a management mechanism for traceability enabling the input/output documentation to be managed in an orderly and traceable manner even over long periods (the development of a large-scale project may require adherence to the approvals processes over many years);
– a cutting-edge IT infrastructure capable of administering online on an interchange platform hundreds of PCs spread out at various sites in the territory and equipped with dozens of technical software packages, constantly updated and running the most recent versions available on the market;
– a rapid and efficient system for purchasing '*engineering services*' enabling the Project Design Team to promptly engage the assistance permitted under article 31, paragraph 8 of Legislative Decree 50/2016 ('*the successful contractor may not subcontract, except for geological, geotechnical and seismological surveys, probes, surveys, measurements and staking and the preparation of specialist and detailed project documents, except for geological reports and the preparation of the project drawings*');
– an archiving system with built-in redundancy enabling the project documentation, whose value often runs in several million euros, to be preserved safely and securely from events that could cause irremediable damage to it.

3 PROJECT DESIGN, LEGISLATIVE CONTEXT AND CONTRACT REQUIREMENTS

Apart from the technical specifications laid down in legislation, a Project Designer about to tackle the Project Implementation Plan must also refer to at least three Codes or 'consolidated texts':

– the Contract Code for Competitive Bidding, Legislative Decree 50/2016 and the current Regulation, Presidential Decree 207/2010 with regard to technical and regulatory aspects;
– the Environment Code, Legislative Decree 152/2006 as regards aspects related to the environment and to the project's compatibility with the territorial context;
– the Safety Code, Legislative Decree 81/2008 for matters related to workplace safety.

In view of this notorious legislative complexity – which, moreover, albeit with different nuances – is common to all the Member States of the European Union, given that the three Codes referred to above are the national transpositions of Community Directives – the definition of the figure of Project Designer is also particularly complex, even though the reference legislation attempts to summarise highly articulated functions in a few concepts:

- article 15, paragraph 12 of Presidential Decree 207/2010 provides that '*all the project documents must be signed by the project designer or by the project designers responsible for them and by the project designer responsible for coordination between the various specialist service providers*', implying that given the existence of a Specialist Project Designer or a Team specialising in each engineering discipline, coordination by an Engineer is required to ensure that the project documents are perfectly coherent within one another;
- article 24, paragraph 5 of Legislative Decree 50/2016 provides that '*regardless of the legal status of the entity engaged to carry out the tasks described in paragraph 6, said tasks must be carried out by professionals entered in the corresponding registers provided for under the current regulations governing their profession, personally liable and identified by name when the bid is submitted, stating their respective professional qualifications*', which means that the tasks of Project Designer may only be performed by Engineers entered in the corresponding professional registers;
- article 254, paragraph 1 of Presidential Decree 207/2010 states that '...*the engineering firms are obliged to appoint at least one technical director, whose duties are to collaborate in laying down the company's strategic guidelines and those governing collaboration and monitoring in relation to the services rendered by the specialists engaged to carry out the design work, who must have a degree in engineering or architecture or in a technical discipline that is relevant to the company's main field of activity, authorised to exercise his or her profession by at least ten years' experience and entered, when he/she takes up his/her post, in the corresponding professional register provided for under current legislation or qualified to exercise the profession in question in accordance with the laws of the countries of the European Union of which the person in question is a citizen. The company shall delegate the technical director or another engineer or architect answerable to the former, holding a graduate degree and qualified to exercise the profession, the task of approving and countersigning the technical documents relating to the services for which the contract has been awarded. The approval and the signature of the project documents triggers the joint and several third-party liability of the technical director or delegate with the engineering company vis-à-vis the commissioning body*', describing the Technical Director's duties, responsibilities and qualifications.

Hence the aforementioned legislation may be summarised by stating that the figure of Project Designer is not '*individual*' but is made up of three joint professions:

- the Specialist Project Designer (PS), who is the 'pure' Technician specialising in a specific field of engineering;
- the Person Responsible for Coordination amongst the Specialist Service Providers (RIPS), who coordinates and ensures that the contributions made by the various Specialist Project Designers are coherent;
- the Technical Director (DT), who approves and countersigns the technical project documents.

The presence of these three figures who, as we have seen above, are precisely defined by the sector regulations that actually describe their professional characteristics, must be expressly requested by the client via clauses in a design agreement, so as to make sure that project documents are compliant with the laws in force and that the respective responsibilities are covered. Similarly, it is fundamental that the signing of the project documents be performed in the correct sequence - PS =>RIPS=>DT – something that can be ensured only via specific software: once the operating procedures have been finalised, an appropriate workflow traces the entire approvals process for each project document or report, from the moment they are initially processed by the designers until they are signed by the DT, who – following on from the signatures appended by the PSs and the RIPS – concludes the authorisation process. Following approval by the DT, the 'graphometric' signatures of the PS, RIPS and DT appear on the cover of the project document and the file is converted into a pdf file with the 'digital' signatures and 'frozen' so that it can no longer be amended, other than by means of a 'revision' process. This change in the status of the project document is also registered by the audit rails and automatically inserted into the list of project documents on the cover of the table.

It is essential that this requisite be stated in the design agreements, at least in the case of large-sized projects, penalty the impossibility of ensuring that the documentation validation process is performed correctly.

The resulting documentation must also ensure the following requisites:

- uniqueness: all the project documentation must be unambiguously codified (without redundancies affecting names or versions);
- traceability: all the documents must be placed in a single archive with a logical structure and rendered accessible so that they can be located as any time as required;
- completeness: the document archive must contain all the documentation relating to the same project;
- standardisation: the project design documentation must be drafted in accordance with unique and clear standards, accepted by everyone.
- updating: the most recent version of the documentation is always available in the system, ensuring the filing of all the versions and revisions at the same time;
- historical traceability: accesses, modifications and revisions relating to each document are registered in the system so as to enable knowledge of its internal evolutionary history.

4 PROJECT ORGANISATION AND RISK MANAGEMENT

The organisational structure proposed under legislation provides that the Specialist Project Designers (PS) should make their own individual contributions to the Project, with the Person Responsible for Coordination amongst the Specialist Service Providers (RIPS) coordinating the technical aspects at the outset and the Technical Director (DT), who collaborates in the task of defining the strategic guidelines and monitors the services rendered by the engineers responsible for design tasks.

Normally, this type of process is managed by setting up an ad hoc Task Force of Specialists for each Design Project, coordinated by an RIPS, with the Technical Director overseeing a number of initiatives.

This organisational structure guarantees maximum productive efficiency, provided the Task Force is of the right size and includes all the professional profiles required for the development of the Project.

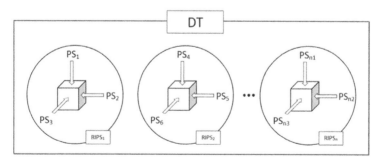

Figure 2.   Diagram for the implementation of a Project with various Task Forces involved in the work.

This type of organization exposes the project to a risk that has a high chance of occurring and that has a large impact on operations: the Task Force centralizes the processes and envisages an ON/OFF type of result, i.e. the objective is fully achieved or dramatically missed. Normally, this organization is not advisable for the management of large-scale projects, for which it is preferable to set up a complex sequence of processes that allows for greater flexibility in managing risks and is more suitable for monitoring progress and the use of resources. In this way, it is possible to spot critical events in advance and find a way to avoid them, reduce their effect or accept the consequences in part or fully, thereby minimising their impact on activities.

When it comes to developing a number of multi-projects alongside one another or managing 'large-scale projects' – i.e. initiatives containing a multiplicity of engineering activities within them – it is advisable to employ a matrix-like structure, an arrangement which, in fact, is usually adopted by engineering companies.

This is precisely the case of the Genoa Gronda, the project of which consists of 10 execution lots that have been developed in parallel:

– Lot 1: works for the preparation of the work sites
– Lots 2/3/4: relating to the tunnels excavated via "drill&blasting" in the Eastern sector
– Lot 5: works at sea for the deposit of excavation materials
– Lot 6: electric and mechanical systems installed in the tunnels and outdoors
– Lot 7: "Genova" cable-stayed viaduct
– Lot 8: TBM-excavated tunnels in the Western sector
– Lot 9: completion works
– Lot 10: "slurry duct" pipeline for the transport of excavation material.

Within the 'matrix', each project only elicits the specialist contributions it requires, solely involving the Offices that are strictly required and leaving the other disciplines unrelated to the purpose of the work on 'standby' (e.g. the PS of the Tunnelling Office is not involved if the Project does not involve any tunnelling). The various contributions are coordinated by the RIPS, who monitors the gradual development of the project documents and their intrinsic coherence. Finally, the DT oversees the project design process.

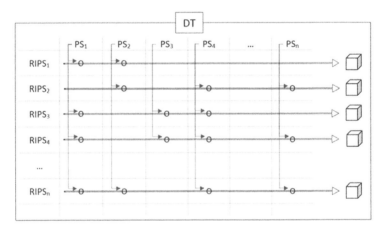

Figure 3.   Diagram showing the functioning of the project design 'Matrix'.

The 'matrix' organization cannot eliminate the risk that parts of the project are not finished in time, but it does allow the Technical Manager to constantly have a clearer picture of the production process that takes on an 'assembly line' format. Among other things, any corrective actions implemented to remedy an error on one of the lines can be immediately applied to the other production lines, thereby maximising the effectiveness of the measures taken.

5   QUALITY SYSTEM AND RISK MANAGEMENT

As is well known, the system certification UNI EN ISO 9001 is a virtually obligatory prerequisite for participating in calls for tenders in design projects, as confirmed in the new Contract Code for Competitive Bidding.

Chapter 8.3 of the regulation provides an effective tool for developing a Project in compliance with the predefined rules of the Quality System, with the aim of eliminating the main causes of design errors, which various studies attribute (in order of importance) to:

- repeated changes in the framework of requirements laid down by the Commissioning Body, which in the long run are not incorporated during the reworking of the project documents;
- insufficient or an absence of coordination between the Engineers involved in the design work;
- lack of comprehension regarding the technical specifications;
- material errors due to a lack of expertise in the drawings and reports.

The steps provided for under legislation for the design process will now be examined one by one. They provide an efficient method to which the Project Designer should refer at all times, and provide an effective "Risk management" system that has been frequently employed during the development of the Gronda Execution Plan.

5.1 Design Planning

A requirement is that the design activities must be planned in advance, through a process that must include:

- scheduling of the various design activities using Gantt diagrams so as to create the baseline curves relating to the distribution over time of the submission of the project documents and the use of Resources;
- preparation of a process for controlling the design progress, indicating the results that should be attained and the drivers that should be monitored;
- identification of the inputs, in terms of the functional requirements, required under the law or under applicable standards;
- identification of the outputs of the design progress, so that they meet the Commissioning Body's requirements.

This 'Design Plan' should be developed employing levels of detail that are compatible with the Project's degree of complexity, using tools enabling the identification at the very least of the following aspects: object and purpose of the activities, technical requirements, development times, responsibility, internal/external Resources that should be employed, internal/external interfaces that should be managed.

5.2 Design input elements

All the input information, at the current design phase, should be archived and shared on an IT platform with the Project Design Team:

- documentation relating to the preceding design levels, with relevant opinions and requirements furnished by the Bodies;
- results of topographical, geognostic and environmental surveys;
- functional and performance requirements stipulated by the Commissioning Body (standards and specifications).

The input elements should be re-examined several times during the Project, to ensure that they are appropriate and complete and to resolve any ambiguous or conflicting requirements.

5.3 Design output elements

This phase, corresponding to the Project's actual development phase, is guided by the RIPS whose main duties include:

- preparation of the Project's structure (according to Specifications, Paragraphs, WBS and Parts of Work);
- technical coordination between the various services rendered by the Specialist Project Designers involved in the drafting of the Project Design;
- the retrieval and transmission of information and/or documentation to all the members of the Design Team.

In order to manage and properly archive the documentation the following items are used:

– the list of project documents, which acts as a road map for the Project, enabling the changes made during the work to be guided and checks on its progress;
– in/out emails to the entire Team, indicating the filing path on an IT platform;
– exchange folders with the specialist departments (from_for_Internal);
– exchange folders with the external suppliers (from_for_External);
– final transfer of the files in the editable/non-editable folders;
– preparation of folders for official deliveries (hard and soft copies).

The project design outputs must be capable of meeting the framework of requirements laid down by the Commissioning Body and must furnish all the information required to produce the product/render the service.

5.4 Review of Design Process

The review of the design process constitutes the implementation of the periodic scrutiny over the organisation provided for in the 'Planning' phase and its aim is to assess the progress of the activities with respect to the predefined baselines and evaluate the risks associated with any departures from the predefined targets.

Hence an evaluation takes place of the actions required to make up for any delays (increase in Resources) or re-orientate the project in order to meet the client's requirements. The Review meetings must be set down in minutes and distributed to the Project Design Team, so that all its Members are aware of the current state of activities.

5.5 Verification of Design Process

The checks conducted on the project design constitute the implementation of the periodic scrutiny of the technical results provided for in the 'Planning' phase and their aim is to assess whether the outcome of the decisions taken has led to a product that is coherent with the original input data.

The tests that should be carried out on the project design concern:

– checks on properties and performance;
– comparison with similar projects implemented in the past;
– scope of the environmental impact;
– checks on compliance with safety requirements.

The Verification meetings must be set down in minutes and distributed to the Project Design Team, so that all its Members are aware of the current state of activities.

5.6 Validation of Design Process

For projects concerning public works governed by the Contract Code for Competitive Bidding, this phase coincides with the provisions of article 26 of Legislative Decree 50/2017: '*the commissioning body, in works contracts, shall check the conformity of the project documents with the documents described in article 23, and their conformity with current legislation*'.

In particular, the checks must examine:

– the completeness of the project design;
– the coherence and completeness of the financial framework in all its aspects;
– whether the chosen design solution can be put out to contract;
– the preconditions for the durability of the work over time;
– the minimisation of the risks of introducing variants and of disputes;
– the chances of completing the work within the stipulated deadlines;
– the safety of the workforce and of the users;

- the appropriateness of the minimum unit prices applied;
- the ability to maintain the work, if required.

Validation consists of the examination of fitness for use, i.e. the activities carried out before delivery or use of the product.

In order to maximise the effectiveness of this sequential process, it was applied to each Lot of the Gronda project and not to the overall project. This was the only way to identify the individual deviations from the technical and time schedule objectives and to intervene with the necessary corrective actions.

6 CONCLUSIONS

We have seen that the definition of Project in the sense of a technical and organisational challenge can be applied aptly to the case of a Project Designer called upon to participate in the implementation of a large-scale infrastructure project.

In such cases the technical competencies of the engineering companies should not be regarded as a test bed and must be unquestionable right from the moment the Commissioning Body awards the contract. However, given that the development of a Project always coincides with a critical path leading to the creation of an infrastructure – the real competitive element refers to the time required to prepare the project and hence to the Project Designer's capacity in managerial terms to set up a working team, to equip it with all the IT tools required and to ensure it works in a context governed by appropriate standards and procedures.

Accelerating the time required for project design – which has become a 'simple' way of bringing forward the commissioning of work due to be undertaken – should now be regarded as one of the main constraints on engineering activities and requires a response characterised by the abundant deployment of technologies and managerial skills. The Project Designer's contractual and risk-management skills therefore are becoming increasingly necessary, and should be applied extensively to the project development phase in order to keep its technical, economic and scheduling variables under control.

Tunnels and Underground Cities: Engineering and Innovation meet Archaeology,
Architecture and Art, Volume 8: Public Communication and Awareness/Risk Management,
Contracts and Financial Aspects – Peila, Viggiani & Celestino (Eds)
© 2020 Taylor & Francis Group, London, ISBN 978-0-367-46873-6

Managing a complex rail construction project: The new Milan-Genoa high speed/high capacity line

G. Morandini, P. Quarantotto, A. Marcenaro, V. Gabrieli & V. Capotosti
Italferr S.p.A., Roma, Italy

ABSTRACT: The *Terzo Valico dei Giovi* (TVG) is a new High Speed/High Capacity railway infrastructure, aiming at developing a significant modal split of freight transport from road to train. The new railway, part of the Rhine-Alps corridor, connects the ports of Genoa with the existing national rail network. It is mainly developed in deep tunnels long more than 27 km (Valico tunnel).

The TVG, has been entrusted in 1992, and to date it is still under construction. Such a long-time frame makes the natural mutations of the external context generate unavoidable impacts in the realization. Furthermore, the assignment is "turn key" to a General Contractor, the first case in Italy, with the Client in the role of both High Supervision and of the Works Management, roles exercised through the Italferr Company. This article describes the features and the critical aspects inherent in the contractual management of a long-duration contract as TVG with a mid-term evaluation about management model adopted for long-term projects.

1 INTRODUCTION

The *Terzo Valico dei Giovi* (TVG) is a rail infrastructure connecting the port of Genoa and the northern existing Italian rail network and represents the Mediterranean end of the Rhine-Alpine Corridor, one of the main corridors of the strategic trans-European transport (TEN-T) network. Its construction is part of the strategy for strengthening the role of rail transport in the development of freight traffic between the large ports of northern Europe and the Mediterranean, and between western and eastern Europe. In particular, the TVG rail line is a key infrastructure for the development and growth of the port of Genoa (Italy's number one container port, with 2.2 million TEUs handled in 2015) and, generally speaking, for handling non-bulk goods; moreover, the infrastructure will also increase the performance capacity of the logistical system, cutting travel time between Milan and Genoa to no more than 90 minutes.

1.1 Technical infrastructure data

The new TVG line is 53 km long, overall, with 37 km of the line running in tunnels in addition to 14 Km of interconnection lines with the existing network (Genoa – Turin and Tortona – Piacenza lines). The project, runs almost entirely in two main twin-tube tunnels with a single track, called the Serravalle and the Valico tunnels.

The Serravalle Tunnels, 7,094 m long, are under construction with a tunnel boring machine (TBM), with one TBM for each tube. The Valico Tunnels, approx. 27,110 m long, are made through four lateral access tunnels, over 1000 m long, ending at the main tunnel with variable-section chambers. Two different tunnelling methods are used: a TBM from north to south in the initial section, about 7 km long, while the remaining stretch of about 20 km, towards Genoa, is excavated using the conventional tunnelling method.

With these technical characteristics, the Valico Tunnel is the longest rail tunnel in Italy and ranks 5th in Europe, as shown in the following Table 1.

Table 1. Europe's longest railway tunnels.

N.	Name	Location	Length in km	Year completed
1	Gotthard Base Tunnel	Switzerland	57.104 km & 57.017 km, 2 tubes	2016
2	Channel Tunnel	France/UK	50.45 km	1994
3	Lötschberg Base Tunnel	Switzerland	34.576 km & 27.2 km, 2 tubes	2007
4	Guadarrama Tunnel	Spain	28.418 km & 28.407 km, 2 tubes	2007
5	Terzo Valico dei Giovi	Italy	27.110 km, 2 tubes	2022
6	Lainzer/Wienerwaldtunnel	Austria	23.844 km, partially 2 tubes	2012
7	Simplon, 2 tubes	Italy/Switzerland	19.8 km	1906/1922
8	Vereina	Switzerland	19.1 km	1999
9	Vaglia	Italy	18.7 km	2009
10	Apennine Base Tunnel	Italy	18.5 km	1934

1.2 *Progress work*

The work is divided into six lots of which four have been currently financed and under construction. The first batch is close to 90%, the others from 60% to 20%. The fifth lot is scheduled for delivery in early 2019, while the sixth must still be financed.

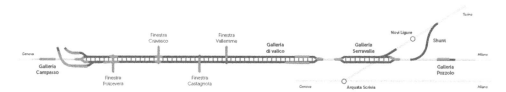

Figure 1. Rail route layout. In green realized at December 2018.

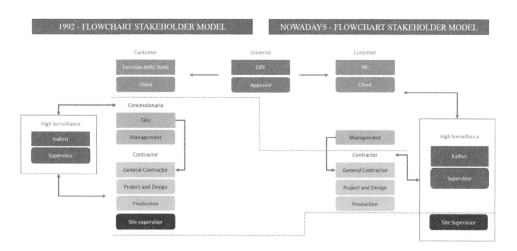

Figure 2. Stakeholder flowchart.

2 TVG PROJECT PLANNING AND DESIGN

2.1 *The role of TVG in the main rail network development strategy*

The TVG covers a role inside the High Speed/High Capacity Italian rail transport network development, started in 1986 by Ministry of Infrastructure and Transport with a specific

4511

Transportation Masterplan (Law 245/1984, DPCM of 10 April 1986) to "*relaunch the role of railways, rebalancing the transport system and facilitating the process of European integration*".

In 2001, the TVG project took on a strategic role within the Europe-wide transportation system as well, when it was included within the "*Tyrrhenian-North European Multimodal Corridor*" as part of the "*Ventimiglia-Genoa-Novara-Milan (Sempione) rail line*", and was therefore included, under Law 443/2001 (the so-called "Objective Law"), in the First Programme of Strategic Projects approved by the CIPE (Inter-departmental Committee for Economic Programming) (resolution no. 121 of 21 December 2001, published in the Italian Official Journal no. 51/2002 S.O.). The project was then presented to the authorities of the regions concerned between 2002 and 2003, within the framework of the so-called *Documenti programmatici di accordo* (agreement programme documents) between the central and regional governments (General Framework Agreement between the Central and Regional Governments), in particular: on 6 March 2002 with the Region of Liguria and 11 April 2003 with the Region of Piemonte.

On 21 April 2004, the European Parliament approved the list of new strategic infrastructure projects, including the Terzo Valico dei Giovi rail line, as part of Corridor 5 with the ports of Liguria. More recently, the Infrastructure has been included in the programme called "Linking Italy" attached by the Ministry of Infrastructure and Transport (MIT) to the 2016 Budget and included in the RFI Programme Contract for 2017-2021, signed on 1 August 2017.

2.2 The actors of the TVG

The TVG actors/responsibilities model is synthetically based on the following points:

– CIPE (Interministerial Committee for Economic Planning): approves and assigns the financing of the project, once received the opinions of the Ministries concerned;
– Ferrovie dello Stato: it is the Client of the work;
– TAV: a company founded specifically for managing planning, design and construction of those high speed-high capacity railway lines;
– General Contractor: responsible company with direct assignment of the design and construction of the works;
– Italferr: engineering company of Ferrovie dello Stato in charge of the High Surveillance (ref 4.1), further also in charge of the Works Supervision.

Subsequently, the evolution of the external context (regulatory, financial and social) made it necessary to update the model as briefly described in the scheme. In particular, the TAV company was incorporated into RFI (a company of Ferrovie dello Stato Italiane), which consequently became the Customer of the work.

2.3 The design stages: Preliminary Design

The Preliminary Design was entrusted by TAV in July 1992 to General Contractor, together with the Environmental Impact Assessment (EIA) and the publication of the Environmental Impact Study (EIS). The Preliminary Design was then presented in 2003, with a positive assessment and certain requirements (to be addressed in the Final Design stage), especially with regard to its environmental compatibility; at the same time, the Environmental Impact Study (EIS) also received a positive response, with requirements by the dedicated EIA Committee. The CIPE – in the light of the positive response obtained – then approved the Preliminary Design (resolution no. 78/2003), for an overall cost of 4,719 M euros – including any contingencies – and decided the start of the relevant road construction and site set-up works for 450 M euros.

2.4 The Final Design

Following the approval of the Preliminary Design, the General Contractor moved on to the Final Design stage and, between September and October 2005, published the *Avviso di avvio del procedimento per l'approvazione del Progetto Definitivo* (Notice of the commencement of the process for the approval of the Final Design), followed by the convening of the preparatory *Conferenza di Servizi* (meeting organised by the competent departments) held on 19 December 2005, with the favourable opinion of the Ministry of the Environment, the Ministry of Culture – DG for Architectural and Landscape Assets, the Ministry of the Interior, the Regions of Liguria and Piemonte, the Provinces of Genoa and Alessandria, 14 local authorities and a further 44 administrations and entities affected by the project works, followed by the transmission of the relevant preliminary report by the Ministry of Infrastructure and Transport to the secretariat of the CIPE.

The Final Design, as approved (cf. CIPE resolution no. 80/2006), provides for the construction of the rail line from the Genoa node, across the provinces of Genoa and Alessandria, along the main Genoa – Milan line to Tortona, and then along the Alessandria – Turin line to Novi Ligure, after which it would join the main existing lines to Milan and Turin.

3 THE CONTRACT DOCUMENTS. THE "INSTRUMENTS" FOR DEALING WITH CRITICAL CONSTRUCTION ASPECTS

3.1 *The 1992 Contract*

In 1992, TAV, the concession-holding company established by Ferrovie dello Stato, for the construction of the High-Speed Lines, awarded the Contract for the execution design and construction of the Milan – Genoa line – including the related infrastructure and interconnections – to the General Contractor Cociv, a consortium set up by several of Italy's leading contractors. In this specific case, TAV decided to apply the international "general contracting" model for the construction of the infrastructure (as well as for the construction of the other HS/HC lines in the same years), which means that the design and construction services are contracted by a single entity, on a turnkey basis, as is generally the case in complex infrastructure projects. Moreover, this specific design-build procurement process was later introduced into the Italian legal system by Law 443/2001 (the so-called "Objective Law"), and the ensuing Legislative Decree 190/2002, setting out the relevant enforcement regulations. Therefore, considering that, at the time of the 1992 Contract, no rules had yet been put into place for regulating this type of procurement process, such as to bind TAV in the definition of the tasks and responsibilities of the General Contractor, the Contract itself, and the related Contract Documents, are the product of extremely atypical arrangements, which also differs, in part, from the current rules governing general contracting agreements. On top of this is the fact that, since the procurement process did not envisage a competitive bidding process involving other bidders, the Contract in question provides for the assignment by the general contractor COCIV of a part of the works (amounting to 40%) to third-party companies. Under the 1992 Contract, therefore, the General Contractor:

– was responsible for designing and building the infrastructure, on a turnkey basis;
– was obliged to assign the technological systems to a nominated subcontractor approved by the Client;
– was obliged to assign at least 40% of the value of the infrastructure to third-party companies approved by the Client.

According to the 1992 Contract, the services for which the General Contractor was responsible, in connection with the construction of the infrastructure, included the designation of the Site Supervisor, subject to approval by TAV. This designation – also confirmed by the Contract Addendum signed in 2011 – was only recently modified, by Contract Amendment 4, signed on 2 May 2017 (see 4.2 below). At the same time, the Contract assigned the High Surveillance functions, on behalf of the Client, regarding the services performed by the Consortium, to Italferr.

Finally, the 1992 Contract assigned the approval of the General Contractor's design, and of the final agreement on the lump sum consideration due to COCIV, to a subsequent contract addendum. In this regard, as a result of the regulatory amendments affecting the design process, the 2011 Contract Addendum contained the approval of the Final Design of the infrastructure – and not of the execution design, as originally provided by the 1992 Contract – with the obligation, for the Consortium, to develop the Execution Design at a later stage.

3.2 *The 2011 Contract Addendum*

The Addendum to the 1992 Contract was signed on 11 November 2011, by RFI S.p.A. – which, in the meantime, had merged with and absorbed TAV S.p.A., effective from 31.12.2010 – and COCIV. The Addendum, which contained the approval, by RFI, of the Final Design of the Line and the agreement of the lump sum payable for its construction "according to the best industry practices and standards", further defined the responsibilities and risks undertaken by the General Contractor (for instance, the Consortium became responsible for (i) adjusting the design, at its own expense and under its own responsibility, and at no extra cost for the Client, to adapt it to any new regulations, (ii) managing relations with any third party, (iii) disposing of any waste products produced or found, etc.. Moreover, the Consortium also became the only party entirely responsible for the health and safety of the workers, pursuant to Legislative Decree 81/2008). Furthermore, the Addendum, introduced after nearly two decades from the conclusion of the 1992 Contract, also necessarily took into account the intervening economic and regulatory developments. The following paragraphs contain an overview of the principal innovations introduced by the Contract Addendum, some of which as a result of the changes in the regulatory, regional and institutional context of which the infrastructure is a part.

3.3 *Lot-based construction*

In 2010, the CIPE (resolution no. 84) – acting in accordance with article 2.232 of Law 191/2009 (2010 Financial Law) – authorised the construction of the TVG line according to a breakdown of the project in six so-called "Construction Lots". Consequently, the Addendum had to take into account this different manner of construction of the Line, providing for specific rules for the single Lots and for dealing with the consequences of the delayed or failed allocation of the financial resources for the Lots after the First Lot.

3.4 *Evolving needs: project's updates and adjustments*

Furthermore, the "Addendum" contains specific rules for dealing with the intervening need to adjust the design, providing that COCIV must make the necessary changes and additions to its Final Design within approx. 5 months, for the following purposes:

- to comply with any new rail interoperability standards and adopt the Level 2 ERTMS signalling system;
- to adapt the design to the rail tunnel safety requirements;
- to adapt the design to the provisions governing the management of the tunnelling spoil materials and to the changed availability of the delivery sites;
- to remediate any contaminated sites;
- to provide for the aboveground/underground utility interference works.

The Addendum, besides providing the schedules for the above adjustments to the Final Design, also sets out the procedures for updating the price, to take the changes into account.

3.5 *Changes to the rules on market opening: the new apportionment of works to third parties*

Regarding the assignment of works to third parties, as a result of the commitment undertaken by the Italian Government towards the EU, to ensure compliance with the principles of free competition, non-discrimination, cost-effectiveness and efficacy in the execution of the Project, the Contract Addendum contains an amendment to the 1992 Contract whereby the General Contractor undertakes to assign 60% – and not 40%, as previously agreed – of the civil and superstructure works to third parties, selected in accordance with the relevant EU public procurement procedures.

3.6 *Execution Design*

Compared to the management of "ordinary" design-bid-build contracts, in which the design and construction stages of a project are clearly separated, in the case of this design-build contract, the Execution Design – the Contract Documents do not envisage a Detailed Design – is developed alongside the construction process. This depends on the "construction" arrangements of the infrastructure, which, as mentioned previously, provides that the TVG line must be built according to Construction Lots, and that each Lot is "activated" by the Parties only if and when RFI receives the relevant financial resources. Until the activation, therefore, all Lot-related activities – including the design process – are put on hold.

Since the purpose of the contract and the related lump sum consideration have been finally established by the Parties in the Contract Addendum (except in the case of project changes during construction), based on the Final Design, and given the responsibilities and independence of and the risks undertaken by the General Contractor, the Contract Documents do not provide for the approval of COCIV's Execution Design by RFI. However, the Design is constantly supervised by RFI, through Italferr.

3.7 *Project Changes during construction*

Given the turnkey arrangement agreed with the General Contractor, the Parties have also agreed – in the Addendum – to restrict the admissibility of any project change during construction to the following specific cases, namely:

- any project change required by the Client RFI;
- the project change arising out of the occurrence of any proven force majeure events.

Any other event is covered by the risks included in the turnkey contract, so the related costs are a covered by the General Contractor.

Any project change proposed by COCIV can be admitted if representing proven needs of project technological improvement, which is intervening since the conclusion of the Addendum, or in the event of important archaeological finds that require significant project change to the Final Design.

3.8 *The instruments for managing critical tunnelling aspects*

3.8.1 *The rules for Technical Changes*
The design and construction of underground works are subject to the observation-based A. DE.CO. model, which provides that the typical tunnelling cross-sections, in the case of the application of conventional tunnelling techniques, must be based on the geological and geomechanical context emerging from the surveys carried out during the construction process. In the case of turnkey contracts, this method may be managed by means of the contractual instrument called "Technical Changes", in relation to the design and construction of tunnels.

In fact, lacking appropriate contract procedures for adjusting the lump sum price to any design improvements made during the tunnelling process, the Client RFI would have been unable to economically "benefit from" the said improvements, which would have been beneficial for the General Contractor alone.

Under the Contract Addendum, any "Technical Change" are subject to the following absolute requirements:

- they must in no way reduce the quality and performance levels of the infrastructure;
- they must not increase the Lump Sum Price.

Subject to these requirements, the Contract Addendum provides that the General Contractor may propose any Technical Change by sending RFI and Italferr the relevant support documents, which are then examined by Italferr, also to assess whether or not each Technical Change, as proposed by the General Contractor is grounded on objectively intervening circumstances, which could not have been foreseen at the date of the Addendum. This assessment process is particularly important, because, depending on its outcome, a different price adjustment system shall apply:

– if a Technical Change is justified, based on objectively intervening circumstances that could not have been foreseen at the date of the Addendum, the (higher or lower) amount entailed by the Change is entered in a special logbook; at the completion of the works relating to each Construction Lot, the relevant Technical Change are aggregated and if the resulting algebraic sum entails a price increase, then nothing is due to the; if, instead, the resulting algebraic sum entails a price decrease, the Lump Sum Price shall be reduced by an amount equal to 65% of the result of the algebraic sum;
– if the Technical Change is not justified by any objectively intervening circumstances that could not have been foreseen at the date of the Addendum (e.g. in the case of design errors), the extra cost entailed by the Change shall be incurred entirely by the General Contractor, while the lower cost shall be entirely subtracted from the Lump Sum Price.

The above considerations highlight the purpose of these rules: on the one hand, to promote any improvements to the tunnelling design – without impairing in any way the quality and performance levels of the infrastructure – in the light of any geological and geo-mechanical data collected during the construction work, and which was unknown or unforeseeable during the Final Design stage; on the other hand, to prevent the General Contractor from benefitting from the introduction of technical solutions that it could/should have foreseen during the final design process.

Considering that the Addendum authorises the general application of Technical Change, in relation to the design and construction of tunnels, any Technical Change effectively proposed and implemented by the Consortium, during construction, did not concern only the typical section of the tunnels, excavated according to the conventional method, but other tunnelling activities as well (such as, for the example, the tunnelling techniques).

To date, a total of 430 Technical Change have been introduced, of which 240 deemed justified by objectively intervening circumstances that could not have been foreseen at the date of the Addendum.

3.8.2 The "asbestos" issue

The specific characteristics of this type of infrastructure (i.e. long tunnels, at a depth of between 300 and 600 m) prevented the overall definition, during the risk planning stage, of the risk of finding asbestos in the tunnelling spoil materials. Therefore, the Addendum has also provided for the possibility of finding asbestos during the tunnelling. This risk was not fully foreseeable by the Parties in the Design stage and, therefore, the Lump Sum Price established in the Addendum could not have included the expenses incurred for handling and disposing of the asbestos containing materials. Therefore, in the Addendum, the Parties agreed that the costs arising out of the discovery of asbestos during the tunnelling, not having been included in the Lump Sum Price, would be determined and paid by RFI on the basis of "Documented Costs", i.e. of the support documents submitted by the General Contractor and stating the costs actually incurred by it as a result of the discovery of asbestos in the spoil materials, namely, contracts, invoices and payment receipts. Therefore, the costs incurred for handling and disposing of the asbestos-containing spoil materials – unlike those relating to all the other spoil materials produced or found during the tunnelling, which are incurred exclusively by COCIV – are effectively reimbursed by the Client.

In this specific case, asbestos fibres were found during the excavation of the Cravasco access tunnel, between km 0+710 and km 0+800 (a section of the tunnel crossing geological formations consisting primarily of grey-black carbonate and clay schists, frequently intercalated with quartz and calcite, and a strongly milonized serpentine inclusion), as well as in the excavation of the Castagnola access tunnel, between km 2+114 and km 2+270 (featuring Monte Figogna metabasalt formations, with a predominance of grey-blue/greenish rocks similar to ophiolite sequences). The higher costs entailed by the management of these findings have been assessed, and approved for payment to COCIV, in accordance with the Documented Cost system.

3.9 The stakeholders

3.9.1 Regional and local authorities
From an administrative point of view, the construction of the new rail line involves two Regions – Liguria and Piemonte – and 14 local authorities. The local authorities take part in the Committees of Mayors (for both regions), coordinated by the Government-appointed Commissioner for the TVG, respectively with the Prefect of Genoa and the Chief of the Infrastructure Department of the Region of Piemonte. Alongside the above mentioned local authorities are those that host the sites identified by COCIV for storing the spoil materials and which, therefore, are also affected by the construction works of the TVG Line.

3.9.2 The environmental protection authorities
An Environmental Observatory has been set up, at the Ministry for the Environment, to monitor the construction of the "Terzo Valico dei Giovi" HS/HC Milan-Genoa railway line and its effect on the local environment, as well as the environmental components with which the infrastructure interferes. The purpose of this Observatory is to optimise the governance of environmental protection, jointly with the other institutional entities responsible for informing the competent local authorities about the progress of the construction work and any relevant critical aspects. Furthermore, the Observatory fully collaborates with the Directorate General for Environmental Assessment, providing and sharing the monitoring data and analyses relating to the various environmental components, and the information feeding the Databases of the Ministry's Environmental Assessment Portal. The members of the Environmental Observatory are designated by the following entities: The Ministries of the Environment and of Infrastructure and Transport; the Higher Institute of Health; the Regional governments of Liguria and Piemonte, ARPA (Regional Agency for Environmental Protection) of Liguria and Piemonte, the Province of Alessandria and the Metropolitan City of Genoa.

3.9.3 The Government Commissioner
In 2009, the President of the Council of Ministers appointed an Extraordinary Commissioner with the task of facilitating the construction of the infrastructure by providing guidelines and support, monitoring the progress of the works and promoting agreements between the various public and private stakeholders. To date, the Government Commissioner – an "independent" figure, appointed to listen to the local communities and to mediate and resolve conflict at the local level, in respect of the construction of the infrastructure – has tackled the following fronts:

- ensuring workplace health and safety, in respect of the asbestos issue, in close collaboration with the Environmental Observatory, of which he is a member since June 2016;
- minimising the impact of the construction sites, by mediating between the parties;
- adapting the compensation measures, already contemplated by the 2006 Final Design, to the new and changed local requirements;
- promoting the development of the province of Alessandria, together with the Piemonte mayors, the provincial government, the regional government and RFI, exploiting the opportunities provided by the new railway line;
- fostering the increase of local employment, based on the application of the so-called "social clause" envisaged by the contract notices published since April 2017;
- providing transparent information about the progress of the construction work through the dedicated media outlets.

4 THE ROLE OF ITALFERR

4.1 High Surveillance

Italferr was assigned High Surveillance tasks, in respect of the Project, from the beginning, in the 1992 Contract. The company's High Surveillance duties are defined by both the Contract Documents stipulated with the Client RFI, and the Contract Documents regulating relations between RFI and Cociv (and the related attachments). Italferr is primarily responsible for verifying the

Execution Design, managing the contract on behalf of RFI, carrying out the preparatory activities, in respect of the Project Change during construction, the Technical Change and any claims, monitoring compliance with the construction schedule and the progress of the works, etc.

4.2 *Supervision of Works*

At a certain point during the execution of the project, the Client RFI decided to strengthen its control over the construction work by taking over the supervision of works function, which, until then, had been the responsibility of the COCIV Consortium. This decision, moreover, was fully consistent with the developments taking place in the regulations governing construction contracts and, in particular, the rules regarding "general contractors", following the introduction of Legislative Decree 50/2016, as a result of which the Client was given both high surveillance and supervision of works duties. Therefore, following the signing of Amendment 4 to the 2011 Addendum, Italferr was also assigned the Supervision of Works duties relating to the construction of the TVG line. In the wake of this assignment, it became necessary to define the boundaries of the supervision duties by Italferr. At the time, in fact, this was absolutely uncharted territory – certainly with regard to railway contracts, and possibly with regard to all works contracts, nationwide – and the very first time the supervision of works activities were performed by a company engaged by the Client, in the presence of a General Contractor, with full responsibility for designing and building the infrastructure, according to the turnkey model.

Of course, the issue at stake was not the technical controls assigned to Italferr, which are an essential part of the supervision of works activities (checking the conformity of the construction to the designs, acceptance of the construction materials, etc.), but certain administrative and accounting functions (for example, issuing of the work progress reports, investigations into the claims and extension applications, subcontractor approval, etc.), due to the fact that the project features a number of contracts, in some of which the Client is not even directly involved. As mentioned above, in fact, the General Contractor itself also acts as Contracting Authority for 60% of the project works, being required to contract out the works to third-party companies through bidding procedures (which, therefore, are the contractors of COCIV and are themselves entitled to subcontract to other companies the works already contracted out to them).

Against this backdrop, and since Italferr is responsible for the Supervision of Works only within the framework of the contract entered into by RFI and COCIV, but not in relation to the contracts concluded by COCIV and its own contractors – the management of which is entirely the responsibility of the General Contractor, in the capacity of Contracting Authority, and which, therefore, can in no way affect the arrangements between General Contractor and RFI (by amending the schedules and the contract price, for example) – the Parties have agreed that Italferr, as Supervisor of Works, will manage the accounting and administrative aspects solely with regard to the RFI-COCIV contract, and not the accounting aspects relating to the contracts awarded by the General Contractor, nor the further administrative aspects (Change, claims, time extensions), in respect of the said contracts. Therefore, while increasing the control exercised by the Client RFI over the construction of the infrastructure, through Italferr, the organisational and management autonomy and freedom of the General Contractor, typical of its role in a turnkey arrangement, has nevertheless been preserved.

The General Contractor itself, by signing Amendment 4 to the Contract Addendum, acknowledged that the Supervision of Works function by Italferr – if respectful of the above mentioned boundaries agreed by the Parties – does not amount to an active interference in the performance and organisation of the works, which, therefore, continue to fall within the exclusive competence and to remain the responsibility of the Consortium.

The Supervision of Works by Italferr is structured in such a manner as to satisfy the needs required by such a large-scale and important project. Being a multi-disciplinary contract (involving civil works, superstructure and technological systems), the Company has identified 5 Supervisors (3 for the civil works and one each for the superstructure and the technological systems), each featuring a team of Construction Site Inspectors, logistically based in the two Regions and coordinated by a Coordinator. The official management documents produced by the Supervision of Works have been entirely digitised, with the creation of a shared cloud space, managed by the head office; furthermore, the Supervision of Works also utilises the IT tools implemented by the company, such as the Construction Platform and the Subcontract Platform, as well as the tools for Claims Management and Accounting (both physical and economic), which interact, compare the data and carry out cross-sector checks.

5 CONCLUSIONS

The purpose of this paper was to provide information about the contract arrangements and management model adopted in relation to the execution of the project, given its specific features, as a result of:

– the lack of binding regulations;
– the awarding of the entire design and building process to a General Contractor, selected without a bidding process.

The adopted contract model is a turnkey arrangement, in which the Lump Sum Price, fixed and invariable throughout the term of the contract, is defined on the basis of the Final Design developed by the General Contractor. Such an arrangement is typically characterised by the considerable breadth of the responsibilities and risks undertaken by the General Contractor, as well as by its organisational independence and freedom, which, however, is significantly limited when it comes to the choice of the manner of execution (direct/outsourced), due to the restrictions imposed by the European Union (as a result of which, the Cociv Consortium was required to subcontract over half the civil works and for the construction of the superstructure to third-party companies, under the European procurement rules).

One of the major problems in this complex context is the need to reconcile the features of a turnkey contract with the enormous amount of risks involved by a railway route that runs mainly inside tunnels, with considerable geological challenges and even the high probability of encountering asbestos. As evidenced in this paper, it has proved impossible to address these specific risks, unlike other more ponderable risks with a lower financial impact, simply by adding a remuneration item in the Lump Sum Price; they require more sophisticated contract tools and mechanisms, which may then be easily replicated and transferred, even in different contexts, to manage these critical aspects, which often recur in projects that require significant amounts of tunnelling.

Last but not least, the experience built up in connection with this project has also highlighted the possibility of reconciling the characteristics of a turnkey contract with stringent controls by the Client, through a client-designated Supervision of Works structure.

REFERENCES

Fiorenzo Ferlaino & Sara Levi Sacerdotti. 2005. *Processi decisionali dell'alta velocità in Italia, Il ruolo del Piemonte nel Corridoio Sud dello Spazio Alpino*. Milano. Franco Angeli

Ministero dell'Ambiente e della tutela del Territorio e del Mare. *http://www.osservatoriambientali.it/online/home/gli-osservatori-ambientali-e-le-linee-avac/la-linea-terzo-valico-dei-giovi/iter-autorizzativo-e-la-delibera-cipe-n.-80-del-29032006.html*

Camera dei Deputati. SILOS Sistema informativo Legge Opere Strategiche *http://silos.infrastrutturestrate giche.it/admin/scheda-pdf.aspx?id=741*

IIC Istituto internazionale delle comunicazioni. *http://www.iicgenova.com/terzo-valico*

Ministero delle Infrastrutture e dei trasporti. Commissario Terzo Valico *http://terzovalico.mit.gov.it/*

Project Management Europa. *https://www.projectmanagementeuropa.com/governance-del-progetto-il-suc cesso-passa-da-qui/*

European Commission, Mobility and Transport *https://ec.europa.eu/transport/themes/infrastructure/rhine-alpine_en*

Terzovalico *https://www.terzovalico.it/*

RFI *http://www.rfi.it LINEE STAZIONI TERRITORIO > Alta Velocità - Alta Capacità > Le direttrici > Milano-Venezia e Terzo Valico*

Commissario di Governo Terzo valico dei Giovi https://commissarioterzovalico.it

Tunnels and Underground Cities: Engineering and Innovation meet Archaeology,
Architecture and Art, Volume 8: Public Communication and Awareness/Risk Management,
Contracts and Financial Aspects – Peila, Viggiani & Celestino (Eds)
© 2020 Taylor & Francis Group, London, ISBN 978-0-367-46873-6

The claims, dispute avoidance and dispute resolution procedure in the new FIDIC Emerald Book

C. Nairac
White & Case LLP, Paris, France

ABSTRACT: The Emerald Book is a new FIDIC form of Contract for Tunnelling and Underground Works. Over the years, throughout its various model conditions of contracts, FIDIC has developed claims and dispute avoidance and resolution procedures that have become a benchmark in the international construction industry. Those procedures underwent a significant evolution in FIDIC's latest edition of its "Rainbow Suite" of contracts that were released in December 2017. The joint FIDIC-ITA task group in charge of drafting the new Emerald Book has chosen to follow suit and adopt the same claims and dispute avoidance and resolution procedures in the Emerald Book. This contribution aims at presenting those procedures.

1   INTRODUCTION

As is well known to all members of the ITA, there is an ever-growing demand for utilizing underground space for infrastructure. The difficulty in predicting underground behaviour and conditions poses unique challenges regarding construction practicability, time and cost. Thus, allocation of underground risks among the stakeholders becomes critical in under-ground construction. To address these unique risks the International Tunnelling and Under-ground Space Association (ITA) and the International Federation of Consulting Engineers (FIDIC) joined forces to draft the new FIDIC Form of Contracts for Underground Works (the "Emer-ald Book"). To accomplish this, the two organizations setup a joint task group (TG10).

The Emerald Book has been modelled on the 2017 FIDIC Yellow Book (Conditions of Contract for Plant & Design Build) but with significant innovations tailored to underground construction. Consistent with FIDIC's philosophy of achieving a fair allocation of risks among the parties, the Emerald Book has been drafted with a view to promoting a balanced risk allocation that is specifically adapted to the risks inherent and unique to underground works.

At the moment of drafting this contribution, the Emerald Book is not yet published. Therefore, there may be differences in wording between the referenced sub-clauses in this paper and the published form of contract.

While a balanced risk allocation reduces the overall cost of the project and the risk of disputes, the uncertainty and risk inherent to underground works mean that projects comprising such works remain, to an even greater degree than other construction projects, prone to claims and disputes.

The philosophy behind a proper claims and dispute resolution procedures in this field is to enable parties to resolve claims and disputes at the earliest possible stage and as close to the work face as possible. In this regard, the ITA has published the 'ITA Contractual Framework Checklist for Subsurface Construction Contracts' (ITA Contractual Framework Checklist for Subsurface Construction Contracts is published by the ITA WG3 and is available at: https://about.ita-aites.org/publications/wg-publications/content/8/working-group-3-contractual-practices) prepared by the ITA-Working Group 3 on Contractual Practices which "identifies the key contractual practice areas that are fundamental for ensuring the success of subsurface construction projects" and specifically highlights the need for early identification of issues requiring resolution, as well as the necessity for providing robust dispute resolution procedures for dealing with them.

Over the years, throughout its various model conditions of contracts, FIDIC has developed claims and dispute avoidance and resolution procedures that have become a benchmark in the international construction industry. Those procedures underwent a significant evolution in FIDIC's latest edition of its "Rainbow Suite" of contracts that were released in December 2017. The joint FIDIC-ITA task group in charge of drafting the new Emerald Book has chosen to follow suit and adopt the same claims and dispute avoidance and resolution procedures in the Emerald Book. This contribution aims at presenting those procedures.

2 CLAIMS AND DISPUTE RESOLUTION PROCEDURE UNDER THE EMERALD BOOK

2.1 *Background*

The claims and dispute resolution procedure in the Emerald Book have generally been adopted from the 2017 FIDIC Rainbow Suite and particularly the 2017 edition of the FIDIC Yellow Book. This procedure is divided into four tiers, as follows: (i) the Engineer's agreement or determination of a Claim (Capitalized terms used in this paper have the meaning ascribed to them in the Emerald Book); (ii) a referral of any resulting Dispute to a Dispute Avoidance/Adjudication Board (DAAB); (iii) an attempt to reach an amicable settlement; and (iv) arbitration.

The 1999 FIDIC Suite contained a similar multi-tier claims and dispute resolution procedure but the procedure under the 2017 FIDIC Suite and the Emerald Book is more prescriptive and complex. For example, unlike the 1999 FIDIC Suite which dealt with Employer and Contractor claims under separate Clauses, under the Emerald Book (and the 2017 FIDIC Suite) Clause 20 now deals with Employer and Contractor Claims and the same procedure applies to both, while a new Clause 21 deals with Disputes and Arbitration. The procedures prescribed under Clauses 20 and 21 are discussed below.

This contribution first discusses and explains the steps involved from the moment a 'Claim' arises under the Emerald Book, up to the time that it may crystalize into a 'Dispute' which is referred to a DAAB and ultimately culminating with arbitration. Before proceeding to explain the importance of each tier and the relevant process under each tier, it is important to note that at all stages the focus is on preventing a Claim from escalating into a Dispute and, if a Dispute has already arisen, on the resolution of such Dispute at an early stage, with arbitration being the option of last resort.

Unlike the 1999 FIDIC books, the 2017 FIDIC Rainbow Suite contains a definition of a 'Claim'. The Emerald Book adopts this definition. A 'Claim' is defined very broadly as 'a request or assertion by one Party to the other Party for an entitlement or relief under any Clause of these Conditions or otherwise in connection with, or arising out of, the Contract or the execution of the Works'(FIDIC Emerald Book, Sub-Clause 1.1.6.). The word 'Claim' covers not only the conventional types of claims related to time and/or money (*i.e.*, Contractor's claims for additional payment and/or Extension of Time and Employer's claims for payment and/or extension of the Defects Notification Period) but also claims concerning other matters, such as the interpretation of a particular provision of the Contract or the validity of a Notice, among others. Additionally, the definition of 'Claim' not only covers claims arising under a Sub-Clause of the Emerald Book but also claims arising under the applicable law. Consequently, the claims procedure pre-scribed in the Emerald Book must be followed regardless of the legal basis of the Claim.

Despite the broad definition of a Claim, compliance with the claims procedure is not required with regard to certain entitlements addressed in Sub-Clause 13.8 of the Emerald Book. This Sub-Clause provides for the adjustment of the Contract Price and/or the Time for Completion as a result of measurement of the performed Excavation and Lining. Such measurement is to be agreed or determined by the Engineer in accordance with Sub-Clauses 3.7 and 13.8 of the Emerald Book and then paid according to lump sums or unit rates and associated charges set out in the Schedule of Rates and Prices. These adjustments to the Contract Price are made as part of the normal payment procedure under the Contract and are therefore not subject to the claims procedure. Similarly, adjustments to the Time for Completion as a result of this measurement are also made without any resort to the claims procedure (FIDIC Emerald Book, Sub-Clause 8.5, last paragraph). These situations should be distinguished from those concerning Unforeseeable physical conditions, including conditions that are outside the scope of possible conditions foreseen in the Geotechnical Baseline Report, under Sub-Clause 4.12, where a Contractor seeking relief in the form of additional time and/or money must comply with the Claims procedure discussed below (FIDIC Emerald Book, Sub-Clause 4.12, third paragraph—the impact on progress and/or cost of any physical conditions that are foreseen in the Geotechnical Baseline Report must be assessed under Sub-Clause 13.8 and not under Sub-Clause 4.12).

Similarly, unlike the 1999 FIDIC books, the 2017 FIDIC Rainbow Suite also contains a definition of 'Dispute'. Some readers may wonder why it is necessary to define what is a 'dispute'–like the elephant in the story, it might be hard to describe but you know it when you see it. The reality is that without a definition there can be uncertainty as to when a dispute has arisen, in particular whether a claim has crystallized into a dispute, and thus whether the dispute can be referred to the dispute resolution tier. The Emerald Book adopts the 2017 FIDIC Rainbow Suite definition, which contains three conditions the fulfillment of which characterize any situation as a 'Dispute' (FIDIC Emerald Book, Sub-Clause 1.1.31). First, one Party must make a claim against the other Party. This claim may be a type of Claim contemplated under the Contract, or any other matter to be determined by the Engineer under the Contract or otherwise (*e.g.* under applicable law). Second, the other Party (or the Engineer acting under Sub-Clause 3.7.2 [*Engineer's Determination*]) rejects the claim in whole or in part. Lastly, the claiming Party does not acquiesce with the rejection of the claim (by giving a Notice of Dissatisfaction under Sub-Clause 3.7.5 [*Dissatisfaction with Engineer's Determination*] or otherwise) (The definition of 'Dispute' also clarifies that a failure by the non-claiming Party (or the Engineer) to oppose or respond to a claim may constitute a Dispute in certain circumstances).

Compared to the claims procedure in the 1999 FIDIC suite, the claims procedure in the Emerald Book is more prescriptive and elaborate in terms of actions by the Parties and/or the Engineer and the time periods within which such actions should be taken.

The 1999 FIDIC Suite dealt with parties' claims in two separate provisions: (i) Sub-Clause 20.1 which sets out the procedure for the Contractor's claims, and (ii) Sub-Clause 2.5 which sets out the procedure for the Employer's claims. Under the 1999 FIDIC Suite the procedure for the Contractor's claims was much more rigorous than the procedure for the Employer's claims. There, the Contractor was required to issue its notice of claim within 28 days after it became aware or should have become aware of the event or circumstance giving rise to the claim, failing which the claim was time-barred. By contrast, there was no reciprocal time period and similar sanction applicable to the Employer's claims. Similarly, under the 1999 FIDIC Suite the Contractor had to submit the full supporting particulars of its claim as part of its fully detailed claim within 42 days after it became aware, or should have become aware, of the relevant event or circumstance. By contrast, there was no time period within which the Employer had to do so.

Following the approach in the 2017 FIDIC editions, the Emerald Book introduces a uniform procedure that both Parties should follow regardless of the Party that makes the Claim. Further, unlike the 1999 FIDIC books, which addressed claims for time and money only, the Emerald Book distinguishes between claims concerning time and/or money (which are dealt with under Sub-Clause 20.2) and claims other than claims for time and/or money (which are dealt with under Sub-Clause 20.1). The latter type of claims may include matters, such as the interpretation of a provision of the Contract, the rectification of an ambiguity or discrepancy found in the Contract documents or a declaration in favour of the claiming Party (Guidance for the Preparation of Particular Conditions to Yellow Book, Second Edition 2017, p. 50). The procedure for both types of Claim is discussed below.

The steps that are required to be followed under the Emerald Book with respect to a 'Claim' and subsequently a 'Dispute', if any, are discussed below.

2.2   *The Engineer's Agreement or Determination of a Claim*

The Engineer must proceed under Sub-Clause 3.7 to agree or determine a Claim. The procedure leading to such agreement or determination is different depending on whether the Claim is related to time and/or money or is a Claim of a different nature.

2.2.1   *Claims procedure for Claims related to time and/or money*

The claims procedure for time and/or money in the Emerald Book is more rigorous compared to the claims procedure for other claims. It includes multiple steps and several time bar provisions. The main steps are as follows:

1. The claiming Party gives a Notice of Claim to the Engineer within 28 days after it became aware, or should have become aware, of the event or circumstance that gives rise to the Claim; the claiming Party's Claim will be time barred if the Party fails to give its Notice of Claim within this period (FIDIC Emerald Book, Sub-Clause 20.2.1);
2. The Engineer gives a Notice to the claiming Party within 14 days after receiving the Notice of Claim if the Engineer considers that the claiming Party has failed to comply with the period of 28 days mentioned above. If the claiming Party disagrees with the Engineer's Notice or considers that there are circumstances that justify late submission of the Notice of Claim, then the claiming Party should include details of such disagreement or why such late submission is justified in its fully detailed Claim. If the Engineer does not give such a Notice, the Notice of Claim will be deemed to be a provisionally valid one (FIDIC Emerald Book, Sub-Clause 20.2.2. The non-claiming Party may disagree with such deemed valid Notice of Claim by giving a Notice to the Engineer. Thereafter, the Engineer must review such disagreement when agreeing or determining the Claim under Sub-Clause 3.7);
3. The claiming Party must keep contemporary records to substantiate its Claim (FIDIC Emerald Book, Sub-Clause 20.2.3);
4. The claiming Party must submit to the Engineer a fully detailed Claim within 84 days after it became aware, or should have become aware, of the event or circumstance giving rise to the Claim, or such other period as may be proposed by the claiming Party and agreed by the Engineer. The fully detailed Claim must include, among other things, a statement of the contractual and/or other legal basis of the Claim. A failure to submit the basis of the Claim within the abovementioned period will invalidate the previously given Notice of Claim (FIDIC Emerald Book, Sub-Clause 20.2.4);
5. The Engineer must give a Notice to the claiming Party within 14 days after the expiry of the time limit under item 4 above if it considers that the claiming Party has failed to submit a statement of the basis of its Claim within such time limit. If the claiming Party disagrees with the Engineer's Notice or considers that there are circumstances that justify late submission of the statement of the basis of the Claim, the claiming Party should include details of such disagreement or why such late submission is justified in its fully detailed Claim. If the Engineer does not give such a Notice, the Notice of Claim will be deemed to be a provisionally valid one (FIDIC Emerald Book, Sub-Clause 20.2.4. The non-claiming Party may disagree with such deemed valid Notice of Claim by giving a Notice to the Engineer. Thereafter, the Engineer must review such disagreement when agreeing or determining the Claim under Sub-Clause 3.7);
6. After receiving the fully detailed Claim (or an interim or final fully detailed Claim, in the case of Claims of continuing effect), the Engineer must proceed under Sub-Clause 3.7 (discussed further below) to agree or determine the Claim. If the Engineer has given a Notice to the claiming Party under item 2 and/or item 5 above, the Engineer must nevertheless proceed under Sub-Clause 3.7. The agreement or determination of the Claim in this case should deal with the question whether the Notice of Claim is a valid one (FIDIC Emerald Book, Sub-Clause 20.2.5);
7. The Engineer may require necessary additional particulars by promptly giving a Notice to the claiming Party and the claiming Party must submit these particulars 'as soon as practicable' after receiving the Notice. In this case, the time period for the Engineer to agree or determine the Claim under Sub-Clause 3.7 commences to run on the date when the

Engineer receives the additional particulars but the Engineer is nevertheless required to give its response on the contractual and/or other legal basis of the claim within the time limit under Sub-Clause 3.7, that is, within the time limit calculated from the date of receiving the fully detailed Claim (FIDIC Emerald Book, Sub-Clause 20.2.5.).

Special rules prescribed under Sub-Clause 20.2.6 apply to Claims that have a continuing effect. Claims of continuing effect are likely to be frequent in contracts concerning tunneling works as the encountered underground conditions entitling the Contractor to relief typically continue for a certain period of time. In cases of Claims of continuing effect, the following process will apply:

1. The submitted fully detailed Claim must be considered as interim;
2. The Engineer must give its response on the basis of the Claim in respect of the first interim fully detailed Claim by giving a Notice to the claiming Party within the time limit for agreement under Sub-Clause 3.7;
3. The claiming Party must submit further interim fully detailed Claims at monthly intervals; and
4. The claiming Party must submit a final fully detailed Claim within 28 days after the end of the effects of the relevant event or circumstance (or such other period as proposed by the claiming Party and agreed by the Engineer) with a total amount of the additional payment and/or time claimed (FIDIC Emerald Book, Sub-Clause 20.2.6).

The claims procedure for time and/or money includes two time bar provisions (those under items 1 and 4 above) and the claiming Party's failure to comply with these provisions will result in forfeiture of the right pertaining to its Claim. The first time bar provision applies to claims by both parties and not only to Contractor's claims as was the case in the 1999 FIDIC Suite. The second time bar provision had no analogue in the 1999 FIDIC Suite (However, a similar time bar provision was contained in Sub-Clause 20.1(c) of the FIDIC Gold Book released in 2008). As discussed below, the Engineer has discretion to disapply these time bar provisions in certain circumstances.

Sub-Clause 3.7, which deals with the Engineer's obligation to agree or determine Parties' Claims, is much more detailed and elaborate compared to the corresponding Sub-Clause 3.5 of the 1999 FIDIC Yellow Book. The latter Sub-Clause provided for a two-stage process where the Engineer was required first to consult with each party in an endeavour to reach an agreement and, in case no agreement was reached, to make a fair determination taking due regard of all relevant circumstances. There were no explicit time periods within which the Engineer was required to act.

The Emerald Book provides that when carrying out its duties under Sub-Clause 3.7 the Engineer must act neutrally between the Parties and shall not be deemed to act for the Employer. The Engineer should first consult with the Parties in an endeavour to reach an agreement on the Claim. The time limit within which the Parties should reach an agreement is 42 days or some other time limit which is proposed by the Engineer and agreed by both Parties and commences on the date when the Engineer receives the fully detailed Claim (or an interim or final fully detailed Claim, in the case of Claims of continuing effect) (FIDIC Emerald Book, Sub-Clause 3.7.3, sub-paragraph (c)). If no agreement is achieved within this time limit or if both Parties advise the Engineer that no agreement can be reached within this time limit, the Engineer shall give a Notice to the Parties and proceed to make a fair determination of the Claim, in accordance with the Contract, taking due regard of all relevant circumstances (FIDIC Emerald Book, Sub-Clauses 3.7.1 and 3.7.2). The Engineer must give its Notice of determination within 42 days or such other time limit as may be proposed by the Engineer and agreed by both Parties after the date corresponding to its obligation to proceed to make a determination (FIDIC Emerald Book, Sub-Clause 3.7.3). Thus, it may take up to 84 days for the Engineer to make a determination on a Party's Claim. If the Engineer does not give a Notice of agreement or determination on the Claim within the relevant time limits, the Engineer is deemed to have rejected the Claim (FIDIC Emerald Book, Sub-Clause 3.7.3).

As part of the process of determining Parties' Claims, the Engineer may decide to disapply the time bar provisions discussed in items 1 and 4 above by taking into account the details included in the fully detailed Claim and other circumstances. Such circumstances may include: (i) whether or to what extent the other party would be prejudiced by the late submission, and (ii) the other Party's prior knowledge of the event or circumstance giving rise to the Claim and/or the legal basis of the Claim, as the case may be (FIDIC Emerald Book, Sub-Clause 20.2.5).

The Engineer's authority to disapply time bar provisions was introduced for the first time in the 2017 FIDIC suite and reflects the enhanced role of the Engineer under the new FIDIC books, including the Emerald Book.

If a Party is dissatisfied with the Engineer's determination, it must give a Notice of Dissatisfaction ("NOD") with the Engineer's determination (or the Engineer's deemed rejection of the Claim) within 28 days after receiving the Notice of the Engineer's determination (or within 28 days after the time limit for determination has expired, in the case of a deemed rejection of the Claim) (FIDIC Emerald Book, Sub-Clause 3.7.5). There was no corresponding time period in the 1999 FIDIC Suite within which a Party was required to express its dissatisfaction with the Engineer's determination. After giving a NOD, a Dispute arises which a Party may refer to the DAAB for a decision. If no NOD is given by either Party within the abovementioned time period, the Engineer's determination becomes final and binding on the Parties and, if not honoured, may be enforced directly in arbitration (FIDIC Emerald Book, Sub-Clauses 3.7.5).

2.2.2 *Claims procedure for Claims unrelated to time and/or money*
The procedure for Claims unrelated to time and/or money is dealt with in the last paragraph of Sub-Clause 20.1 of the Emerald Book. This procedure is more lenient and there are no strict time limits within which the claiming Party must give a Notice of Claim and/or submit a fully detailed Claim.

This procedure requires the claiming Party to give a Notice to the Engineer after which the Engineer will proceed under Sub-Clause 3.7 to consult with the Parties in an endeavour to reach an agreement, and thereafter if no agreement has been reached then the Engineer must make a 'fair' determination of the Claim. The procedure presupposes that the requested or asserted entitlement or relief has already been made to the other Party or the Engineer who has disagreed with it (either by way of explicitly rejecting it or by failing to respond with-in a reasonable time). The Notice of such Claim must be given as soon as practicable after the claiming Party becomes aware of such disagreement (or deemed disagreement) and must include details of the claiming Party's case and the other Party's or the Engineer's disagreement (or deemed disagreement). Like in the procedure for Claims for time and/or money, a Party that is dissatisfied with the Engineer's determination (or the Engineer's deemed rejection of the Claim) must serve a NOD in which case a Dispute will arise which may be referred to the DAAB for a decision.

3 DISPUTE AVOIDANCE

Like the 2017 FIDIC Suite, under the Emerald Book either Party may refer any Dispute to the DAAB for a decision. As discussed below, under the Emerald Book the DAAB is a standing body which is put in place at the outset of a Project and whose members are appointed from a list of members that the Parties have agreed on under the Contract Data. While no specific qualification is necessary or required for a person to be nominated, given the peculiar nature of underground works the Parties may be well advised to nominate persons who may have experience in decision making on excavation, support, ground treatment, etc., in similar subsurface contexts, paired with a fair knowledge of the FIDIC contractual system.

Further, in line with the 2017 FIDIC Suite, under the Emerald Book the Parties can now jointly request in writing the DAAB to provide informal assistance with a view to avoiding a Dispute— and the DAAB may also invite the Parties to make such a request. Previously, the DAB could only provide an 'opinion' when requested by the Parties. Such informal assistance can take place at any time including during any meeting, Site visit or otherwise in the presence of both Parties unless otherwise agreed. That said, the Parties are not bound to act on any advice given or received during such informal meetings. Similarly, the DAAB is also not bound by any views provided by it orally or in writing during any future Dispute resolution procedure or decision.

According to Rule 3 of the DAAB Procedural Rules contained in the Emerald Book any Site visit(s) required to be undertaken by the DAAB must be carried out at intervals of not less than 70 days, and not exceeding 140 days, unless otherwise agreed jointly by the Parties and the DAAB or in case of special circumstances that require Site visits to be carried out at shorter intervals.

It is recommended that the Parties consider organizing Site visits at key moments during the evolution of the project such as after the beginning of a new activity (excavation or lining of a given section, for example). Additionally, the Parties must provide the DAAB with copies of all documents that the DAAB may request, such as progress reports, the initial and/or revised programme, the Contractor's Statements, certificate issued by the Engineer, relevant Notices, and other relevant communications between the parties, to enable the DAAB to remain informed about the evolution of the Project (FIDIC Emerald Book, Rule 4.3 of the DAAB Procedural Rules). Even if such documents are not requested, regularly or at all, by the DAAB, they may do so of their own volition, on a monthly basis for example, without waiting for a request from the DAAB asking them to do so.

Lastly, apart from permitting the DAAB and the Parties to engage in informal discussions for avoiding a Dispute, unlike the 2017 FIDIC Suite, the Emerald Book introduces two new concepts—a Risk Register and a Risk Management Plan. The Risk Register is a register identifying relevant risks and the Party/Parties best able to manage or control those risks. The Risk Management Plan is the Contractor's roadmap for dealing with risks identified in the Risk Register, should they arise, and is required to be prepared, maintained, revised and updated from time to time in accordance with the relevant provisions of the Emerald Book (*See* FIDIC Emerald Book, Sub-Clause 1.17).

These two concepts have been introduced specifically to deal with the vagaries of underground constructions works, on the one hand, while continuing to foster Dispute avoidance on the other. This is because the Risk Register and Risk Management Plan have been designed as tools which enable the Parties to specify and identify relevant risks at the time of entering into the Contract and thereafter, by allowing the Parties (and the Engineer) to alert each other at the earliest possible opportunity, including by way of advance warning notices under Sub-Clause 8.4, in case of any known or probable future events or circumstances that are likely to jeopardize the timely completion of the Works and/or a Milestone, or adversely affect the work of the Contractor's Personnel and/or have the effect of increasing the Contract Price. Therefore, by providing for a Risk Register, Risk Management Plan and Advance Warning, the Emerald Book allows contracting Parties to take mitigating steps in order to avoid Disputes, and allows the DAAB to be better informed of the relevant risks and thus to be in a better position to perform its dispute avoidance duties. Of course, it is fully possible that despite these measures, a Dispute may arise between the Parties; however these tools are designed to reduce as much as practicably possible the circumstances that may give rise to a Dispute.

4 DISPUTE RESOLUTION PROCEDURE UNDER THE EMERALD BOOK

If a Parties is dissatisfied with an Engineer's determination with respect to a Claim (whether a Claim for time and/or money or a Claim unrelated to time and/or money), then as stated above it must issue a NOD within 28 days after receiving the Engineer's Notice of the determination (or within 28 days after the time limit for such determination has expired). The dispute resolution procedure under the Emerald Book is set out in Clause 21 [Disputes and Arbitration] and primarily consists of three steps. They are: (i) Reference of a Dispute to a DAAB and Obtaining DAAB's Decision [Sub-Clause 21.4]; (ii) Amicable Settlement [Sub-Clause 21.5]; and (iii) Arbitration [Sub-Clause 21.6].

Under the 2017 FIDIC Suite the Dispute Avoidance/Adjudication Board (DAAB) is the successor to the Dispute Adjudication Board (DAB) that was provided for under the 1999 FIDIC Suite. However, there is one key difference. Under the 2017 FIDIC Suite significant emphasis has been placed on the Dispute "avoidance" function of the board, hence the DAB as it was called under the 1999 FIDIC Suite now being referred to as the "DAAB" under the

2017 FIDIC Suite. The Emerald Book adopts the provisions of the 2017 FIDIC Suite relating to the DAAB. These provisions are briefly discussed below.

4.1.1 *Standing Board*

Under the 1999 FIDIC Suite, depending on the specific FIDIC book the DAB was either an ad-hoc entity constituted after a Dispute had already arisen, or a standing body to be constituted at the very beginning of any Contract with a view to facilitate dispute avoidance. Consistent with the focus on dispute avoidance, the 2017 FIDIC Suite, and the Emerald Book follows suit and the DAAB is to be put in place at the outset of the project.

4.1.2 *Constitution of DAAB (Appointment of Members & Timing): Sub-Clause 21.1*

The DAAB must be constituted jointly by the Parties. Like the 2017 FIDIC Suite, the Emerald Book also introduces a 'list system' in order to enable the Parties to constitute a DAAB. The list is contained in the Contract Data, and requires the Employer and the Contractor to each nominate 3 members each to the list. The purpose for doing so is that at the time of appointing the DAAB member(s), the Parties must only nominate members from the list, except under exceptional circumstances such as a listed nominee's inability or unwillingness to act. While no specific qualification is necessary or required for a person to be nominated, given the peculiar nature of underground works the Parties may be well advised to nominate persons who may have experience in decision making on excavation, support, ground treatment, etc., in similar subsurface contexts, paired with a fair knowledge of the FIDIC contractual system. The provisions concerning the remuneration and termination of DAAB member(s) are set-out in Sub-Clause 21.1.The DAAB must be constituted within the time stated in the Contract Data, and if no time is stated, then within 28 days from the date on which the Contractor receives the Letter of Acceptance.

In case of failure to appoint DAAB member(s) or a replacement member or refusal to cooperate in their appointment, a default-mechanism will apply and accordingly a three-member DAAB will be appointed.

4.1.3 *Challenge Procedure: Rules 10 and 11 of the DAAB Procedural Rules*

A new challenge procedure has been introduced under the 2017 FIDIC Suite which has also been adopted under the Emerald Book. Under this challenge procedure, any challenge to a DAAB member's ability to discharge its functions is to be decided by the International Chamber of Commerce. The specific details of the challenge procedure are set out in Rules 10 and 11 of the DAAB Procedural Rules which are provided as an appendix to the Emerald Book.

4.1.4 *Reference of a Dispute to the DAAB and Obtaining DAAB's Decision: Sub-Clause 21.4.1 and 21.4.3*

A Dispute must be referred to the DAAB within 42 days from giving or receiving a Notice of Dissatisfaction with the Engineer's determination under Sub-Clause 3.7 [*Agreement or Determination*], if the subject-matter of the Dispute is covered under Sub-Clause 3.7, failing which the Notice of Dissatisfaction will be invalid and deemed to have lapsed. Previously, under the 1999 FIDIC Suite there was no time-limit for referring disputes to the DAB.

The DAAB must give its decision on the Dispute within 84 days after receiving the reference or by such other period that may be proposed by the DAAB and agreed by both Parties.

Where the DAAB orders a payment to be made by one Party to the other it may, at the request of the paying Party, order the payee to provide security for repayment of such amount, if there are reasonable grounds to believe that the payee will be unable to repay in the event that the decision is reversed in arbitration. Previously, under the 1999 FIDIC Suite there were no explicit provisions allowing this.

Finally, any applicable limitation period is automatically interrupted once a Dispute has been referred to a DAAB, unless prohibited by law.

4.1.5 *Dissatisfaction with DAAB's decision: Sub-Clause 21.4.4*

If either Party is dissatisfied with the decision of the DAAB then it must give a NOD to the other Party and the Engineer within 28 days after receiving the DAAB's decision. If the DAAB has failed to give its decision within the 84-day period (or the period otherwise agreed to by the Parties) stated in Sub-Clause 21.4.3, then in such a case the NOD must be given within 28 days after the period stated under Sub-Clause 21.4.3 has expired. Arbitration proceedings under Sub-Clause 21.6 cannot be commenced unless an NOD in respect of that Dispute has been given according to the requirements of Sub-Clause 21.4.4. An arbitration can only be commenced without giving a NOD for Dispute(s) that arise under the last paragraph of Sub-Clause 3.7.5 [*Dissatisfaction with Engineer's Determination*], or Sub-Clause 21.7 [*Failure to Comply with DAAB's Decision*] or Sub-Clause 21.8 [*No DAAB in Place*]

4.1.6 *Compliance with DAAB Decisions: Sub-Clause 21.7*

DAAB decisions which are 'binding' or 'final and binding' (that is, any DAAB decision whether or not a NOD has been served) can be enforced directly through arbitration without the need for first referring them to a DAAB or amicable settlement procedure. This is a notable difference from the 1999 FIDIC Suite which explicitly provided for the opportunity of enforcement of final and binding DAB decisions only (1999 FIDIC Yellow Book, Sub-Clause 20.7.).

4.1.7 *Amicable Settlement: Sub-Clause 21.5*

In order to promote the resolution of any disputes by way of amicable settlement rather than arbitration, arbitration can only be commenced on or after the 28th day, once a NOD has been given (*See* 1999 FIDIC Yellow Book, Sub-Clause 20.5. Previously, an arbitration could be commenced on or after the 56th day, once a NOD had been given.).

4.1.8 *Arbitration: Sub-Clause 21.6*

In the event that a Party is dissatisfied with the DAAB's decision, it may refer the Dispute to arbitration. However, prior to pursuing arbitration the Parties are required to settle the Dispute amicably (see above). There is however, no sanction for not doing so, and either Party may refer the matter to arbitration after the 28-day amicable settlement period has lapsed.

The arbitrator(s) are now expressly authorized to take into account a Party's failure to cooperate in constituting the DAAB when allocating the costs of arbitration.

5 INDICATIVE PROCEDURAL TIMELINE UNDER FIDIC EMERALD BOOK

Table 1. Indicative procedural timeline under FIDIC Emerald Book.

Number of Days*	Particulars	Number of Days from Notice of Claim**
84	Submission of a Fully Detailed Claim	—
up to 84	Engineer's Agreement or Determination	168
28	NOD against Engineer's Determination	196
42	Referral to DAAB	238 (around 8 months)
84	DAAB Decision	322
28	NOD against DAAB Decision	350
28	Attempt at Amicable Settlement	378 (around 12 months)
—	Arbitration	—

* allocated for each step
** calculated cumulatively

6 CONCLUSION

To conclude, the Emerald Book adopts the claims and dispute avoidance and resolution procedures provided under the 2017 FIDIC Suite. While these procedures are similar to those prescribed under the 1999 FIDIC Suite, they have undergone a significant evolution and have become more prescriptive and complex. This contribution has endeavoured to briefly present these procedures in a simple and easy-to-follow manner so that the reader may have a basic understanding of these procedures.

The claims and dispute resolution procedures set out in the Emerald Book consist of four-tiers that begin with the Engineer's determination of a Claim under Sub-Clause 3.7. A Party's dissatisfaction with an Engineer's determination of a Claim will result in a Dispute, which the claiming Party may refer to the DAAB.

The DAAB has been designed to be a standing body, which may assist the Parties in the avoidance of Disputes. Regardless of the Dispute avoidance function of the DAAB, should a Dispute arises a Party must refer the Dispute to the DAAB. Once the DAAB renders a decision, if either Party or both Parties are dissatisfied with it they may seek to resolve the Dispute through arbitration, provided that a Notice of Dissatisfaction has been served against the DAAB decision within 28 days after receiving the decision. However, prior to going to arbitration, the Parties must mandatorily seek to reach an amicable settlement within 28-days from which a Notice of Dissatisfaction had been issued by a Party in respect of the DAAB decision.

As this contribution endeavours to highlight, the four tiers in the claims and dispute resolution procedures (that is, the Engineer's determination of a Claim, dispute adjudication by a DAAB, an attempt to reach an amicable settlement, and arbitration) have their own peculiarities which must be followed, and the rigorous pursuit and exhaustion of each step of the four tiers is a prerequisite for proceeding with the next step, thus, promoting the resolution of Claims and Disputes as early and close to the work face as possible.

ACKNOWLEDGEMENT

This contribution is based upon the work of the FIDIC Task Group 10 "New Form of Contract for Tunneling and Underground Works". The author wishes to thank FIDIC, the ITA and his colleagues Gösta Ericson (IC Consultants, Lund, Sweden), Hannes Ertl (D2 Consultants, Linz, Austria), James Maclure (Independent Consultant, Durham, United Kingdom), Andres Marulanda (Ingetec, Bogotà, Columbia), Matthias Neuenschwander (Neuenschwander Consulting Engineers Ltd, Bellinzona, Switzerland) and Martin Smith (Matrics Consult Ltd., Seoul, Republic of Korea) for their important contributions.

Caveat: at the moment of writing of this article, the FIDIC Emerald Book is still under review. Part of the content may therefore be in contrast with the published Form of Contract. The reader should always consult the published FIDIC Form of Contract for Underground Works.

REFERENCES

FIDIC, "Conditions of Contract for Underground Works", 2019
FIDIC, "*Conditions of Contract for Pland & Design Build*", Second Edition 2017
FIDIC, "*Conditions of Contract for Design, Build and Operate Projects*, 2008
FIDIC, "*Guidance for the Preparation of Particular Conditions to Yellow Book*", Second Edition 2017

Tunnels and Underground Cities: Engineering and Innovation meet Archaeology,
Architecture and Art, Volume 8: Public Communication and Awareness/Risk Management,
Contracts and Financial Aspects – Peila, Viggiani & Celestino (Eds)
© 2020 Taylor & Francis Group, London, ISBN 978-0-367-46873-6

Measuring the excavation and lining in the Emerald Book

M. Neuenschwander
Neuenschwander Consulting Engineers Ltd., Bellinzona, Switzerland

A. Marulanda
Ingetec, Bogotà, Colombia

ABSTRACT: One of the key features of the new FIDIC Form of Contracts for Underground Works (the "Emerald Book") is the remeasurement of time and of quantities for excavation, support and lining of the works. It is an important aspect of the balanced allocation of risk related to the subsurface conditions, as in this way, the effectively encountered ground is taken into account. In the tender documents, the relevant quantities are presented in specific schedules for time related, quantity related and value related cost items. These items are the used by the Contractor and the Engineer in order to establish the realized works on a regular basis, allowing for a timely issuing of the necessary statements, as well as for the control of cost evolution in the project. The present contribution is one of several regarding the Emerald Book, and treats the remeasurement process and tools regarding time and quantity related cost items, as well as the documents that must be prepared by the Employer prior to issuing the tender documents during his procurement process.

1   INTRODUCTION

As is well known to all members of the tunneling industry, there is an ever-growing demand for utilizing underground space for infrastructure. The difficulty in predicting underground behaviour and conditions poses unique challenges regarding construction practicability, time and cost. Thus, allocation of underground risks among the stakeholders becomes critical in underground construction. To address these unique risks the International Tunnelling and Underground Space Association (ITA) and the International Federation of Consulting Engineers (FIDIC) joined forces to draft the new FIDIC Form of Contracts for Underground Works (the "**Emerald Book**" (see Reference [1]). In order to accomplish this, the two organizations setup a joint task group (TG 10).

The Emerald Book has been modelled on the 2017 FIDIC Yellow Book (Conditions of Contract for Plant & Design Build) (see Reference [2]) but with significant innovations tailored to underground construction. Consistent with FIDIC's philosophy of achieving a fair allocation of risks among the parties, the Emerald Book has been drafted with a view to promoting a balanced risk allocation that is specifically adapted to the risks inherent and unique to Underground Works.

The FIDIC "Rainbow Suite of Contracts" provides standard forms of contract for different types of risk sharing models, amongst which the "Red Book" (see Reference [3]), a Form of Contract for Construction Works designed by the Employer, the "Yellow Book", a Form of Contract for Plant and for Construction Works, designed by the Contractor, and the "Silver Book" (see Reference [4]), a Form for EPC Contracts, where all the risk of Engineering, Procurement and Construction is taken by the Contractor against a lump sum price.

Two salient differences between the Yellow and the Red Book lie in the responsibility for design and in the mode of payment of the Contractor. While the first difference appears in the titles ("designed by the Employer" vs. "designed by the Contractor"), the second difference requires a comment.

2 MEASUREMENT UNDER THE FIDIC RED AND YELLOW BOOKS

Under the Red Book, the Contract Price is based upon a Bill of Quantities, where the Employer lists the payment items and estimates the quantities for each item, and where the Contractor offers a unit rate for each item. The Contractor is the paid upon remeasurement of the quantities of the different items according to what he has actually realized. The General Conditions of Contract provide, amongst others, a mechanism for the case where an item is completely missing in the Schedule, and one for the case where the remeasurement of the quantities realized of an item show a remarkable difference against the estimated quantity. The unit rates typically include all charges that the Contractor has to sustain in order to realize the Works (including charges related to the time of construction, like i.e. costs for supervision, plus a profit of his choice). The items will usually cover all parts of the Permanent Works in more or less detail.

Under the Yellow Book, the Contract Price is a Lump Sum that, again, includes all charges the Contractor has to sustain in order to realize the Works, plus a profit of his choice. It is considered that, as the Contractor is in charge of the design of the Works, he should bear the quantity-related risks.

In both Forms, the General Conditions of Contract provide mechanisms for claims by the Contractor, in particular for the cases of Unforeseeable Physical Conditions.

3 THE CONTRACT PRICE IN UNDERGROUND WORKS

Compared to other kinds of Works, underground construction faces a main uncertainty due to the fact that in order to realize the Works, the necessary space must first be created by excavation in a ground mass that is neither known perfectly in its characteristics, nor in its reaction to excavation. It is therefore difficult to assess the effort (in terms of energy, time and money) it will take to excavate and to finish the Works.

3.1 *Risk allocation under the Emerald Book*

A balanced risk allocation is central to Underground Works contracts (this is also why FIDIC explicitly discourages the use of the Silver Book for sub-surface works). As a rule, the most important risks in Underground Works are related to the quality of the ground to be excavated and supported during construction (there are other important risks related to i.e. access points, available space for installation etc.). The Emerald Book specifies that the Geotechnical Baseline Report GBR shall be the only contractual source of allocation of the risk related to the sub-surface physical conditions: within the limits described in the GBR, the risk of production rates and related cost is allocated to the Contractor, while outside these limits it is considered to be in a situation of Unforeseeable Physical Conditions under Sub-Clause 4.12.

3.2 *Measurement in a Design-Build approach*

If the Contractor is responsible for the design of the Works he should logically bear the risks related to the construction method he selects and to the quantities: unless a situation of Unforeseeable Physical Condition occurs, the Contractor's remuneration and allowance for time to perform the Works should be fixed (except cost escalation according to the Contract). This is the main incentive to the Contractor for finding cost-effective solutions, while satisfying the fitness for purpose required by this type of Contract.

3.3 *Cost uncertainty in Underground Works*

In Underground Works, the greatest uncertainty is related to the sub-surface conditions, and in particular to

- the difficulty in excavating the necessary space for the Works, including but not limited to natural or man-made obstacles, water inflow etc.,

- the reaction of the surrounding ground mass to the excavation of this space,
- the effort to create the provisional and final support and lining of the surrounds of the Works,
- and last but not least, the handling, transportation and disposal of the excavated ground (spoil), according to the nature of the ground, the excavation method, available space, recycling possibilities and legal environment.

The last of these factors will, as a rule, influence cost only, while the other three will have a strong influence on the cost of the Works as well as on the time for completing the underground excavation and lining and therefore in many cases on the Time for Completion of the Works.

Time is an important cost driver (not only) in underground works, because an important part of the charges that the Contractor sustains are independent from the actual performance of the Works and depend only on the necessary time, such as i.e. the devaluation of and interest rates for expensive equipment, or the site supervision and management. The uncertainty of the time available for excavation and lining due to the uncertainty of the sub-surface physical conditions leads therefore to a similar uncertainty regarding the time-related charges.

3.4 *Adjustment of the Contract Price*

Following the logic of the Employer carrying the risk of the sub-surface physical conditions, in relation with the contractually agreed construction methodology, and the Contractor carrying the risk of production rates and cost in any situation within the limits specified in the GBR, the Contract Price should be adjusted according to the difference between the agreed sub-surface physical conditions and the conditions as encountered during the Works.

On the other hand, no adjustment should be made for any difference between the Contractor's offered production rates and the real performance in ground conditions as described in the GBR.

The adjustment of the Contract Price should take into account the cost for performing the Works as such (quantity-related charges) and the cost related to the adjustment of Time for Completion (time related charges). Of course, an adjustment may mean an increase or a reduction of the Contract Price as compared to the Agreed Contract Amount: as a principle, if the sub-surface physical conditions as encountered lead to easier realization of the excavation, support and lining of the Works than what was contractually agreed, the Contract Price should be reduced, while in the opposite case it should be increased.

The adjustment is possible in two ways: either through a change in a lump sum by the claims procedure as foreseen by i.e. the FIDIC Yellow Book, or through measurement. For the Emerald Book it was decided that a measurement provision for the Excavation and Lining works and for the time related charge items necessary for these works was the most effective way of managing the adjustment.

No adjustment of the Contract Price should be made for any part of the works that is not subject to the risks related to sub-surface physical conditions: this portion of the contract price is a lump sum, and as such it shall not be adjusted through measurement, but may only be varied following a Contractor's Claim and/or a DAAB award.

4   MEASUREMENT UNDER THE EMERALD BOOK

4.1 *The concept of measurement and the different cost items*

The Emerald Book distinguishes between those parts of the Works that are subject to the risk related to the sub-surface physical conditions, and those parts that are not. The underground excavation and temporary support and the final lining are deemed to be subject to these risks, while all other parts of the Works are deemed not to be subject to these risks. Therefore, the underground excavation, support and lining Works should be remeasured, while all other Works should be remunerated through the lump sum component of the Contract Price. This is consistent with the logics of a Design-Build Contract. For the purpose of measurement, the following categories have been postulated in the Emerald Book (Sub-Clause 13.8 "Measurement of Underground Works and Adjustment of Time for Completion"):

- Fixed rate items
- Quantity-related rate items
- Time-related rate items

4.2 Fixed rate items

All those items that are necessary for the performance of the underground excavation, support and lining Works, but that are independent from the variation of the sub-surface physical conditions, may be considered in the exclusive risk sphere of the Contractor's, and may therefore be paid with lump sums as fixed rate items. Typically, these items include the transportation to site, setting up, dismantling and evacuation of the Contractor's equipment, the construction of the site infrastructure and the making available of equipment and of site supervision and management for the time estimated by the Contractor for performing the works.

In the Bill of Quantities, the fixed rate items may be described in detail ("*113/861.101 Electrohydraulic Drilling rig for drill&blast excavation with three drilling arms and working platform, maximum working height 6.75m, covering an excavation face of 120m², with a minimum drilling length of 5 m, including transportation to site, setting up, availability for the entire duration as estimated by the Contractor for completing the excavation of Section X, dismantling and evacuation*"), or in grouped items ("*113/121.111 Complete equipment for excavation and support of Section 1, according to Drawing XY in the Employer's Requirements, including transportation to site, setting up, availability for the entire duration as estimated by the Contractor for completing the excavation of Section X, dismantling and evacuation*").

4.3 Quantity-related rate items

The performance of excavation, support and lining Works, including all necessary ancillary activities such as i.e., the drainage of water seepage, the drilling of probe holes, the handling of the spoils etc., shall be remunerated through "quantity-related rate items". These are unit rate items that are paid according to the rate offered by the Contractor and agreed under the contract, with a price that shall be paid for each measured unit that was performed, irrespective of the real effort and/or time it took the Contractor to perform it. The items and units shall be as per the Bill of Quantities prepared by the Employer in his tender documents, and each item shall be completed with the respective rate by the Contractor.

In the BoQ, the quantity related rate items may be described and measured in detail (i.e., "*261/123.456 Excavation in support class 3A according to drawing XY attached to the Geotechnical Baseline Report, including transportation of spoil to Contractor's disposal area. Measurement according to theoretical cross section on drawing and measured tunnel length in situ. Unit: m³, Quantity: 125'000*"; "*261/234.567 Supply to excavation face and setting up of expanding rock bolts with adherence on the entire bolt length. Minimum yield load per bolt: 100 kN, bolt length: 4m. Including anchor plate 20x20cm. Unit: piece (pce)), Quantity: 5000*") or as grouped items (i.e., "*261/111.211 Excavation and temporary support of tunnel cross section according to drawing XY attached to the GBR, including all necessary measures and activities, measurement in m in the tunnel axis of excavated and supported tunnel. Unit: m, Quantity: 2'500.*")

4.4 Time-related rate items

Time-related rate items are used for the remuneration of charges supported by the Contractor that don't depend on the quantity of performed Works, but on the time required. Typical time-related charges are the supervision and management on site, the running and maintenance of the construction yard, the workshops and the rail system (if any), and (without limitation) the availability of Contractor's key equipment not covered by the Fixes rates for the time lapse estimated by the Contractor in his tender.

While during the Contractor's estimated time these charges shall be covered by fixed rate items, the same charges due to adjustment and/or extension of time related to risks that are not allocated to the Contractor, depend on the length of this adjustment and should be paid throuth time-related rate items.

The time-related rate items shall be measured in days (Calendar days) or weeks. They shall be related to the respective fixed rate items, and shall refer to the same (i.e. *"113/891.861 Extended availability of items 113/861.101 to 861.191. Unit: day, Quantity: 120"; "113/891.862 Reduced availability of items 113/861.101 to 861.191. Unit: day, Quantity: 90 (negative rate)"*).

Only the differences between the estimated and the measured amounts in the Baseline Schedules may give rise to the adjustment of time for completion of the Works, a Section or any other Milestone. Further, in order for an adjustment to be applicable, the respective part of the Works must be on the critical path of the same Milestone. For example, it is possible that the measurement in a Baseline Schedule leads to extended time for completion of the Milestone "End, Excavation of tunnel Section XY", but does not lead to extended time for completion of the Works, because the Excavation of tunnel Section XY is not on the critical path for completion of the Works. This is to be considered when adjusting and/or extending time.

4.4.1 Adjustment and extension of time

In order to measure any quantity for time related charge items, the respective time for completion must first be adjusted and/or extended. As pointed out above, if the sub-surface physical conditions are better than expended (i.e. if the amount of rock requiring light support is higher than expected and the amount of rock requiring heavy support is lower than expected), this will lead to a shortening of the time available to the Contractor as measured in the Baseline Schedule (see Reference [5]). Vice versa, if the sub-surface physical conditions are worse than expended (i.e. if the amount of rock requiring light support is lower than expected and the amount of rock requiring heavy support is higher than expected), this will lead to a lengthening of the time available to the Contractor as measured in the Baseline Schedule. To the total of adjusted time, extensions of time under Sub-clause 4.12 shall be added.

4.4.2 Measurement of quantity of time related rate items

Time related rate items shall be measured in units of days of extension and/or reduction of the Time for Completion of the respective Milestone. For example, if the Time for Completion of the Milestone "End, Excavation of tunnel Section XY" is extended according to the measurement in its Baseline Schedule by 45 days, then the measurement for all the time related rate items for the same Milestone should also be 45 days.

4.5 Value-related charges

Some charges that the Contractor must support are value-related, such as i.e. the premium for construction insurance, or interest on retention money, etc. In order for these charges to be properly taken into account, they should be added as percentage rates to all items in the Bill of Quantities. In this way it is assured that the Contractor's cost will be covered according to the contractually agreed payment plan and to the remeasured quantities of Works performed, as the case may be.

4.6 Adjustment of the Contract Price

The Contract Price shall be adjusted according to the results of the measurement of all items in the BoQ and according to any awarded claim of the Contractor's. As with all measured contracts, the adjusted Contract Price may result higher or lower than the Accepted Contract Amount.

4.7 Validity of Unit Rates

The rates and prices shall remain fixed irrespective of the actual quantities measured. This is a major difference against the measurement principle as postulated in the Red Book (both the 1999 and the 2017 Editions), where in the case of a difference between the estimated and the measured quantities of more than 10%, a change in the rate or price may be asked for.

The difference between the Emerald and the Red Book lies in particular in the distinction between the time-related and the quantity-related rate items. Because the time related charges are covered irrespectively of the actual quantity of work performed, the cost supported by the Contractor for the performance of a quantity-related charge item varies much less due to the actual quantity performed than in a situation where the time-related charges must be included in the unit rates.

4.8 *Absence of unit rates*

In the case of absence of unit rates (i.e., if a new type of rock bolt is instructed by the Engineer, for which there is no unit rate in the BoQ), there are two possibilities: either a new unit rate is negotiated between the parties, in analogy with similar items wherever possible, or the particular part of the works shall be remunerated through day-work rates.

4.9 *Payment in exceptional situations*

As the Yellow Book, 2017 Edition, the Emerald Book includes two provisions regarding exceptional situations: Sub-Clause 4.12 "Unforeseeable Physical Conditions" (UPC) and Clause 18 "Exceptional Events". In both cases, the Contractor is entitled to reimbursement of the Cost (and Extension of Time, if any) he has suffered (meaning that he is entitled to reimbursement of the entire Cost but not to Profit). The particularity of the Emerald Book resides in the treatment of time-related cost the Contractor may have suffered under Sub-Clause 4.1 (Unforeseeable Physical Conditions): in case of such an event relating to Excavation and Lining Works, the Cost "shall be determined gy reference to the time-related rate items provided in the corresponding Bill of Quantities, to the extent that unit rates exist and are comparable, deducting profit as defined in Sub-Clause 1.1.27 [Cost plus Profit]". This provision relieves the Contractor from the burden of the proof of the cost he suffered per unit of time under the said UPC, and allows the parties to simply use the time-related rate items and the duration of time loss that was agreed or determined.

The same provision does not apply to Clause 18 "Exceptional Events". In this case it was the opinion of the authors of the Emerald Book that a full claim by the Contractor would be justified.

4.10 *Treatment of a particular case: geological overbreak*

How can particular cases be treated under the provisions of Sub-Clause 13.8 "Measurement of Underground Works and Adjustment of Time for Completion and of Contract Price", such as i.e. geological overbreak?

Geological overbreak per definition is a void outside the theoretical excavation line which is due neither to poor workmanship, nor to the excavation methodology, but to the nature of the ground mass. The cost generated by overbreak is typically caused by the need of evacuating more psoil than originally estimated, and by the need to backfill the extra void with concrete, sprayed or vibrated. Of course, overbreak may cause quantity-related cost and time-related cost. According to the logics of the Emerald Book, the risk regarding geological overbreak cannot be assigned entirely to the Contractor. As with other natural phenomena such as, i.e., flood), this risk to the Contractor should be capped. The way of limiting the Contractor's risk depends on the legal environment and the culture of the Employer's. In the following, two possibilities amongst several are presented:

One possibility is to assign the risk of geological overbreak in percentage rates of the theoretical excavation surface, taking into account the fact that in order to create the minimum necessary space for construction of the Lining, some overbreak will be inevitable. There might therefore be three categories of overbreak: a), the inevitable (technical) overbreak to be included in the rates and prices for excavation. The percentage of overbreak to include in the excavation rates should be defined by the Employer (i.e. 5%, 10% or x%, according to the Employer's estimation). Outside this surface of $(1+x\%)$ x (theoretical excavation surface), another percentage may be allocated to the Contractor as a risk, i.e. the additional void ranging from 10% (included in the unit rates) to 15% will be in the Contractor's risk sphere, while the handling and backfilling of all geological overbreak beyond 15% shall be remunerated through especially created quantity-related unit rate items.

The other possibility presented here is based upon distances from the theoretical excavation line.

In the tender documents, the Employer shall specify the distance from the theoretical excavation line that is the limit of the Contractor's risk sphere for geological overbreak (boundary line G).

In his tender, the Contractor shall define up to which distance from the theoretical excavation line any overbreak shall be considered "technical" and therefore included in the unit rates for excavation.

Any overbreak that remains inside the boundary line "G" shall be considered in the risk sphere of the Contractor's, meaning that the Contractor shall have no entitlement for either Extension of Time for Completion or additional remuneration due to this overbreak. For any overbreak with a boundary outside the "G" line, the volume for which there is an entitlement for spoil handling and backfill, if any, is measured as indicated in Figure 1.

An estimation of the total volume may also be specified in the Baseline Schedule, so that the incidence of the hindrances due to the overbreak on production rates may be contractually agreed beforehand, and the entitlement to adjustment of Time for Completion be a mere question of measurement without the need for a claim.

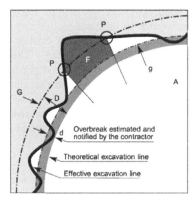

A: Theoretical excavated surface area

D: Distance between the theoretical excavation line and the boundary line G

d: Distance between the theoretical excavation line and the boundary line g

g: Up to this line, the overbreak is calculated into the excavation price

G: Boundary line, dependent on D

F: Surface area for which the geological overbreak will be paid for

P: Point of intersection between the boundary line G and the effective excavation line

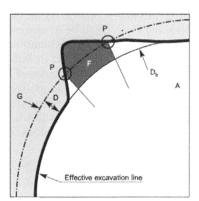

A: Theoretical excavated surface area

D: Distance between the effective excavation line and the boundary line G

D_b: Bore diameter with worn tool

G: Boundary line, dependent on D

F: Surface area for which the geological overbreak will be paid for

P: Point of intersection between the boundary line G and the effective excavation line

Figure 1. Definition of geological overbreak in rock excavation for conventional (Drill & Blast) and for mechanical (TBM) excavation. From: Swiss Code SN 507198, "General Conditions for Underground Construction" (see Reference [6]).

5 CONCLUSIONS

The Emerald Book combines the Design-Build approach with a balanced allocation of the risk related to sub-surface physical conditions. In particular, the definition of these conditions and of the relevant ground categories, including natural and man-made obstacles and hindrances, is proposed by the Employer in his tender documents (namely in the Geotechnical Baseline Report and in the Baseline Schedules). The production rates and the rates and prices for excavation, support and lining according to these ground categories are offered by the Contractor in his tender. Both the description of the ground categories with their estimated amounts and the production rates are contractually agreed. For all parts of the Works except Underground Excavation (including support and ancillary measures) and Lining, the Price shall be an agreed Lump Sum Amount.

The Underground Excavation and Lining shall be measured by the Contractor according to a Bill of Quantities. The Engineer shall determine whether the measurement by the Contractor applies to works that have been necessarily performed or not, and shall only certify payment and the adjustment of the Contract Price, if any, for those works that were necessary. In case of disagreement, the Contractor shall be entitled to claim the difference, and to address the Dispute Avoidance and Adjudication Board.

Because a high portion of the charges that the Contractor supports is time-related (such as, i.e. the devaluation and interest rates on equipment, the site supervision and management etc), the BoQ shall include Fixed rate items, Quantity-related rate items and Time-related rate items. The time-related rate items allow for a measurement and payment of the Contractor based upon the adjustment of time according to the differences between the contractually agreed and the encountered quantities of different ground categories. Thus, the Quantity-related charges become independent of the time-related elements, which greatly reduces their dependency on the actual amount of items produced. The Quantity-related rates may therefore remain unchanged irrespectively of the measured quantities.

Finally, there is no precise prescription on how the Bill of Quantities should be drafted, nor on what measurement rules should apply. These are specific to each project and to each legal and cultural environment, and should be defined in the Particular Conditions of Contract.

ACKNOWLEDGEMENT

This contribution is based upon the work of the FIDIC Task Group 10 "New Form of Contract for Tunneling and Underground Works". The authors wish to thank FIDIC, the ITA and their colleagues Gösta Ericson (IC Consultants, Lund, Sweden), Hannes Ertl (D2 Consultants, Linz, Austria), James Maclure (Independent Consultant, Durham, United Kingdom), Charles Nairac (White & Case LLP, Paris, France) and Martin Smith (Matrics Consult Ltd., Seoul, Republic of Korea) for their important contributions.

Caveat: at the moment of writing of this article, the FIDIC Emerald Book is still under review. Part of the content may therefore be in contrast with the published Form of Contract. The reader should always consult the published FIDIC Form of Contract for Underground Works.

REFERENCES

FIDIC, "*Conditions of Contract for Underground Works*", 2018
FIDIC, "*Conditions of Contract for Pland & Design Build*", Second Edition 2017
FIDIC, "*Conditions of Contract for Construction*", Second Edition 2017
FIDIC, "*Conditions of Contract for EPC/Turnkey Projects*", *Second Edition 2017*
Marulanda A. and Neuenschwander, M., "*Contractual Time for Completion Adjustment in the FIDIC Emerald Book*", ITA, WTC2019 Naples
Swiss Standards Association SNV, "*General Conditions for Underground Construction*"- *Swiss Code 507198*, SIA Zurich 2007

Tunnels and Underground Cities: Engineering and Innovation meet Archaeology,
Architecture and Art, Volume 8: Public Communication and Awareness/Risk Management,
Contracts and Financial Aspects – Peila, Viggiani & Celestino (Eds)
© 2020 Taylor & Francis Group, London, ISBN 978-0-367-46873-6

Regulatory framework and railway safety approval procedures in a bi-national context – the example of the Montcenis base tunnel

P. Poti
TELT sas, Turin, Italy

A. Chabert
TELT sas, Bourget du Lac, France

ABSTRACT: The Montcenis base tunnel between Italy and France is part of a series of 20+ km-long railway tunnels that have recently replaced, or are replacing, the older mountain pass crossings. These long tunnels require a specific look at the aspects of railway safety and accident management. For cross-border tunnels, the granting of safety certifications is complicated by the juxtaposition of multiple national authorities. This paper begins with a short description of the procedures followed for the Channel and Brenner links, which are similar to the future Montcenis base tunnel. The particular Italian-French bi-national context characterizing this latter is then presented under three aspects: the definition of a reference regulatory framework, its harmonization, and the initiatives followed to obtain the "Clearance for Development and Completion of the Project" from the respective National Agencies, as well as, upon completion of all work, the "Clearance for the Commercial Commissioning" of the infrastructure.

1 THE MAIN EUROPEAN CROSS-BORDER TUNNELS

1.1 Introduction

We live in a context where Europe is equipping itself with railway corridors aimed at speeding up the connections between major cities, facilitating the exchanges of goods over different countries, and shifting most of freight traffic onto railways. This in turn will decrease gas emissions and improve road safety. The large base tunnels are a core element of this strategy, as they are needed to reach the performance objectives expected for these fast connections.

Some of these tunnels are already operational (Channel since 1994, Lötschberg since 2007, Gotthard since 2016), while others will become so in the near future (Koralm in 2022, Brenner in 2025, Moncenis in 2030). All of them have required a careful consideration of the subjects of railway safety and accident management.

For cross-border tunnels, the approval process in relation to railway safety has been further complicated since the early 2000s by the contemporary presence of several competent national authorisation bodies, in particular the National Safety Authorities for railways (NSA).

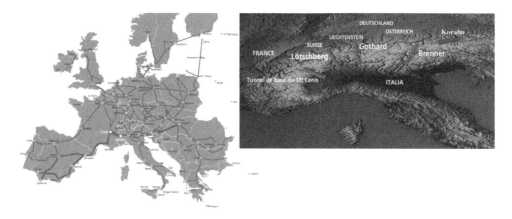

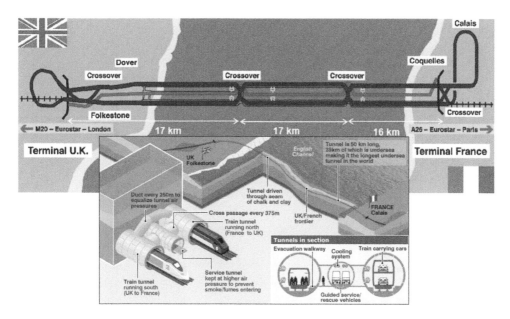

Figure 1. Main railway corridors and relative base tunnels.

The 50.4-km long Channel Tunnel links the city of Calais in France with the city of Folkestone in England. It consists of two main outer tubes, where the trains travel, and a third service tunnel in the middle for the various maintenance and safety services.

Figure 2. Channel tunnel diagram.

Cross-connections between the railway tubes are located at every third of the tunnel, allowing for greater flexibility of operations during breakdown situations. The maximum speed of the trains is 160 km/h for passenger convoys and the tracks are controlled via the French TVM 430 standard signalling system. In scenarios requiring passenger evacuation, the service tunnel between the two railway tubes serves as a safety site. The damaged tube is then isolated from the undamaged one with the help of shutters and watertight doors, protecting the passengers and other trains.

The tunnel came into service in 1994 and, since the NSAs did not exist at the time, the design and construction stages were supervised by an *ad-hoc* Intergovernmental Commission (IGC), whose members were drawn from the French government, the UK government, British Rail and SNCF.

The IGC consisted of 5 sector committees: finance, design, operation and maintenance, safety/rescue (firefighters, police, healthcare personnel, etc.) and security.

In 1993, the testing and commissioning stages were approved by the IGC and consequently the French and UK governments granted the "clearance for operation".

1.2 Brenner Base Tunnel

The Brenner link covers the 64 km from Tulfes to Fortezza and it includes a 55-km-long tunnel connecting the village of Fortezza in Italy with the city of Innsbruck in Austria. It consists of two main tubes using ERTMS level 2 as the signalling system. Passenger trains will be able to run at a maximum project speed of 250 km/h and freight trains at 120 km/h.

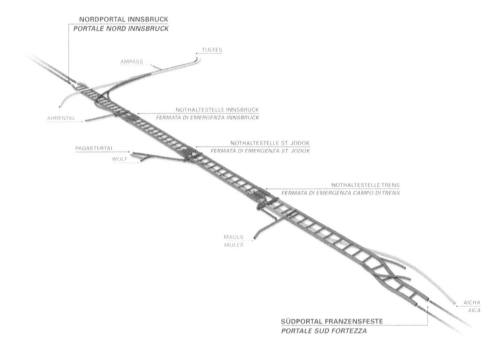

Figure 3.   Brenner tunnel diagram.

Since September 2017, BBT has been engaged in developing a "Cross-border operation regulation" with the help of RFI, ÖBB, the ITA/AUS railway safety Authorities and ERA (with an advisory role). The Regulation will apply the safety principles for operation established by the Authorities and it will be based on the functional specifications already defined in the Definitive Plan of the Brenner base tunnel. This includes the very important "Reversing" function, an operation mode of the ERTMS system that authorises the reversal of the signalling and the backward motion of the trains in safe conditions. This will permit to reverse the train flux in case of emergencies or irregular events.

The set of functional specifications and the definition of the cross-border regulation will also enable the definition of procedures for the safe management of accidents during operation.

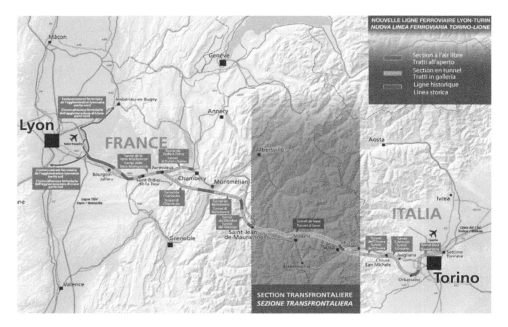

Figure 4. The new Turin-Lyon railway line.

2   THE MONTCENIS BASE TUNNEL

2.1   *Main design characteristics*

The New Turin-Lyon Line consists of three parts: the French part (under SNCF-Reseau management); the Italian-French joint part (international section); and the Italian part (under RFI management). As shown in Figure 4, the international section includes the cross-border section that spans from St-Jean-de-Maurienne to Susa/Bussoleno and whose design, construction, and management is in the hands of the French- Italian company TELT (Tunnel Euralpin Lyon Turin).

This section consists of: an open-air area in St.-Jean-de-Maurienne (3.7 km) that includes the new passenger station of St.-Jean-de-Maurienne, the safety site and the connection to the existing French line; the Montcenis base tunnel (57.5 km); an open-air area in the Susa Valley (2.7 km) that includes the new international passenger station and the Susa safety site; the connection tunnel (2.1 km) to the existing Bussoleno railway track; and an open-air area for the linkage with the existing line in Bussoleno (0.9 km). The cross-border part of the joint section is therefore 66.9 km long in total and it includes 2 outside safety sites (Saint-Jean-de-Maurienne and Susa), 3 underground safety sites (La Praz, Modane and Clarea, accessible from the outside via access tunnels) and an additional tunnel for access by rescue services in Saint Martin la Porte. The two tunnels (base and interconnection) consist of two single track tubes connected via corridors built every 333 m (reduced to 50 m in the underground safety sites of the base tunnel). Along both tunnels, the cross-section of the current section consists of a service and evacuation walkway (no more than 1.20 m wide – from the side of the second tube), a rail traffic track, and a maintenance walkway on the outside side (see Figure 5).

In particular, the base tunnel has a lighting system and smoke extraction systems that can be activated in the event of an accident, as well as detectors, a liquid collection system, and a fire protection network. The New Turin-Lyon Line will be a mixed passenger and freight traffic line designed with a nominal track speed of 250 km/h. The following train categories will be able to run on the line:

• High-speed passenger trains (HS): maximum operating speed 220 km/h on the Saint-Jean-de-Maurienne–Susa section;

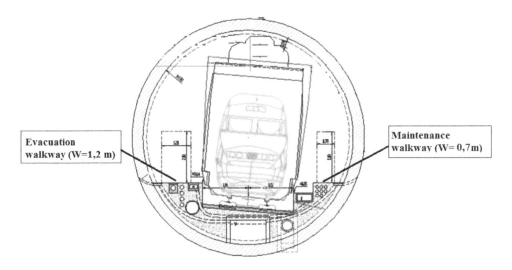

Figure 5. Tunnel typological cross-section.

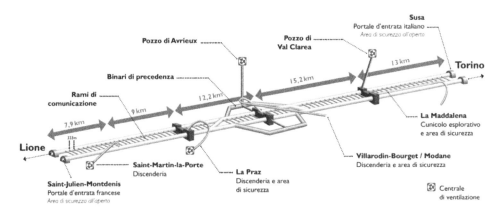

Figure 6. Diagram of the 57.5 km of the Base Tunnel and its main connected works.

- High-profile Railway Trains (AFGG) and Modalohr Railway Trains (AFM): maximum operating speed 120 km/h.
- Conventional freight trains (M), maximum length of each train 750 m. Maximum operating speed: 100 or 120 km/h depending on the category.

3   THE RAILWAY SAFETY APPROVAL PROCEDURES

3.1   *The Bodies involved in railway safety*

A massive cross-border work, such as the Montcenis base tunnel, has seen, since its early design stages, the intervention of a series of competent "players" in the field of railway safety. In particular, TELT, as Public Sponsor, is responsible for concluding and overseeing the execution of the contracts required by the design, creation, and operation of the cross-border section of the work, as well as for obtaining all the authorisations concerning the safety of the railway operation. In accordance with Directive 2001/14/EC, which provides for a specific infrastructure manager for the cross-border section, the "Public Sponsor" was officially created on 23/02/2015 and it is called 'Tunnel Euralpin Lyon Turin (TELT) SAS.

Additionally, the Intergovernmental Commission, set up by an agreement between Italy and France dating from 15 January 1996, is charged with approving the project and with proposing to the two governments the specifications for the final works, the methods of construction, and their financing, as well as the conditions of operation.

To accomplish these tasks as defined in the aforementioned agreement, the Intergovernmental Commission has decided to set up a Safety Committee to assist it with decisions concerning the technical safety of the work during the design, construction and management stages.

This Safety Committee consists of experts in the following sectors:

• Infrastructure safety and traffic in the railway sector,
• Civil safety and rescue.

The National Safety Authorities for railways (NSA) of the two countries are also represented in the Safety Committee.

All safety certification requests for the cross-border section are addressed by the Sponsor (TELT) to the National Safety Authorities for railways who, while competent for their own national territory, coordinate and express a joint decision on the basis of a 2014 protocol.

The Designated Body (DeBo) is responsible for instituting the verification procedure of structural subsystems when national standards are applied. This procedure is preliminary to the presentation of the safety files to the NSAs.

The notified body (NoBo) is responsible for assessing the conformity of the subsystems with the TSIs (Technical Specifications for Interoperability) and the applicable regulatory provisions, and it certifies the EC declaration of conformity and suitability for use of the components. This activity must also be conducted prior to the presentation of the safety files to the NSAs.

The assessment body (AsBo), in accordance with Regulation (EU) No. 402/2013 ("Common safety method for risk evaluation"), is responsible for the independent assessment of the correct application of the risk management process and for the results obtained.

TELT has envisaged appointing a single, joint body to ensure the roles of DeBo, NoBo and AsBo. This in order to avoid duplications in implementing the DeBo and NoBo missions, in accordance with Article 6 of Regulation (EU) No. 402/2013. The DeBo/NoBo/AsBo was identified by TELT as the Grouping Belgorail – RINA.

3.2   *Reference regulatory framework and consequent international harmonisation*

Over the years, a complex work of regulatory harmonisation has been conducted under the supervision of the Italian-French Intergovernmental Commission (IGC). The goal was to arrive at a common set of standards for the cross-border portion of the railway work.

These standards were first divided into two macro-categories: general standards, not strictly railway-related (for example the standards on geology, environment and non-railway infrastructure such as roads), and railway-specific standards. Within this second macro-category, the standards have been grouped into three sections, namely: "joint" standards, valid for both high-speed lines and conventional/existing lines, specific standards for high-speed lines, and specific standards for conventional lines. The regulatory framework also aims to apply technical recommendations issued by International Associations over very specific topics, such as guidelines and best practice recommendations. From the regulatory viewpoint, the standards and regulations for the international section are classified according to the following order of priority:

1. EU directives and TSI standards come first.
2. Failing that, the safety criteria established by the IGC take precedence over national standards. The IGC may lay down rules more stringent than the European directives and TSI standards, except for rolling stock.
3. In the absence of European directives, TSI standards, or IGC criteria, the standard applied is whichever national standard (Italian or French) is the strictest, subject to verification of the consistency of all provisions.

Organisation:

Hierarchisation:

1° level	«European Direttives» + TSI
2° level	«CIG Criteria»
3° level	Most restrictive «National standard»

Territoriality principle for specific work

Figure 7.   Structure of the regulatory framework.

It is clear that compliance with the TSIs is a necessary condition for the safe integration of the international section into the trans-European rail system, to which it will be connected through both the French and Italian sections. On the contrary, for all "non-line" works strictly linked to their geographical location (for example, in the case of technical buildings), the territoriality principle applies, i.e. the legislation of the country of origin applies (see Figure 7).

In the context of the regulatory framework described above, and in application of French Decree 2006–1279 on the traffic and interoperability of the railway system, TELT had to draft, already at the time of issue of the Definitive Plan, a Preliminary Safety Dossier (DPS: Dossier Préliminaire de Sécurité), subject to the approval by the French Rail Safety Authority (EPSF: Établissement Public de Sécurité Ferroviaire).

In compliance with the agreement between the two NSAs of 2014, the DPS was also sent for analysis to the Italian National Railway Safety Authority (ANSF: Agenzia Nazionale italiana di Sicurezza Ferroviaria).

In short, during the course of the year 2017, the DPS for the cross-border section was first subject to an independent investigation by the NoBo/DeBo, and then to the joint investigation by the two NSAs (EPSF and ANSF).

3.3   *Obtaining the "clearance for development" for the construction of the work and, upon completion, the "clearance for commercial commissioning" of the infrastructure*

The first version of the DPS was issued in March 2017, covering an overall summary of the full TELT Project. It was subjected to compliance checks with the Technical Specifications for Interoperability (NoBo) and national standards (DeBo) by the independent validator identified by TELT for the roles of both NoBo and DeBo. This investigation ended in early July 2017 with a favourable opinion on the compliance of the infrastructure portion of the DPS (Rousse 2018).

On 13 July 2017, the reviewed version of the DPS and the NoBo/DeBo compliance report were sent to the NSAs to begin the investigation aimed at obtaining the "clearance for development" for the part relating to the infrastructure.

The special Italian-French bi-national context characterising the Base Tunnel involved a close exchange of information, clarifications and requests for additional documents, in particular:

- The forwarding by TELT of more than 110 design documents in reply to requests for support/clarification;
- The definition of the reference regulatory framework applicable to the infrastructure, including aspects regarding its Italian-French harmonisation, and the approval of its scope and completeness by the NoBo/DeBo;
- Demonstration of compliance with the Rail Traffic Regulations (ANSF Decree No. 4/2012);

- Redefinition of some procedures to manage tunnel accidents and train reversing;
- Drafting of a risk analysis in accordance with Ministerial Decree 28/10/2005 "Safety of Railway Tunnels".

On 2 May 2018, TELT received formal approval of the infrastructure portion of the DPS by the EPSF, pursuant to French Decree 2006–1279 of 19 October 2006 on the safety of rail traffic and the interoperability of new railway lines. This approval formally allowed the beginning of civil works in the base tunnel in France.

As previously stated, the investigation for the DPS Infra was carried out jointly by the Authorities of the two countries on the basis of the 2014 protocol, although at this stage of the design process it was an exclusively "French" obligation.

The technological part (Energy and Command/Control/Signalling) of the DPS should be developed later (TELT plans to issue it by 2023). It will take the form of sector-specific DPSs which will then follow the same approval process: with the independent NoBo/DeBO validator first, and then with the NSAs, thus clearing the start of technological works in the tunnel.

Once construction of the infrastructure is completed, TELT, as infrastructure manager, will have then to obtain from each relevant National Authority an authorisation for railway activation with territorial validity, which will certify its compliance with a regulatory framework structured on European and national directives (see Figure 8).

In particular, the Italian Authority will issue the Commissioning Authorisation (AMIS: Autorizzazione alla Messa in Servizio) for the section in the Italian territory, and the French Authority will issue the Commercial Operation Authorisation (AMEC: Autorisation à la Mise en Exploitation Commerciale) for the section in the French territory.

These authorisations require first the approval of a Safety File (TELT plans to issue it by 2028) which will describe in detail the whole work and everything related to the safe management of railway operations. In particular, the plan will detail the emergency procedures adopted, the management procedures for critical issues, the interoperability constituents used in accordance with the TSIs, the testing and inspection programmes, the maintenance levels of the work, the minimum operating conditions, and the tasks and functions of the personnel employed.

To further complicate the approval procedures, according to the so-called One-Stop-Shop initiative, the EU Agency for Railways (ERA) will become, from 2019 onwards, a European-level Authority with the power to:

- Issue single EU-wide safety certificates to railway undertakings;
- Issue single EU-wide vehicle operation certifications;
- Grant pre-approval for ERTMS infrastructure.

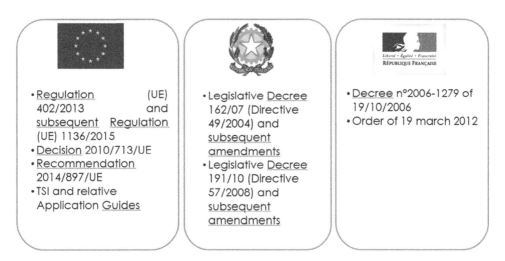

Figure 8. Reference regulatory framework for the commissioning authorisation procedures.

Clearly, TELT, as infrastructure manager, will be involved exclusively in the pre-approval of the Command/Control/Signalling sub-system (based on ERTMS), but it will nonetheless be required to verify the certificates and authorisations necessary for railway undertakings to run in its area of competence.

4 CONCLUSIONS

A cross-border project structured as a long railway tunnel requires, since the early design stages, to respect a series of complicated approval steps which aim to ensure safety of the railway and its compliance with the interoperability requirements described in the TSIs.

To date, the local entities with the authority to issue development approvals are the National Safety Authorities for railways, but, as of 2019, they will also be supported by the European Agency (ERA), which will give advance notice on the aspects related to ERTMS.

In the case of the Montcenis base tunnel, at the end of the design stage, the National Safety Authorities have jointly expressed a "clearance for development" in relation to its "Infrastructure" portion and, in the future, they will have to do the same for the "Energy" and "Command/Control/Signalling" systems.

The Commissioning is therefore the final act in this complicated process by which the infrastructure managers brought a structural element into operating status, after providing proof of compliance with all the essential requirements and therefore with all applicable regulations.

REFERENCES

Brino, L. 2012. La Nuova Linea Torino Lione: gli aspetti tecnici di un grande progetto transfrontaliero. *Gallerie e Grandi Opere Sotterranee* 102: 11–15.

Brino, L. et al 2012. Analisi degli schemi e dei costi della sicurezza nella concezione di un tunnel ferroviario lungo e profondo. *Gallerie e Grandi Opere Sotterranee* 104: 21–31.

Marzoni, M., Tatarelli, L. 2017. Quadro di riferimento per i procedimenti amministrativi di competenza ANSF. In *Riunione "Comitato di Sicurezza Torino-Lione", Torino, 29 November 2017*. Turin.

Poti P. & Brino, L. 2009. Nuova linea Torino-Lione - Impianti Elettrici. *Ingegneria Ferroviaria* 5: 439–453.

Poti P. & Brino, L. 2011. Necessità e concezione del sistema di raffreddamento sia in fase di scavo che di esercizio nel tunnel di base della nuova linea ferroviaria Torino-Lione. *Ingegneria Ferroviaria* 6: 551–564.

Poti P., Brino, L., Foresta, M. & Glarey L 2015. Il nuovo tunnel di base del Moncenisio: sicurezza e nuove tecnologie in un ambito pluri-normativo. In *Le Grandi Infrastrutture e la funzione strategica dei trafori alpini, Proc. intern. symp., Roma, 22–23 October 2015*. Rome: FASTIGI.

Rousse, F 2018. Compliance Report according to Directive 2008/57/EC on railway interoperability. *Exchange of letters between the Directors of TELT and of the EPSF*. Le Bourget du Lac.

Tunnels and Underground Cities: Engineering and Innovation meet Archaeology,
Architecture and Art, Volume 8: Public Communication and Awareness/Risk Management,
Contracts and Financial Aspects – Peila, Viggiani & Celestino (Eds)
© 2020 Taylor & Francis Group, London, ISBN 978-0-367-46873-6

A review on the risk analysis and construction claims in Tehran underground projects

S.M. Pourhashemi
Tehran Metro and Suburban Railway Group of Companies, Tehran, Iran

S. Tarigh Azali
Line 6 Project, Tehran Metro and Suburban Railway Group of Companies, Tehran, Iran

M. Ghorbani
Tehran Behro Consulting Engineers Co., Tehran, Iran

E. Khorasani
University of Tehran, Tehran, Iran

ABSTRACT: In this paper, the main risk groups which are specific to projects related to underground works, such as: difficult or unclear geotechnical and environmental conditions and constraints; proper selection of design, construction and monitoring methods and tunnelling equipment; safety hazards in working confined underground environment; onerous contractual conditions and unbalanced risk allocation, were discussed. This research presents the results of a study of the types, causes, and frequency of claims in a variety of underground projects (metro lines, road tunnels and utility tunnels) which were constructed in densely populated urban area in soft ground using different tunnelling methods in Tehran, Iran. The data were analyzed and the results of this analysis along with recommendations on how to reduce/prevent claims in tunnelling projects are then presented.

1   INTRODUCTION

In 2006, tunnelling projects became uninsurable due to a tremendous increase of loss ratio of 500%. It could give the impression that insurance had been the cheapest risk management tool. As a result, risk management became an integral part of most underground construction projects during the late 1990s. Since April 2003, international guidelines on tunnelling risk management had been established showing how risk management may be utilized throughout the project's phases of design, tendering and contract negotiation and construction (Eskesen et al., 2004). Further, the insurance industry issued the joint code of practice for risk management of tunnel works in 2006 that is now being used worldwide and effective in risk sharing and encouraging best practice of risk management procedure in tunnelling (ITIG, 2006).

Due to the increasing complexity of underground infrastructure, the industry progressively extends the risk management to the complete lifespan of underground projects (design, construction, operation).

Due to the inherent uncertainties, including ground and groundwater conditions, there might be significant cost overrun and delay risks as well as environmental risks. Proper geotechnical exploration is fundamental for appropriate planning of the underground operation, and for reducing the risks to a manageable level. There can be various risk factors underground related to geotechnical conditions, which are specific to the actual projects, and may affect the proper selection of design, monitoring, construction methods and equipment.

To facilitate the application of FIDIC red book contractual framework in tunnelling contracts, required modifications are presented. In this study the risks and constraints in urban tunnelling will be discussed both technically and contractually.

Construction claims are considered by many project participants to be one of the most disruptive and unpleasant events of a project (Ho and Liu, 2004). According to Vidogah and Ndekugri (Vidogah and Ndekugri, 1997), however, claims are becoming a way of life and, indeed, an indispensable part of modern contract systems. In general, claims are common in construction projects and can happen as a result of several reasons that can contribute to delaying a project and/or increasing its costs. Finishing a project on schedule is a difficult task to accomplish in the uncertain, complex, multiparty, and dynamic environment of construction projects (Kartam, 1999). To enhance the chances of success, contractors submitting claims must closely follow the steps stipulated in the contract conditions, provide a breakdown of alleged additional costs and time, and present sufficient documentation (Kululanga et al., 2001). On the other hand, project owners need to follow an overall comprehensive step-by-step procedure for tracking and managing the claims submitted by contractors (Abdul-Malak et al., 2002; Singh and Sakamoto, 2001; Scott, 1997).

Once a claim has been presented, the owner and contractor can come to an agreement concerning the claim and, thereby, create a change order or a modification, or they may disagree and create a construction contract dispute. Analyzing the various types and causes of claims is an important task to resolving these claims (Ren, 2003; Janney et al., 1996). Since project participants are becoming more aware of the high costs and risks associated with claims and their litigation, the construction industry needs to develop methodologies and techniques to reduce or prevent claims. Even though construction claims are frequent and their resolution is difficult, many times legal advice is not sought because it is not available or because it is expensive (Barrie and Paulson, 1992; Diekman and Nelson, 1985).

Several attempts were made in the literature to study the types of construction claims and determine their main causes and ways of avoiding them. Scott (1997) conducted a survey to investigate the causes and mechanisms that are used to prepare and evaluate delay claims in United Kingdom. Zaneldin (2006) studied information of 124 claims related to different construction projects in Dubai and Abu Dhabi Emirates in UAE. Based on collected data they classified different types of claims in construction projects in UAE into six main types: (1) contract ambiguity claims; (2) delay claims; (3) acceleration claims; (4) changes claims; (5) extra-work claims; and (6) different site condition claims. Hartman and Snelgrove (1996) evaluated the effectiveness of written contract language to communicate risk apportionment between contracting parties.

There is still, to a great extent, a lack of information related to the causes of construction claims in Tehran and the ways to prevent or minimize them. This research, therefore, presents the results of a study of the types, causes, frequency of construction claims and methods used to resolve claims using a data collected for claims related to different underground projects in Tehran, Iran. The data are analyzed to identify problem areas and recommendations to reduce claims in urban tunnelling projects are then presented.

2   RISK ASSESSMENT AND MANAGEMENT

Construction projects are not risk free and always associated with some degree of risks. Risk can be managed, minimised, shared, transferred or accepted but it should not be ignored.

Risk Management is continuous and systematic cyclical process for identifying, analysis, evaluation, monitoring and the planning and implementation of appropriate risk-prevention measures and harm reduction. Therefore, the risk control is part of the project life cycle from project initiation to project completion (Eskesen et al., 2004).

Risk Management involves the systematic processes as listed below:

a) Identifying hazards and associated risks, through Risk Assessments, that impact on a project's outcome in terms of costs and programme, including those to third parties;
b) Quantifying risks including their programme and cost implications;

c) Identifying pro-active actions planned to eliminate or mitigate the risks;
d) Identifying methods to be utilised for the control of risks;
e) Allocating risks to the various parties to the Contract.

Hazards in design and construction of underground projects may be attributed to following categories:

Ground: Unforeseen subsurface conditions (e.g. unforeseen faults, weaknesses zones, subsurface obstacles); Unexpected System Behaviour, (e.g. overstressing, damage); Ground failure (e.g. collapse of excavation face, excessive ground surface settlement); Groundwater (e.g. unforeseen water inflow, unforeseen water chemistry, unforeseen water level, wells).

Construction: Failure of construction equipment; Interruption of access to site; Failure of power supply; Problems with material quality.

Project implementation: Change of guidelines; Environmental regulations; Safety standards.

Hazards shall be identified and evaluated on a project-specific basis and their consequent risks shall be identified and quantified by Risk Assessments through all stages of a project (Project Development Stage, Construction Contract Procurement Stage, Design Stage, Construction Stage and operational stage for any stipulated maintenance period).

Systematic monitoring as an important part of risk management can prevent catastrophic failures especially in urban tunnelling projects. Technical monitoring of tunnels provides methods and applications of measurements to control and monitor the construction phase of projects and its effects on the structures itself and the surrounding ground. The monitoring phase does not end when the tunnel is completed, but continues during its whole life to constantly monitor safety when it is in service (Lunardi, 2008).

The main causes of risks in tunnelling projects especially in urban areas are as follow: 1) Unexpected ground conditions/lack of data; 2) Unforeseen water level, unforeseen water inflow and wells; 3) Old sewers with unknown locations above the tunnel route; 4) Temporary works; 5) Fire; 6) Air or gas explosion; 7) Poor workmanship and quality control; and 8) Flood.

Risks to third parties from underground construction can be classified in physical, economic and societal terms. Control of the physical risk is governed by the health and safety legislation applicable to construction works. The economic risk is managed through the contractual arrangements between the constructor, the insurance provider and the affected infrastructure owners. Societal risk is more difficult to define and manage as it can include a wide range of issues including wide spread detriment and socio-political issues.

Urban projects introduce additional risks to tunnelling work due to the density of the existing infrastructure and the spread of the population. Construction methodologies may need to be adjusted to suit local environmental restrictions and working space can be difficult to locate and safeguard. The close presence of possibly aging and unfavourably sited infrastructure can introduce hazards that are not met on rural sites.

The main risk types of the construction projects are: juridical risks, scheduling risks, financial risks, technical risks, management risks, and surroundings' risk.

The risk policy is the necessary statement from the client/employer about the extent and kinds of risks that he is willing to take and accept in pursuant to its objectives. Therefore prior to the beginning of a risk analysis the client has to determine the financial efforts that he is willing to take to avoid or minimize risks or what kinds of risks he will be willing to accept subject to the costs. The risk policy is one of the most important parts of the risk analysis and the following risk management. The project management needs the specification of the risk policy to react accurately to potential and occurred hazards and chances. The main types of the risks are explained with more detail at following subsections.

3 URBAN TUNNELLING CONSTRAINTS

Tunnelling in urban areas has some particular geotechnical, structural and environmental characteristics and constraints compared to rural area tunnelling as listed below:

• Potential interferences with buildings at the surface, underground utilities and other surface and sub-surface structures.

- Poor or inadequate site investigations due to a lack of permission or to the occupation of the surface.
- Urban tunnelling is generally carried out at a shallow depth which often consists of soft ground, with non-uniform, infinitely variable; and perhaps incorporating man-made features of varying age.
- Presence of water table(s) above the tunnel.
- Ground settlements induced by tunnelling with potential damage to the existing structures (buildings, bridges, etc.) and utilities above the tunnel.
- An extensive and redundant geotechnical–structural –environmental monitoring plan is required, which needs extra and direct money input and also additional human efforts.
- Limitations for instrument's installation at the tunnel route, on the buildings and other surface structures.
- During tunnelling in urban areas, an appropriate plan for the temporary diversion of the traffic, an accurate planning of worksite areas, a particular attention to control of dust and noise emissions and a special care for safety issues are necessary.
- The political and physical environment of larger cities makes decisions on heavy construction projects not always as clear an issue as in rural areas. Urban tunnelling has to be considered not as three-dimensional, not even as four dimensional including time, but as five-dimensional including politics.

3.1 *Design challenges*

Engineering design and construction of underground openings, requires in situ stress, rock and soil strength and deformation parameters, pore water pressure and etc., as input parameters. These parameters are essential for stability analysis and ground support system design for these geo-structures. Without them, the process of engineering design is not possible.

The fundamental objective of the design process is that of achieving a design where the risk of failure or damage to the tunnel works or to a third party from all reasonably foreseeable causes, and including health and safety considerations, is extremely remote during the construction and the design life of the tunnel works. High consequence but low frequency events that could affect the works or a third party shall also be considered.

The design process shall include an assessment of the impact of construction on third party infrastructure. In this respect, the Designer shall assemble as far as reasonably practicable all available records of foundations and other structures/artificial obstructions which could affect and/or be affected by the tunnel works.

Experienced and correct interpretation of the Environmental Assessment Report (EAR), Geotechnical Baseline Report (GBR) and related data, analysis and models is first and foremost at the selection of design, tunnelling and monitoring methods. Incorrect interpretations of these reports can create various risks with serious consequences as described above. There are several well proven design methods for calculating and modelling the ground behaviour and quantifying the effects, pressures and expected forces, which will be carried by the underground structures. For establishing the proper design input by using the most realistic and practical assumptions and rational safety factors, can reduce the risk of underground construction preventing structural or ground failure.

Where appropriate, the design shall detail excavation/support sequences and identify appropriate monitoring measures during the works for the range of anticipated ground and groundwater conditions and shall also include for the provision of contingency measures.

Considering huge amount of pedestrian and vehicle use and frequent traffic congestion in urban tunnels (metro lines and road tunnels) and safety problems, it is required to design a proper ventilation system, fire detection and protection systems, power supply (in normal and emergency operation), lighting systems, water proofing system, surface and underground water drainage systems, traffic control and tunnel systems, an appropriate pedestrian egress in case of emergencies and tunnel surveillance (traffic incident detection system, closed circuit television system, environmental monitoring system and other sensors). In urban tunnels, a variety of fire protection equipment for the purpose of derivers' safety and fire prevention have to be installed, and optimal fire-fighting scenario should be established. The use of concrete pavement for urban road tunnels (> 1000 m) is highly recommended to reduce risks. The many strengths of concrete pavements have been illustrated from the perspective of the three mainstays of sustainable development, consisting: 1) Environment; 2) Economy; 3) social importance.

3.2 *Availability of geotechnical information*

The Owner/Prime Consultant select a reputable and experienced geotechnical company to assemble a geotechnical exploration plan. Analysis of previous soils report, research of geological history and investigation of historical land use can provide additional information. Thorough laboratory works provide detailed data of physical and chemical properties, contaminations, gas contents of the soil, rock, groundwater samples. Geotechnical Baseline Report (GBR) shall be prepared with analysis; comprehensive interpretation of all data and illustrations by profiles, sections and 3D models and must provide recommendations and basic parameters for design, summarizing the potential risk elements.

Exploration of geological conditions is an essential component of effectively managing underground space use in urban jurisdiction. It provides essential information about the regional geological conditions and constraints. Applying geophysical exploration methods without disturbing or drilling the unused land can indicate unknown or unexpected findings with archeological, historical nature. It can trigger actual explorations for providing information about ancient use of the land well before the implementation of underground use. Beside of gathering facts on geological and archeological events, the mapping of the underground use and facility inventory is very important.

3.3 *Environmental constraints and mitigation measures*

Investigation, understanding and monitoring of the natural and urban constraints and the potential impacts for the broader environment are very important. To minimize adverse environmental effects a proper Environmental Assessment Report (EAR) shall be prepared and approved by the authorities. The report shall be studied and followed at design and construction for risk prevention and mitigation. Vibration, noise, air quality, accessibility for emergency vehicles, protection of archaeological sites and findings - these are usually the main issues in urban environment. Risk mitigation in the natural environment includes the protection of natural surface and ground waters from pollution and loss, Special features of the flora and fauna shall also be protected.

3.4 *Construction challenges*

For selection of methods and equipment for tunnelling and sub-surface works, the knowledge of typical geological formations and detailed soil and rock characteristics, groundwater and gas conditions along the working area is fundamental. The basic methods can be either construction with cut-and-cover process, or if the geological, functional conditions require a deeper tunnel alignment in variable soils, tunnel boring machines TBM offer the best solutions. Selecting TBM with the required parameters can reduce drastically the risk of tunnelling.

Instrumentation, monitoring, data analysis and evaluation is one of the most important steps to control and mitigate the risks in subsurface construction. The proper design, selection and assembling of the best system shall be based on the understanding and following the detailed geotechnical conditions along the alignment of the future tunnel. Various monitoring systems, sensitive instruments, data loggers, fast data processing and software designs, capable of real-time analysis and activating built-in indicator signals and alarms are developed and available for tunnelling and underground construction. The real time data monitoring system can provide continuous analysis and information of soil, rock and environmental conditions and the effectiveness of the construction. It makes possible immediate evaluation of data, and if required, the modification of operating parameters to match the actual conditions and prevent undesirable events by the designed and custom made techniques of required signals and alarms for the overall project operation. The benefits are two-fold: the system can monitor all the events and conditions as the tunnel construction progresses, with various indicators and alarm levels, and it helps preventing simple errors, failures or catastrophic events. This is a very effective tool for risk prevention.

With particular regard to tunnel works in urban areas and where Third Party equipment or structures are at risk, Method Statements shall clearly identify 'trigger levels' at which contingency action shall be taken. The Method Statements shall clearly identify the reporting roles and responsibilities and what actions are to be taken and by whom at each trigger level. contingency measures may include: increased monitoring frequency, ground treatment, additional support measures, modifications to the excavation/support sequencing.

Other construction challenges are: 1) Lack of new technologies; 2) Lack of Inexperienced (Unskilled) foremen; 3) Lack of qualified project management systems; 4) Lack of machinery; 5) Quality control by contractor; and 6) Life lines.

3.5 Contractual challenges in Urban tunnelling

Contract is the basis for reciprocal handling and the ruling concerning rights and duties between the Contract Parties (technical-, financial-, legal- and organizational-wise) and Risk is a measurement of the probability that a potential hazard/threat actually happens and causes damage (technically, financially, legally, economically).

The clear and mutually satisfactory legal approach for handling the potential risks in tender and contract documents are very important for every project, especially for projects involving works underground. At all of the conventional and advanced contract and project delivery methods the key challenging issues are: contractual relationships, risk allocations and compensation. The three most important contract terms are: the *Scope* (the work or construction services what the client shall receive), the *Price* (how much is the total amount, which shall be paid for the contractor's services), and the *Schedule* (the time of completion, when the work shall be available for the Client's use). These terms must be clear, exact and indisputable. If one of these terms is missing or unclear, the proposal/tender shall be declared incomplete, or the contract is invalid. In underground or tunnelling projects, all the three basic elements can be influenced by the major risk factors discussed above. They must be subject to thorough risk assessment.

Table 1. The risk profile of the contract parties based on the type of the construction contracts.

Contract parties	Risk profile
Contractor	UP: Quantities, geology, construction time; LS: Quantities, geology + rock/soil class distribution, construction time, price escalation; EPC: Quantities, geology, construction time, price escalation, design, approvals, reductions to client; PPP: similar to EPC + right of way + financing portion, compensations; BOT: Full risk included, construction permit, financing and operation.
Client/Owner	UP: Quantities, geology, construction time, price escalation, finance, design, approvals, right of way, compensations. LS: Construction time, price escalation, financing, approvals, design, right of way, etc. EPC: Approvals, Financing, controlling, operation, etc. PPP: Controlling + financing portion, right of way, operation; BOT: Toll agreement (power: PPA (Power Purchase Agreements)), controlling.
Designer/ Consultant	UP: Quantities, geology, design, supervision, Design approval, right of way, compensations, etc. LS: Design approval, design, right of way, supervision, controlling, etc. EPC: right of way, design approval, controlling (construction time, contract, etc.) PPP: Controlling, right of way, approvals, Technical Consultant of the Client BOT: Technical Consultant of the Client (Project Management Consultant)
Investor	UP: Project technical + commercial feasible, Duration and amount of finance; LS: Project technical + commercial feasible, Duration of financing, Qualification Contractor; EPC: Feasible, economical design, operational capability, Qualification Contractor; PPP: Feasible, qualification + solvency acting parties, Operation, Refinancing by operation; BOT: Feasible, qualification + solvency of acting parties, refinancing conditions + Business Plan feasible, etc.

UP: Unit Price Contract; LS: Lump Sum Contract; EPC Contract: Engineering- procurement- Construction; PPP Contract: Public- Private- Partnership; BOT Contract: Built- Operation- Transfer;

Typical contract terms and conditions impacting underground construction includes: The terms of General and Supplementary Contract Conditions; The Financial Conditions, Payment Terms, Evaluation of Changes; The terms related Pre-selected, Pre-purchased Equipment, Materials and Services; Terms related to Schedule, Claims and Penalties, Disputes and Resolution Methods; Terms related to Cooperation; Terms of Joint Risk Sharing.

The risk profile of the contract parties based on the type of the construction contracts are listed in Table 1.

4 ADAPTING THE FIDIC STANDARD CONTRACTUAL FRAMEWORK FOR URBAN TUNNELS

FIDIC (Federation International Des Ingenieurs Conseils) is a standardized model contract for international construction contracts.

FIDIC red book contractual framework refers to works designed by the Employer and constructed by the Contractor and is widely used in surface civil works contracts worldwide. Nonetheless some clauses deserve to be revised to allow their correct application to underground works. In particular, the geologic risk sharing principles are not integrated in FIDIC standard contract. On the other hand, about risks other than geology related ones, the FIDIC framework affects some of them completely to one party without a real risk sharing philosophy. Some clauses have thus to be modified or more detailed.

To facilitate the application of FIDIC in tunnelling contracts the following modifications are recommended (Russo, 2014, Marulanda, 2014):

- Include a more specific differing site conditions clause;
- Use a Geotechnical Baseline Report (GBR) as part of the contract documents to establish a contractual baseline for subsurface risk;
- Implement a ground classification system that properly reflects the cost of excavating and supporting particular conditions;
- Use a unit price contract payment system for items that affected by differing site conditions;
- Specify the detailed difference among temporary and final works;
- Including a well-defined BOQ to the contract;
- Foresee a unit rate revision at bigger rates than FIDIC's principles;
- Foresee a fair risk sharing in consideration of custom's procedures and power supply and on all of the events beyond contractor's control in case the contractor may demonstrate that he acts with diligence.

5 QUALITATIVE AND QUANTITATIVE RISK ANALYSIS

As basis for discussions and negotiations with insurance companies and to provide the Client with a reliable financial overview over all possible and likely costs a risk analysis shall be prepared for the construction of the tunnelling projects. Purpose of this analysis is the estimation of the impact of risks in the project regarding the given uncertainties (hazards and chances). This analysis shall be evaluated based on quantitative and qualitative evaluation methods.

Prior to the execution of the qualitative risk analysis, all project relevant information (general situation, project area, geology & hydrogeology) with more or less high risk potential shall be collected, described and evaluated. Based on the above mentioned information the following risk scenarios shall be evaluated. Those risks were assessed and evaluated by the qualitative risk analysis method for Tehran underground projects as listed below:

General: 1) Land acquisition, right of ways; 2) Protest of residents; 3) Planning and building permission; 4) Ensuring financing resources; and 5) Political change.

Geology: 1) Unsatisfactory exploration of geological and hydrological conditions; 2) Contamination of the soil; 3) Over-breaks, < 20 m^3; 4) Over-breaks, > 20 m^3; 5) Collapse till surface; 6) Face collapse, caused by water-inflow; 7) Boulders within soil formation; 8) Water inflow at the face; 9) Gas detection; 10) Natural caverns; 11) Fault zones; 12) Approach of fault zones; and 13) Water and gravel lenses, water inflow.

Open Cut: 1) Danger of damage, permanent structures; 2) Danger of instable slopes, collapses.

Tunnel drive: 1) Instability of the face; 2) Instability of intersections during intermediate excavation stages; 3) Instability of large excavation profiles; 4) Non-compliance with settlement limitations; 5) Settlements: Damage of door frames; 6) Settlements: Little cracks in bearing walls; 7) Settlements: Considerable cracks in bearing walls; 8) Settlements: Safety problems with brick walls; 9) Settlements: Instability; 10) Settlements: Collapse of buildings; 11) Settlements: Leaky sewers/pipes; 12) Settlements: Destroyed sewers/pipes; 13) Sudden changes in geologic sequence.

Local/special scenarios: 1) Preservation of evidence; 2) Water ingress; 3) Damages of buildings because of settlements after tunnel construction; 4) Damages of sewage, other utilities, and streets because of settlements; 5) Private water wells; 6) Private pit latrines; 7) Old waste deposits; 8) Earthquake incidents; 9) Munitions/explosives form army area; 10) Ruins, old houses; 11) Contractor – Insufficient technical equipment (shotcrete, etc); 12) Contractor – Insufficient spare part storage; 13) Contractor – Insufficient geodetical survey; 14) Contractor – Insufficient know-how in present geology; and 15) Contractor – Insufficient risk management.

Based on the previously executed qualitative analysis the quantitative risk analysis shall be down. All prices shall be continuously reassessed during the continuation of the project with the execution of a risk management.

6 SCOPE OF CLAIM STUDY

Information of claims related to different tunnelling projects in Tehran were collected from 35 different entities (10 contractors, 18 consultants, and 7 owners). The 35 owners and firms were asked to provide information related to types of claims, causes of claims, and frequency of each type and cause by filling a questionnaire, in which they choose one of five possible options for the frequency of each type and cause of claims: (1) never; (2) rare; (3) average; (4) frequent; and (5) very frequent. The entities also were asked to provide information about the method used to resolve each claim. The data were then analyzed and a detailed analysis of the data is shown in the following section.

6.1 *Data analysis*

The data collected represent various types of projects. The types of projects included metro lines, urban road tunnels and utility tunnels. The distribution of the types of projects for metro and road tunnel projects are 50 % and 43 %, respectively. The data collected were analyzed and a detailed discussion of the analysis is shown in the following subsections.

6.2 *Types of claims and their frequency*

The data received indicated that the types of claims in Tehran underground projects can be classified into six main types: (1) contract ambiguity claims; (2) delay claims; (3) acceleration claims; (4) changes claims; (5) extra-work claims; and (6) different site condition claims. To provide a realistic idea about the frequency of each type of claims, firms were asked to fill a questionnaire in which they choose one of five possible options for the frequency of each type of claims, as mentioned in the previous section. A weight in a scale from 0 to 4 was given for each of the five frequencies with a weight of 0 for "never, 1 for "rare", 2 for "average", 3 for "frequent" and 4 for "very frequent". No weight was given when no response was provided. The frequencies for each type of claims received are listed in Table 2. Responses for the "changes" type of claims, for example, indicated that, 0 firms responded as "never", 14 responded as "rare", 11 responded as "average", 10 responded as "frequent", and 0 responded as "very frequent". Data of Table 2 were analyzed and a weighted average was calculated for each type of claims as follows:

$$\text{Weighted Average} = \Sigma\, W_i \cdot X_i / N \qquad (1)$$

where W_i is the weight assigned to the ith option; X_i is the number of respondents who selected the ith option; and N is the total number of respondents 35 in this study). To better

Table 2. Frequency of each type of claims.

Types of claims	No response	Never	Rare	Average	Frequent	Very frequent	Importance index (%)	Rank
Delay claims	-	0	6	15	14	0	55.71	1
Extra-work claims	-	0	7	18	10	0	52.14	2
Changes claims	-	0	14	11	10	0	47.14	3
Contract ambiguity claims	3	0	14	7	7	4	46.43	4
Different site conditions Claims	-	0	15	16	0	4	45.00	5
Acceleration claims	-	6	17	7	5	0	32.86	6

understand the importance of each type of claims, an importance index percentage was then calculated as follows:

$$Importance\ Index = Weighted\ Average \times 100/4 \qquad (2)$$

The importance index values for each type of claims are shown in Table 2. The results of this analysis indicate that "delay" claims are the most frequent type of claims. This type of claims was ranked first with an importance index of 55.71 %. "Extra-work" claims were ranked second with an importance index of 52.14 % while "acceleration" claims were ranked last with an importance index of 32.86%. The ranks of all types of claims are listed in the Table 2.

6.3 Causes of claims and their frequency

The data received indicated that there are 41 possible causes of claims. Similar to what is explained in the previous subsection for types of claims, firms were asked to choose one of five possible options for the frequency of each cause of claims: never, rare, average, frequent, and very frequent with a weight for each in a scale from 0 to 4. A weighted average was calculated using Eq. (1) for each cause of claims and the importance index percentage was then calculated using Eq. (2), as shown in Table 3. The results of this analysis indicate that "Scheduling errors" are the most frequent cause of claims with an importance index of 76.79 % while "Subsurface problems" was ranked second with an importance index of 66.07 %. "Accidents" cause of claims was ranked last with an importance index of 11.61 %. The ranks of all causes of claims are listed in Table 3.

6.4 Methods used to resolve claims in Tehran underground projects

Disputing parties in any construction project may resolve and settle any claim using the normal resolution channels including: (1) negotiation; (2) mediation; (3) arbitration; or (4) litigation. According to this study, the majority of construction claims in Tehran underground projects are resolved using negotiation and mediation and none of claims were resolved using litigation which confirms that firms are quite reluctant to go for litigation because of the long time and high costs and risks associated with this method of resolution.

7 RECOMMENDATIONS

Recommendations on how to prevent/reduce claims and how to deal with such claims in case they happen:

- Allow reasonable time for the design team to produce clear and complete contract documents with no or minimum errors and discrepancies.
- Establish efficient quality control techniques and mechanisms that can be used during the design process to minimize errors, mismatches, and discrepancies in contract documents.
- Have a clearly written contract with no ambiguity.
- Have a third party to read contract documents before the bidding stage.

Table 3. Ranking of each cause of claims based on their frequencies.

The main causes of contractual claims	Importance index (%)	Rank
Scheduling errors	76.79	1
Subsurface problems	66.07	2
Delay in payments by owner	65.71	3
Contractor financial problems	65.71	3
Changes in material and labor costs	65.71	3
Subcontracting problems	60.00	4
Delay caused by owner	59.29	5
Inadequate site investigations	58.93	6
Low price of contract due to high competition	56.43	7
Variations in quantities	56.43	7
Ground settlements induced by tunnelling with potential damage to the existing structures (buildings, bridges, etc.) and utilities	56.43	7
Poorly written contracts	56.25	8
Execution errors	55.36	9
Delay caused by Contractor	55.00	10
Groundwater (unforeseen water inflow, unforeseen water chemistry, unforeseen water level, wells)	55.00	10
Delay in access to site and working area (Owner)	53.57	11
Estimating errors	53.57	11
Protest of residents in project area	52.86	12
Design errors or omissions	51.79	13
Land acquisition, right of ways (Owner)	47.86	14
Specifications and drawings inconsistencies	47.32	15
Bad quality of contractor's work	47.14	16
Environmental regulations	47.14	16
Oral change orders by owner	46.43	17
Bad communication between parties	46.43	17
Delays due to shortages in labor force or goods caused by governmental action or exogenous causes (Sanctions)	45.71	18
Contractor is not well organized	44.29	19
Safety standards	44.29	19
Insufficient risk management by Contractor	43.57	20
Change or variation orders	40.71	21
Owner personality	39.29	22
Planning and construction permission	39.29	22
Delays induced by the Engineer in instruction, tests, etc	39.29	22
Planning errors	38.57	23
Delay in power and water supply (Owner)	36.43	24
Government regulations	27.86	25
Political change	25.71	26
Archaeology preservation and eventual works (Owner)	19.29	27
Suspension of work	16.07	28
Termination of work	13.39	29
Accidents	11.61	30

- Have signed change orders before starting doing these changes on site.
- Low site investigation effort (i.e. cost) resulted in an increase in the level of claims. There was a pronounced reduction in the cost of contractual claims, and the contractual claims continued to reduce in cost as the level of exploration increased. The limit of 0.6 meters of borehole length per meter of tunnel is recommended which is roughly equivalent to an exploration cost of 1% to 1.5% of project cost.
- Use special contracting provisions and practices that have been used successfully on past projects. Useful information can be found in ASCE booklet (1991), which is about avoiding and resolving disputes during construction.
- Develop cooperative and problem solving attitudes on projects through a risk-sharing philosophy and by establishing trust among partners (e.g., the owner and the contractor). This concept is known in the literature as partnering.

8 CONCLUSIONS

This study can be used to identify several problem areas in the underground construction process in Iran. Steps should be taken to clarify any issues or conflicts that may arise in these common problem areas. One of the common problem areas is the "Delay" type of claims which, according to this study, was the most frequent type of claims and needs special consideration. "Extra-work" type of claims came second and "Acceleration claims" was ranked last. It can also be concluded from this study that "scheduling errors" are the most frequent cause of claims while "subsurface problems" was ranked second. "accidents" were ranked last, indicating that it is the least frequent cause of claims.

According to the results of this study, it is recommended that special consideration should be given to contract clauses dealing with change orders, disputes, variations and extra works conditions, and delay. The best means to cope with risk of construction claims is to reduce or avoid them altogether.

REFERENCES

Abdul-Malak, M.A., El-Saadi, M.M. & Abou-Zeid, M.G. 2002. Process model for administrating construction claims. *Journal of Management in Engineering*;18(2): 84–94.
The International Tunnelling Insurance Group. *A code of practice for risk management of tunnel works.* 2006.
ASCE technical committee on contracting practices of the underground technology research council. 1991. Avoiding and resolving disputes during construction: successful practices and guidelines. New York: ASCE.
Barrie, D.S. & Paulson, B.C. 1992. *Professional construction management.* New York: McGraw-Hill.
Diekman J, Nelson E. Construction claims: frequency and severity. *Journal of Construction Engineering and Management* 1985;111(1): 74–81.
Eskesen, S.D., Tengborg, P., Kampmann, J. & Holst Veicherts, T. 2004. Guidelines for tunnelling risk management: International Tunnelling Association, Working Group No. 2. *Tunnelling and Underground Space Technology*, 19, p.217–237.
Fédération Internationale de Ingénieurs de Conseils (FIDIC). 1999. *Conditions of contract for works of civil Engineering construction- Red Book, 1st Ed.*, Lausanne, Switzerland: Blackwell Science Inc.
Hartman, F. & Snelgrove, P. 1996. Risk allocation in lump-sum contracts – concept of latent dispute. *Journal of Construction Engineering and Management*;122(3): 291–6.
Ho, SP. & Liu, LY. 2004. Analytical model for analyzing construction claims and opportunistic bidding. *Journal of Construction Engineering and Management*;130(1): 94–104.
Janney, J.R., Vince, C.R. & Madsen, J.D. 1996. Claims analysis from risk retention professional liability group. *Journal of Performance of Constructed Facilities*;10(3): 115–22.
Kartam, S. 1999. Generic methodology for analyzing delay claims. *Journal of Construction Engineering and Management*; 125(6):409–19.
Kululanga, G.K., Kuotcha, W., McCaffer, R. & Edum-Fotwe, F. 2001. Construction contractors' claim process framework. Journal of Construction Engineering and Management;127(4):309–14.
Lunardi, P. 2008. *Design and construction of tunnels, Analysis of controlled deformation in rocks and soils (ADECO-RS).* Springer. 587 p.
Marulanda, A. 2014. Adapting the FIDIC Standard Forms of Contract for Underground Construction Projects *World Tunnel Congress*, Foz do Iguaçu, Brazil.
Ren, Z., Anumba, C. & Ugwu, O. 2003. Multiagent system for construction claims negotiation. *Journal of Computing in Civil Engineering*;17(3):180–8.
Russo, M. 2014. An analysis of FIDIC red book contractual framework applied to underground works. *World Tunnel Congress*, Foz do Iguaçu, Brazil.
Singh, A. & Sakamoto, I. 2001. Multiple claims in construction law: educational case study. *Journal of Professional Issues in Engineering Education and Practice*;127(3):122–9.
Scott, S. 1997. Delay claims in UK contracts. *Journal of Construction Engineering and Management*; 123(3):238–44.
Vidogah, W. & Ndekugri, I. 1997. Improving management of claims: contractors' perspective. *Journal of Management in Engineering*;13(5): 37–44.
Zaneldin, A. K. 2006. Construction claims in United Arab Emirates: Types, causes, and frequency. *International Journal of Project Management*. 24, 453–459.

*Tunnels and Underground Cities: Engineering and Innovation meet Archaeology,
Architecture and Art, Volume 8: Public Communication and Awareness/Risk Management,
Contracts and Financial Aspects – Peila, Viggiani & Celestino (Eds)
© 2020 Taylor & Francis Group, London, ISBN 978-0-367-46873-6*

Integrating risk management in underground works: The French experience and AFTES Recommendations

M. Pré
Setec, Paris, France & AFTES Technical Committee

E. Chiriotti
Incas Partners, Paris, France & AFTES Technical Committee

G. Hamaide
Cetu, Lyon, France, & AFTES Technical Committee

J. Piraud
Setec, Paris, France & AFTES Technical Committee

ABSTRACT: The formal integration of risk management into underground works began in France in the early 2000s. A dedicated AFTES Working Group focused on the issue of geotechnical risk assessment and management. It also worked on a catalogue of risk categories to be addressed in construction contracts, as well as on the improvement of contractual practices. At the same time, official guidelines on public procurement have made the implementation of Risk Management Plans (RMPs) mandatory in France since 2013. Over the last 10 years, around 20 contracts have been implemented for underground projects, and Owners, Engineering companies and Contractors are now familiar with the approach. Since 2017, an AFTES Committee has been created with the scope to collect, organize and share among all professionals the feedback of the use of risk management in such projects. AFTES is convinced that the sharing of this feedback will be useful for the international community.

1 THE AFTES APPROACH WITHIN THE INTERNATIONAL TRENDS

The principles adopted by AFTES (French Association of Tunnels and Underground Space) for the management of risks in underground works fit into a broader international trend of needs, marked by the growing importance of controlling risks with respect to the contractual budget, the construction schedule and deadlines, and the safety of works. At the international level, the main steps associated to these growing needs have been marked by the following events:

- creation by ITA of Working Group WG3 (Contractual Practices), and issuing in 1995 of a first position paper on the Contractual Sharing of Risk in Underground Construction: ITA views, followed in 2011 by the guideline The ITA Contractual Framework checklist for Subsurface Construction Projects which address the allocation of risk and emphasize how "in some countries formal risk management procedures have become mandatory for obtaining insurance for underground projects";
- the publication in 2004 of first Guidelines for Tunnelling Risk Management, prepared by ITA Working Group WG2 (Research);
- in 2006, issuing by ITIG (International Tunneling Insurance Group) of the "Code of Practice for Risk Assessment in Tunnelling Works", then reviewed in 2012;
- assertion by ITA that only an "objective and equitable risk sharing mechanism could lead to control over the final cost of projects" (see WG19, 2013).

AFTES have worked mainly within the framework of French public works contracts, which are characterized by the following roles and responsibilities among the actors:

– Preliminary and Detailed Design are realized by a specialized engineering company (the Design Consultant) directly contracted by the Owner. The Design Consultant also prepares the technical part of the Tender Documents and cooperates with the Owner for the preparation of the contractual and administrative part of the Tender Documents;
– Following the awarding of the construction contract, the Contractor is responsible for the preparation of the Execution Design, under the validation of the Design Consultant, who ensures also the role of the Engineer on behalf of the Owner and the technical supervision of the construction.

Other contractual frameworks (FIDIC, NEC, etc.) or national regulations (where the Design Consultant and the Engineer could not be the same entity) could require the AFTES Recommendations to be adapted. However, AFTES believes in the importance of sharing the French experience.

The AFTES approach towards risk management has been marked by the following steps:

– an increased consciousness regarding the importance of analysing the causes of abnormal final cost increases affecting underground works in the years 1970–2000 (see § 2);
– definition of general principles to be respected in order to improve the trend (see §3);
– preparation by AFTES Working Group GT32 of three technical Recommendations on management of uncertainties and risks (cf. § 4); this task was also based on other Recommendations issued by Working Groups GT25 (Contractual Practice) and GT16 (Tunnelling-induced effects).

2   1970–2000: OBSERVATION AND ANALYSIS OF NEGATIVE TRENDS

In the last decades of the 20th century, a number of negative trends in the implementation, construction and management of underground works were observed:

– the announced amount of works was very often exceeded;
– the deadlines were rarely respected, especially due to contractual disputes during construction, geological accidents, or misjudged environmental constrains;
– it was increasingly hard to find insuring company accepting to insure contracts for underground work;
– Contractors' claims during construction got a major importance, to the detriment of a fair technical dialogue among the parties for following and optimizing the construction.

For obvious reasons, to cite specific examples is not possible, but such negative behaviours have had serious consequences in a number of recent cases in France:

– Several large tunnels (up to 10 km in length) were put out to tender without the inclusion of the prospective geological section or the geotechnical synthesis report;
– In Paris itself, the constant refusal of some Owners to assume the consequences of geological hazards has led contractors to stop responding to some calls for tenders; the relations were so strained that some contractors, having to face unexpected geotechnical difficulties, waited (even hoped) for the occurrence of disorders in order to be able to put a high price on the corrective measures;
– For several large projects in France, an insufficient analysis of the risks in unfamiliar grounds - consequence of a bad connection between geologists and engineers - led to a multiplication by 3 of the costs and time of execution;
– For engineering consultations for small tunnels, it became frequent that the cost of geotechnical investigations was left to the candidates, which financially favored the most reckless, that is to say those who accepted to put the Owner at risk.

The extent of such negative trends was so important that, in the view of politicians and in the common understanding of public opinion, underground works started to gain a bad reputation and to be considered as a danger to be avoided.

At the same time, the empowerment of jurists and lawyers in the awarding and management of contracts was taking place, often without the minimum required technical background, at the expense of experienced engineers.

AFTES endeavored to analyse in detail the causes of these anomalous trends. They could be assigned to an aberrant scheme that was affecting the Owner-Contractor relationship through contracts:

– Owners willing to impose Contractors to assume all the consequences of geological-related project risks;
– Owners preferring to withhold detailed information about the geological nature or properties of the ground, being afraid to open the way for claims through too detailed records and data;
– some adventurous Contractors offering abnormally low prices, and systematically seeking for cash return through claims carefully identified, planned and prepared in advance;
– Contractors playing fair games but being penalized on the offered price because of the excessive provisions included, due to the amount of risk they are made responsible for in tender documents.

Several French and international comparative studies outline an inverse relationship between cost of geotechnical investigations and extra cost of construction in comparison with cost estimated during the design phase, as shown on Figure 1 (Robert & Humbert, 2017).

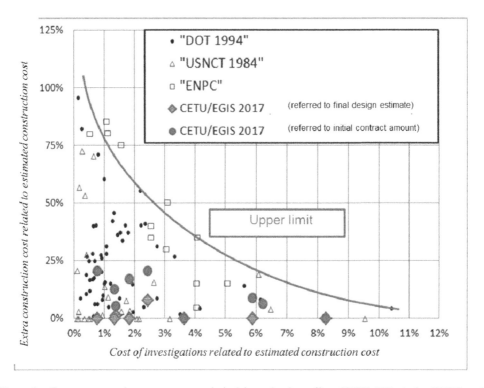

Figure 1. Extra construction cost vs geotechnical investigation effort (DOT: UK study; ENPC and CETU: French studies).

3   AFTES REACTION IN THE EARLY 2000S

AFTES considered that the contractual relationship between Owners and Contractors should not be polluted by manoeuvres or hidden agendas. Due to the growing challenges of the underground works (tunnelling beneath sensitive buildings, under the sea, with high overburden, etc.), it was important to set up *procedures promoting virtuous behaviour, in such a way that technical optimization of work could become again the major concern of stakeholders.*

Hence, in agreement with the Public Works French Authority, AFTES decided to start focusing and promoting the following *behaviour principles*:

- To recognize that uncertainties on the geotechnical conditions, on the existing buildings and on the environmental context should be considered as part of underground work, and consequently be identified and characterized at the earliest stages of a project and with utmost care;
- To fairly inform Contractors of these uncertainties, and to encourage them to provide means to overcome the consequences and agree with them of an adapted mode of remuneration;
- To clearly define in advance to the commencement of the works the responsibilities of the various stakeholders as regards the risks inherent to the works.

These three fundamental behaviour principles, initiated in France the approach to the *management of risks before their occurrence.*

These principles resulted in three AFTES technical Recommendations published in 2003, 2012 and 2016 on management of technical risks in underground works. Their aim was to help re-establishing a fair balance of responsibilities between Owners and Contractors, and to re-create the contractual conditions allowing to focus on what is important in underground works: the constant focus on technical solutions to anticipate the problems, to prevent difficulties and to overcome real unexpected situations.

This new direction promoted by AFTES required changing the attitude of the major players which was neither easy nor obvious, since in complete antithesis with the usual, established habits. The way forward in this matter must be endorsed by the Owners themselves, so that the Contractors are put in the condition to agree playing the game without concerns of not being remunerated for their efforts, and they can reasonably limit provisions for risks in their financial offers.

With regard to the French official texts, it is important to remind that as early as 1992, the Ministry of Equipment had instituted the obligation to include in the tender documents a *Geological, Hydrogeological and Geotechnical Synthesis Report*, with contractual value (see CCTG fasc. 69). In addition, in 2013, a new version of this document introduced the obligation to include a *Risk Management Plan to* the tender documents.

4   APPROACH PROPOSED BY AFTES

We consider that the most original contributions of AFTES in the approach of assessing and manage risks in underground works have been successively:

- the clear definition of the content and of the legal nature of booklets A, B and C of the geotechnical file, which form the backbone of tender documents (cf. Recommendation GT32-1, 2003);
- the transposition of ISO 31000 (Risk Management) to underground works, by defining a strict process of identification, then analysis and evaluation of risks (cf. Recommendation GT32-2, 2012);
- the extension of this process to geotechnics in a broad sense, including all the uncertainties related to neighbouring buildings and environment (cf. Recommendation GT32-3, 2016).

4.1   Booklets A, B and C of the Geotechnical Report

In 2003, AFTES recommends distinguishing three separate geotechnical documents in the Tender Documents, the technical and contractual scopes of which are different:

- The collection of factual data (booklet A) which were used to establish the project. These data, of variable origin and reliability, are provided for information only; they include the results of surveys and tests, geological and geophysical surveys, lessons learned from neighbouring works, etc.
- The Geotechnical Synthesis Report (booklet B), which gives the Designer's (and hence the Owner's) reasoned interpretation on the data in Booklet A (including associated uncertainties), and which constitute the contractual ground on which the Contractor presents his offer. It includes a booklet B1 on geological, hydrogeological and geotechnical conditions, and a booklet B2 on the sensitivity of neighbouring buildings and structures. It has to be completed by a detailed geological and geotechnical profile, duly informed so that residual uncertainties are clearly identified in the space, and by drawings summarizing the sensitivity of neighbouring buildings and structures;
- Constructive proposals by the Design Consultant (booklet C), which outlines his vision in terms of construction methods, as well as the degrees of freedom given to the Contractor to optimize the construction methods, based upon his own technology and experience.

AFTES strongly discourage all attitudes having the objective not to communicate to Contractors all the available geotechnical information, or to present this information in an excessively pessimistic way in order to protect the Owner or the Design Consultant. In addition, asking Contractors to assume the overall risks related to ground uncertainties leads to failure, as it most often ultimately leads to increase costs and duration of the underground works. On the contrary, adequate investigations, accurate detailed design, rigorous tender documents, balanced contractual responsibilities towards risks and a fair contractual management of risks (share and remuneration) are the bases for Contractors to confidently estimate the risks they incurs, thus reducing hidden provisions. Such provisions often penalize those Contractors who are the most aware of their responsibilities.

4.2   Risk analysis process and Risk Management Plan

AFTES considers that geotechnical uncertainties and risks affecting projects must be *identified, represented and evaluated* as early and objectively as possible. Similarly, the constructive methods to overcome them in case of occurrence and their method of remuneration must be outlined in the Tender Documents, and then further completed and validated by both parties (Owner and Contractor) before signing the construction contract, so that the management of risk occurrence is anticipated and impacts on project costs and schedule are controlled and reduced.

In recommendation GT32-2 (2012), AFTES suggests to apply a relatively strict risk management approach to uncertainties related to subsoil, which is directly inspired by ISO 31000 (2008) and based on a rigorous terminology; this approach presents three main steps (see Figure 2):

- the establishment of a synthesis regarding geological, hydrogeological and geotechnical knowledge, featuring a compilation of useful data from booklet A, including a critical analysis of reliability of data, and a final "synthesis of knowledge";
- the assessment of geotechnical risks incurred from identified uncertainties, with respect to their probability of occurrence and their potential impact on project objectives (cost, time, safety, environmental impact, image, etc.);
- the treatment of geotechnical risks, identifying means to reduce the risk to an acceptable level with respect to the Owner's criteria.

AFTES recommends that at the stage of Tender Documents preparation, and in addition to the booklet B which gives a contractual geotechnical interpretation, a Risk Management Plan (RMP) shall be established, consisting of: a policy for risk assessment; a Residual Risk Register; strict

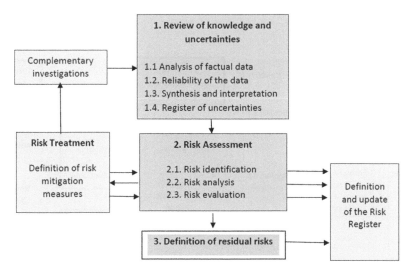

Figure 2. Flow chart defining AFTES approach for risk management.

requirements on the organisation related to risk management; a dedicated bill of quantities associated to the corrective measures for the potential occurrence of risks, with associated unit prices.

The Risk Register must be established in the initial phases of the project, and then updated iteratively in all subsequent phases. At the preparation of the tendering documents, it is split into two: the Technical Risk Register which tracks all the risks fully covered by the technical solution developed in the detailed design (it is provided for information only); the Residual Risk Register which covers those risks that are not covered by the technical solution. The latter can be modified during the contracting phase in order to consider any new construction method as possibly proposed by the Contractor.

What distinguishes the French approach is the fact that corrective measures are predefined and associated to estimated quantities. The Contractor offers his unit prices on this dedicated BOQ.

4.3 *Process extended to risks related to neighbouring buildings/structures and environment*

The AFTES recommendation issued in 2016 extends the scope of previous texts to two other sources of risk that may affect underground works:

- Risks related to existing buildings and neighbouring structures, their interface with the underground works and the tunnelling-induced effects (settlement and vibration);
- Risks related to the effects of the work on humans and on the natural environment: noise, dust, spoiling, constraints related to neighbouring works, impacts on ground and surface water, various perturbations on fauna and flora. These effects are often difficult to quantify in advance and their acceptability by local residents is never well known in advance.

All these risks come from an external context to the actors of the works. They do not include risks of the internal context, which are related to methods and means of the Contractor and are clearly his responsibility, whether for preventive or for corrective measures. They do not include non-technical risks which the Owner may face: political and administrative risks, program and interface risks, etc.

AFTES considers that Tender Documents based on these Recommendations will allow the Owner:

- to receive robust offers from Contractors, by giving them a sufficient vision on the remuneration mode, so that these offers reflect the right price and limit financial drifts during works;

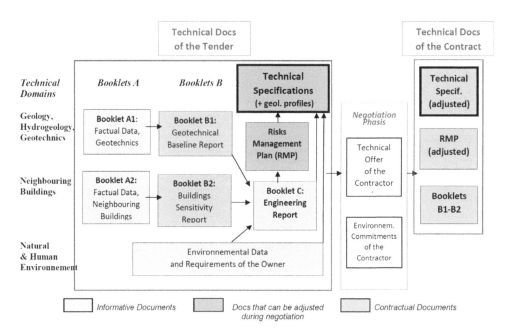

Figure 3. Organization of the technical documents of the Call for Tender and of the Contract. When reading from left to right, the diagram reflects the general timing for writing the documents, while the rows indicate the technical connections.

– to facilitate a rapid decision-making process in the event of a risk occurrence during construction, by providing for control and remuneration mechanisms that are well understood by the stakeholders.

The articulation of the technical parts of the Tender Documents which is recommended by AFTES is shown on Figure 3. It can be noted the similar approach in treating geotechnical-related risks on one hand and risks related to neighbouring buildings and structures on another hand, including data and uncertainties. In both cases a non-contractual, factual booklet A and a contractual, interpretative booklet B are prepared. The presentation of environmental data and constraints is less formalized because there is a multitude of cases. It is up to the Design Consultant to summarize all the data by defining in booklet C – which is by definition unique – a method of execution compatible with all the constraints of the site.

Based on booklet C, the two main parts of the construction contract from a contractual point of view are:

– the Technical Specifications, as imposed on the Contractor (consistency of works, materials to be used, imposed methods, environmental criteria to be respected, etc.);
– the Risk Management Plan (see § 4.2), intended to provide in advance a framework for the technical and financial conditions to be applied in the event of a risk occurrence.

AFTES considers that these two parts could be amended and/or completed during the contract negotiation phase, if relevant, by considering elements issued from the Technical Methodology of the Contractor, in order for example to promote innovation and/or optimization of the project.

5 FINANCIAL ASPECTS RELATED TO RISK MANAGEMENT

Improving the framework of underground works contracts depends on the choice of the most appropriate remuneration methods. In France, a strong tradition exists of using unit prices associated to a bill of quantities. In this type of contract, the Contractor is remunerated based on the quantities actually implemented, most often measured on as-built drawings. In the

French contract system, these drawings are based on the execution design which is prepared by the Contractor, then checked and validated by the Design Consultant/Engineer on behalf of the Owner. The quantities are linked to the Unit Price Schedule and can vary only in a limited range compared to the quantities established at the detailed design stage and which served as the basis for the awarding of the construction contract.

In the French approach, two families of risks are distinguished:

– "ordinary" risks, corresponding to limited variations in quantities, which are automatically covered and remunerated by the technical solution implemented, provided that the Engineer can confirm these variations; an advantage of this type of remuneration is that variations in quantity can be equally negative, and thus resulting in an optimisation of the project;
– "remarkable" risks, which are those for which the corrective measures to be implemented are out of the scope given above, either by their nature or by the exceptional quantities they imply.

Each of these remarkable risks is associated with a section of the contractual Risk Register, and their corrective measures in case of occurrence are associated to a specific, so-called Risk Price Schedule, which will be used to pay for these measures in the event of a proven occurrence. Compared to previous practices, the advantage is that the prices are offered by the Contractor in the competitive context of the bidding, and no longer result from a negotiation under the pressure of the event.

For these prices to be the least speculative possible, it was necessary to introduce a mechanism where, at the bidding stage, fictitious quantities resulting from an arbitrary risks scenario that may be encountered during the works are attached to this Risks Price List. The bids are then judged based on the sum of the prices of the reference project and of the risk scenario.

This mechanism is still perfectible, and it is also one of the subjects of the Observatory (see next section) to examine how it performs in practice and how it can be improved. The approach will work better if the identification of risks has been carefully established. In other words, this mechanism is in no way an alternative to a lack of prior knowledge.

Another approach to mitigate the identified uncertainties could be, for the Owner, to use a lump sum remuneration, being it a global value for a certain works' package, or – in the case of linear tunnels – the price per unit length of tunnel. This practice, which has been observed in the past, is to be absolutely banned because leading straight to litigation. Lump-sums should be limited to those works that are very well defined, which is quite difficult to have in underground works.

At the end, the remuneration mechanisms associated to the occurrence of risks are a fairly natural extension of the traditional mode of remuneration of construction works in the French market. They have proved to be especially well suited to the context of urban tunnels projects, which have seen a very significant increase in recent years. Indeed, these projects are characterized by a large number of interfaces, each of them leading to a lot of uncertainties:

– Internal interfaces, for example between stations and tunnels jobsites, with overlapping critical paths;
– External interfaces, the least of them not being the reaction of the public to the potential or real nuisances that it has to undergo.

It therefore appears appropriate in this context to provide a reference project (the detailed design), capable of reducing the impact of both geotechnical and neighbouring uncertainties and accompanied by an inventory of potential risks and a predefined determination of their mode of remuneration.

6   THE AFTES OBSERVATORY OF CONTACTUAL PRACTICES

AFTES set up an Observatory of Contractual Practices. The objective is to collect the feedback from practical cases in which the Recommendations mentioned above have been implemented, in order to identify areas for improvement, if any. A special focus will be given to the integration of risk management in the contracts of the Design Consultants who are in charge to develop the detailed design and tender documents, and then of the supervision of the works, including the systematic technical and contractual follow-up of risk management during construction.

Nowadays, the major ongoing underground projects are driven by a tight and challenging schedule, and this contractual environment has an impact on risk management.

The approach of the Observatory will be started interviewing the major project Owners who have – or have recently had – the responsibility of major underground works in France, to get the first feedback from their past or current experiences, to check how the RMP fits into their usual approaches in contracting design services and construction works, and if adaptations have been done compared with the AFTES best practice Recommendations.

The interviews will focus on the approach chosen to prepare works contracts and risk sharing, to choose the contractors, and to execute the works of the contract by integrating a formalized risk management procedure that needs to be followed-up. Also, the organizations set up by the project Owners to make the right decisions, in the interest of the project, when important difficulties occur will be discussed in this Observatory.

Since risk management during construction is ensured by the Contractors and by the Design Consultant/Engineer who intervene daily on site, they will be also interviewed with respect to specific projects to understand how they position themselves within the framework of the new construction contracts including the formal RMP.

Then, the lessons learned will be shared with all the actors of the technical AFTES community identifying the practices that work well, those that have to be adjusted and those that are to be proscribed. The sharing will be done respecting the confidentiality of the interviews.

AFTES will then examine the need to correct its Recommendations and/or issue new Recommendations.

The projects focused by the Observatory are urban tunnels at low or medium depth in very sensitive environments where the tunnelling-induced on existing urbanisation is a strong constraint, and long and deep tunnels where the geotechnical risks are increased by the difficulty to obtain an accurate ground model from the investigations.

7 CONCLUSIONS

The efforts of many years of practice of Owners, Contactors and Designers in France on risk management for underground works have collectively led AFTES to prepare a set of documents reflecting a coherent and comprehensive approach to risks.

AFTES are also organised to collect and share the feedback on the implementation of this approach.

The confrontation with other experiences at the international level will be of great interest to the whole community of underground works.

REFERENCES

AFTES, GT32, 2004, Consideration of Geotechnical Risks in Tender Documents for Tunnel Projects. AFTES Recom. N. GT32.R1F1, Tunnels & Espace Souterrain, N. 185
AFTES, GT32, 2012, Characterization of Geological, Hydrogeological and Geotechnical Uncertainties and Risks. AFTES Recom. N. GT32.R2A1. Tunnels & Espace Souterrain, N. 232
AFTES, GT32, 2016, Taking into Account Technical Risks Construction Contracts for Underground Works. AFTES Recom. N. GT32.R3F1, Tunnels & Espace Souterrain, N. 258 [English translation under preparation]
CETU, 2013, CCTG Application Guide Underground Work (Issue 69)
ISO 31000: 2008, Risk Management, International Standard
ITA-AITES, 2004, Guidelines for Tunnelling Risk Management. International Tunnelling Association, Working Group N.2. Tunnelling & Underground Space Techn., N.19
ITIG, 2012, A Code of Practice for Risk Management of Tunnel Works. Recom. of the International Tunnelling Insurance Group
Robert A. & Humbert E., 2017, Risk management in tunnelling: there are still efforts to be made! Proc. of AFTES Int. Congress, Paris, 2017

Tunnels and Underground Cities: Engineering and Innovation meet Archaeology,
Architecture and Art, Volume 8: Public Communication and Awareness/Risk Management,
Contracts and Financial Aspects – Peila, Viggiani & Celestino (Eds)
© 2020 Taylor & Francis Group, London, ISBN 978-0-367-46873-6

A model for fair compensation of construction costs in TBM tunneling: A novel contribution

N. Radončić
Amberg Engineering AG, Innsbruck, Austria

W. Purrer
CCC Purrer, Innsbruck, Austria

K. Pichler
BBT SE, Innsbruck, Austria

ABSTRACT: TBM tunneling represents state-of-the-art excavation method for long alpine tunnels. This kind of tunnel construction is always associated with a considerable amount of uncertainty, having two major impacts on the TBM advance: various adverse occurrences at the tunnel face and imposing reductions of the achievable performance. The reduction is caused by the need to advance with sub-optimal operation parameters and the increased inspection and maintenance efforts. The second influence is given by the usage of additional measures in order to enable a safe TBM advance. The additional time required by encountering circumstances as described above can be hardly anticipated during the design phase. The model presented in this publication uses a very simple solution for the aforementioned issues: the construction activity during TBM advance is divided into three categories: regular advance, hindered advance and event-driven advance stop. A clear set of delimiting criteria is defined in the paper allowing a simple and objective identification.

1 INTRODUCTION

TBM tunneling represents state-of-the-art tunneling method for construction of long alpine tunnels, due to the intrinsic advantages of obtaining an almost finished tunnel by the end of the advance (in case of shield tunneling with pre-cast concrete segments), lower requirements on ventilation and fully mechanized, high-capacity logistics. The drawbacks, when compared to conventional tunnels, are also well known: possibly high performance is traded for lower flexibility, larger requirements on the ground investigation and risk of long standstills in case of TBM becoming trapped.

TBMs equipped with a hard rock cutterhead and without an active face support are generally used in alpine conditions. The desired performance is obtained only in case of a stable face, where intended regular chipping occurs, and stable extrados, where the filling of the annular gap is not hindered by debris in the annular gap. In case such conditions are not given, for instance in a jointed rock mass with high uniaxial compressive strength, this results in continuous dynamic loading of every disc, impact upon impact, and brittle damage to the discs (Figure 1).

In case of weak ground (tectonic faults), large displacements occur frequently and support pressure is mobilized already in the shield area, leading to an increased thrust force demand (Ramoni, 2010) and possible reduction of the performance. In larger tectonic faults perpendicular to the alignment, high volume overbreaks ahead of the face (Figure 2) can also be frequently observed, leading either to situations of "infinite mucking" or to cutterhead blockade. Both require substantial efforts on ground stabilization before an advance restart is attempted.

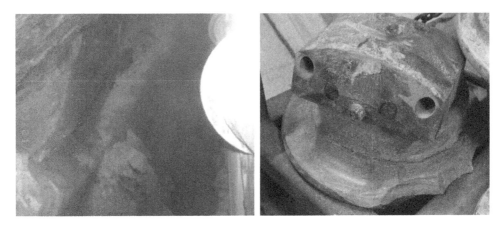

Figure 1. Left: Block rock mass. Right: damaged disc cutter, after operation in such conditions.

Figure 2. Laserscan of an overbreak please note the finger shield of the hard-rock TBM inside.

2 CONTRACTUAL CONDITIONS: STATE-OF-THE-ART IN MIDDLE EUROPE

The normative documents in Austria (ÖNORM 2203-2) and Switzerland (SIA 118/198) adhere to the desire for clear and fair separation of responsibility spheres in underground construction since more then 20 years. In their current and valid versions, they clearly assign the responsibility with regard to risks associated with the encountered ground conditions to the owner. In order to incorporate this into the daily site processes, the contracts have a flexible billing "mechanism" and enable straightforward billing of possibly changing conditions. Both standards split the contractual prices into material and time-dependent costs. Both informations have to be offered by the contractor in his tender bid: the contractor is obliged to estimate and enter his required time demand for each activity. The final payment is thus the result of the quantity of works and the time the contractor has offered for each activity. The described methodology has proven itself indispensable in conventional tunneling repeatedlyin the past decades.

However, in case of TBM tunneling, the occurrences described in the introduction chapter lead to major problems and cost overruns, since the required time cannot be estimated in the tender phase. This holds true even in case of excellent ground investigation: even if the location and the quantity of the stretches featuring blocky ground or massive overbreaks is known, the associated performance prediction for tunneling in such circumstances in not possible. The reduction of the performance parameters (thrust, penetration, cutterhead rotation rate) is a product of interaction between observed tool damage, conveyor belt loading, TBM heading stability and subsequent further face instabilities due to overload during excavation. The situation in case of additional stabilization measures, or – in an extreme case – excavating auxiliary bypass tunnels to free the trapped TBM is the same: neither the amount of material required for stabilization injections nor the duration of the entire endeavor can be anticipated in a reliable manner. These circumstances and the contractual problems frequently accompanying them have been the starting point to develop a novel billing model for TBM advance.

3 NOVEL BILLING MODEL

3.1 *Goals*

The new billing model has the following goals in mind:

1. Improving the contractual fairness and demanding only things from the contractor which can be really determined;
2. Improvement of transparence and flexibility for site implementation. Occurrences not described by the contract and the underlying geological prognosis must be coverable without additional mediation support and/or expertise by external experts;
3. Using TBM machine data as an integral part of the billing, due to the high information density.

3.2 *Activity classification*

The first step towards clear contractual structure is the clear classification of all technically possible activities on the TBM into three distinct categories:

1. Regular measures, composed of activities which are conducted always during the advance. These can be: advancing of the TBM for one stroke, regular inspection of key components, erection of one ring and backfill of the annular gap (in case of a shield TBM), rock bolt and steel arch installation (in case of hard rock, TBM type) etc.
2. Additional measures, composed of measures which are conducted on demand and based on the observed system behavior. These measures do not require installation of any additional equipment on the TBM (everything required must be already installed) and no additional, specialist personnel. Examples are shield skin lubrication by continuous bentonite pumping, silicate foam injections to fill overbreaks and cavities, steel connections and/or additional steel arches for increased capacity of steel segments, etc.
3. Special measures encompass such activities which require additional equipment installation on the TBM and/or specialist personnel, and are conducted only in case of major and extraordinary events and/or situations associated with high risk. Typical examples would be pipe roof umbrella installation with subsequent grouting, drillings for systematic water table drawdown campaigns etc.

The technical tender documents must clearly separate the TBM equipment into these three groups and give them clear assignment along the reasoning presented above. In addition, bill of quantities and time-dependent cost tables must accommodate the same structure as well.

The first principle of the new billing system is: the contractor is obliged to deliver contractually binding "time required" estimates only for activities belonging to the "regular measures". Only in exceptional cases the contractor can estimate beforehand the required times for activities belonging to "additional measures", while "special measures" are explicitly realized as non-calculable. A fair compensation for such activity is therefore found by splitting the tunneling works into regular advance (comprising regular measures), hindered advance and "advance halt".

3.3 Definition of "regular advance"

Regular advance represents the envisioned majority of the TBM operation – if this is not case, either the ground model is severely wrong or an inadequate TBM is being used. It is defined by following criteria:

1. Only regular measures (see above) are being used.
2. A stable face is present, with tight contact between the disc cutters and the face, is systematically present. No face instabilities are occurring, leading to either premature damage to the discs and/or cutterhead or requiring reduction of operation parameters (thrust, cutterhead rotation, etc.).
3. The flow of material is constant and no reduction of operation parameters is required in order to prevent conveyorbelt overload;
4. The mucked material does not cause damage to the conveyor belt immediately after the cutterhead;
5. The shield friction is so low that the capacity of the hydraulics system does not pose a limiting factor for the performance;
6. No premature damage to the disc cutters is occurring, only abrasive wear is observable;

The stability of the face is validated by the following criteria:

1. Visual inspection of the face during cutterhead inspection;
2. Usage of cutterhead cameras (Schuller et al., 2015, Gaich & Pötsch, 2016);
3. Usage of disc cutter monitoring systems (Entacher & Galler, 2013)
4. Usage of TBM data to determine the stability of face (Radoncic et al., 2014).

The consequence of the above criteria is that the performance of the TBM becomes solely dependent on the penetration and contractor's logistics in case of "regular advance". Both of them can be calculated in the bidding phase and can be handled in a straightforward manner: in case a large variability and/or uncertainty with regard to the intact rock strength are present, several penetration classes, with different time-dependent costs attached, can be defined to address this. The contractor's logistics remain entirely in the responsibility sphere of the contractor, and all time dependent costs based on such delays are contractor's responsibility entirely.

3.4 Hindered advance

A "hindered" advance is present if the face (or, to a lesser extent: extrados) behavior forces a change of machine parameters towards reduced performance, or additional measures are conducted in parallel to the advance. If this occurs, the past 250 m (or any other length, as individually specified by contract) of advance are taken as a reference to determine the "would be" average performance. The reasoning is simple: the contractor reached a certain performance in this area, and all the intrinsic influences are incorporated (productivity, logistics, net penetration etc.) in it. This average performance is used to determine the hypothetical time required to advance through the area where the deviation from the regular advance is present. Subtraction of this time from the total time required for the advance through the area where hindered advance is present, yields "time difference" – TD – which has to be paid to the contractor in addition to the normal time-dependent cost.

The above reasoning is best presented when examining the exemplary chart, showing time, chainage and associated activities (Figure 3). It can be seen that from chainage 17.50 to chainage 30.0, the TBM stroke takes longer, and the advance rate does not reach the one observed before. The time difference "TD" awarded to the contractor is determined by extrapolating the past performance and subtracting it from the total time.

Analogous procedure is to be used when the stroke time is not affected, however additional measures (in this case: drillings for bedding improvement) are conducted systematically during the advance (Figure 4). The number of drillings and the time required to fully inject the loosened rock mass with grout or foam can vary from ring to ring, and the depicted determination over the entire stretch represents the simplest and fairest solution.

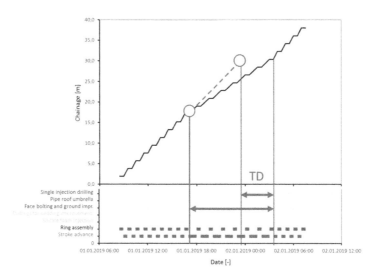

Figure 3.   Exemplary time-advance plot showing the area of hindered advance due to face instabilities and the usage of the past performance to determine the additional time awarded to the contractor.

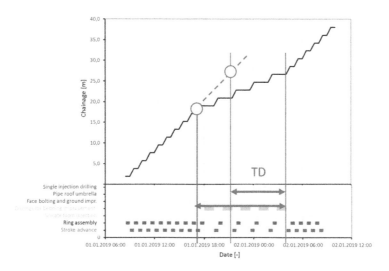

Figure 4.   Exemplary time-advance plot showing the area of hindered advance due to additional measures and the usage of the past performance to determine the additional time awarded to the contractor.

3.5   *Advance halt due to an unexpected event*

In case of major unforeseen events (such as a major overbreak and cutterhead blockade), the contractor is awarded on the strict cost plus fee basis. The reasons for this regulation are as follows:

- The entire duration in case of such circumstances is usually long and unforeseeable, and the "regular advance" duration becomes negligible;
- Usually numerous activities are conducted simultaneously (for instance: face stabilization grouting while drilling the pipe roof umbrella) and the determination of the critical path is complex;
- The activities are determined at different locations on the TBM, but in the same tunnel area. Best example is a passage through a major overbreak area, where the face and the

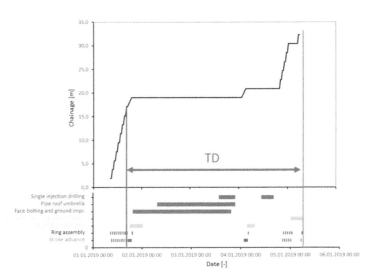

Figure 5. Exemplary time-advance plot showing the activities and billing in case of advance stop due to an unexpected, major event.

roof are stabilized and the advance is restarted. However, the annular gap in this area may require additional grouting injections, conducted from various positions on the backup trailer. Therefore, a single event at a constricted, singular location affects the advance over the length of the entire TBM with its backup trailer;

- Additional equipment and/or personnel are required for the execution of the requested measures ("special measures") and their respective time-dependent costs can not be foreseen and offered in the tender bid.

The contractor is awarded, apart from the actual material cost, time dependent costs for personnel (salaries) and equipment (depreciation) in the actual quantity. The personnel present at the site and the additional equipment used need to be logged meticulously by the site supervision to provide the basis for the accounting.

3.6 *Additional regulations*

The basis for the presented billing model is given by a clear differentiation between a regular advance and circumstances departing from it. In order to enable the presented model to work in real life conditions, additional regulations are required:

1. The contractor is obliged to define the required time for the activities resulting in longer standstills associated with the logistics. These are: main surveying campaign, conveyor belt extension, power and water extension, scheduled cutterhead and TBM overhauls etc. These will be smeared over the entire advance length and accounted for in the determination of the achieved past performance. Simply put, if a conveyor belt extension has been conducted in the past 250 m, not the entire duration of this can be associated with the 250 m.
2. If the past 250 m (or any other length used for determination of the performance in "regular advance") are also affected by the occurrence of hindered advance or major events, then the affected stretch is to be ignored and the entire length is to be increased by the same quantity. Simply put: if a hindered advance was present over 30 m in the past 250 m, then this part of the tunnel is to be ignored in the determination of the performance, and 30 m are to be added to the frame used for performance determination.
3. The contractor is obliged to offer his loan prices in the tender bid. When a major event occurs and cost and fee accounting is used, the contractor will be paid the true personnel

strength present with the offered loan prices. In this way, speculations regarding loans and personnel quantity are prevented.

4. Site supervision is required to take a proactive role, working together with the contractor from the beginning and logging all times and quantities meticulously.

4 CONCLUSIONS

The authors are fully aware that the presented models have certain drawbacks: the contractor may be inclined to reduce the productivity deliberately when knowing that the advance is currently not classified as "regular advance". The site supervision has the task of working closely with the contractor and maintaining high productivity in all circumstances. Furthermore, tunnel construction needs to be understood as a partnership endeavor, where fairness and transparency are of utmost importance. The presented model reduces the potential for speculation considerably, and ensures that reasonable prices are offered both for services within the scope of "regular advance" and for the ones outside of it. Only by a meaningful offer without speculation the contractor can be sure to earn money, because almost all unforeseen occurrences are covered by the presented model.

The authors have been involved with two tenders in Austria where such billing methodology has been used, and the presented philosophy is also being discussed within the Austrian standards committee, tasked with an update of the ÖNORM 2203-2 standard. We sincerely hope that the presented model will enforce meaningful prices in the tender phase and allow a generally fair and transparent financial rewarding for contractor's services, reducing the need for mediation and protracted discussions during construction.

REFERENCES

Ramoni, M. 2010. On the feasibility of TBM drives in squeezing ground and the risk of shield jamming. Veröffentlichungen des Instituts für Geotechnik (IGT) der ETH Zürich, Doctoral Thesis, Zürich, ETH, 2010.

SIA 118/198. 2004. Allgemeine Bedingungen für Untertagbau - Allgemeine Vertragsbedingungen zur Norm SIA 198 Untertagbau – Ausführung.

Austrian Standards Committee. 2005. ÖNORM 2203-2: Untertagebauarbeiten - Werkvertragsnorm - Teil 2: Kontinuierlicher Vortrieb

Gaich, A. & Pötsch, M. 2016. 3D images for digital tunnel face documentation at TBM headings – Application at Koralmtunnel lot KAT2. Geomechanics and Tunnelling 9 (2016), No. 3, pp. 210-221.

Schuller, E., Galler, R., Barwart, S. & Wenighofer, R. 2015. The transparent face – development work to solve problems in mechanized hard rock tunnelling. In: Geomechanics & Tunnelling 8/3, pp 200-210. 2015 Elsevier.

Entacher, R. & Galler, R. 2013. Development of a disc cutter force and face monitoring system for mechanized tunnelling. In: Geomechanics & Tunnelling 6/6, pp 725-731. 2013 Elsevier.

Radončić, N., Hein, M. & Moritz, B. 2014. Determination of the system behaviour based on data analysis of a hard rock shield TBM. In: Geomechanics and Tunnelling 7/5, pp. 565-576. 2014 Elsevier.

Tunnels and Underground Cities: Engineering and Innovation meet Archaeology,
Architecture and Art, Volume 8: Public Communication and Awareness/Risk Management,
Contracts and Financial Aspects – Peila, Viggiani & Celestino (Eds)
© 2020 Taylor & Francis Group, London, ISBN 978-0-367-46873-6

The new Bözberg Tunnel (Switzerland) – risk management and auxiliary measures for the TBM break-through

M. Ramoni, R. Gallus & R. Iten
Basler & Hofmann AG, Zurich, Switzerland

ABSTRACT: The new Bözberg Tunnel is a double-track railway tunnel. He is roughly 2.7 km long and thus the longest of the tunnels, which are being excavated or enlarged for the realization of the so-called "4-metre corridor" on the Gotthard artery through Switzerland. The excavation methods have been chosen according to the expected geology: conventional excavation in soil and by means of a single shielded TBM in (swelling) rock. After a short general overview of the project, the paper focuses on the tunnel section which runs in rock. Particular emphasis is put on the TBM break-through, which – in spite of the very low over-burden of 1–2 m of sound rock – occurred successfully in November 2017 close to the existing railway line without interference of the railway operation. For this purpose, the implementation of auxiliary measures (e.g. triple pipe umbrella), a continuous monitoring and exhaustive risk management were necessary.

1 INTRODUCTION

At the time being, the Swiss Federal Railways are upgrading the Gotthard route – which can be seen as a centerpiece of the European rail-freight corridor between Rotterdam and Genoa – to further increase the volume of transalpine freight carried by rail. The realization of the so-called "4-metre corridor" (Figure 1) between Basel (at the German border) and Chiasso (at the Italian border), which will allow as of 2020 semitrailers with a headroom of 4 m and a width of up to 2.6 m to be carried, requires the enlargement of about 20 tunnels and more than 80 adaptions to platforms, traction current systems, signaling installations and overpasses (Swiss Federal Railways 2018). The investment costs are estimated to be CHF 710 millions (Zieger et al. 2018a).

One of the tunnels, which does not satisfy the requirements with respect to the clearance profile, is the 2.5 km long Bözberg Tunnel, which was opened in August 1875. Based on various studies, for this tunnel the Swiss Federal Railways decided to not refurbish the existing tunnel but to excavate a new double-track tunnel running parallel to the existing one. After opening to commercial railway operation of the new tunnel in 2020, the existing tunnel will be converted to a service and escape tunnel, which will be connected to the new tunnel by means of five cross-passages at a spacing of maximum 500 m. With this solution, tunnel safety is increased and a time-expensing enlargement of the existing tunnel under continued railway operation can be avoided. In this respect, it is worth mentioning that such a refurbishment would have led to a considerable loss of capacity on the Bözberg line and to a delayed opening of the 4-metre corridor. The 2.7 km long new Bözberg Tunnel (hereafter referred to as "the tunnel") is the biggest single project of the 4-metre corridor. The investment costs are esti-mated at CHF 350 millions (Zieger et al. 2018b).

The Swiss Federal Railways tendered the construction works of the tunnel in September 2014 as an overall performance model according to the model "General Contractor Plus (GU+)". According to the tender documents, the GU+ provides the tunnelling works (tunnel and cross-passages), the construction of both portals including two short sections of cut-and-cover

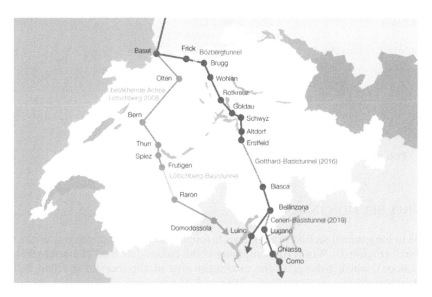

Figure 1. The 4-metre corridor across Switzerland along the Gotthard route (red line; picture: courtesy of Swiss Federal Railways).

tunnel, the implementation of the dewatering system, the installation of the electrical equipment as well as of the rail system and the conversion of the existing tunnel into a service and rescue tunnel. Furthermore, the GU+ undertakes all the design works for the construction project and provides the building structure documentation. During construction, the Swiss Federal Railways (client) are supported and represented by an Owner's Engineer. The Swiss Federal Railways awarded the contract (with a volume of CHF 145 millions) to Implenia Switzerland in November 2015. On behalf of the contractor, the design works are carried out by the following joint-venture of Swiss engineers: Amberg Engineering AG (lead designer), Basler & Hofmann AG (responsible for the tunnel section in rock), F Preisig AG and Heierli AG.

After a short general overview of the project (Section 2), the paper focuses firstly on the tunnel section in rock (Section 3), which has been excavated by means of a single shielded TBM. Subsequently (Section 4), the paper pays particular attention to the TBM breakthrough, for which the implementation of auxiliary measures, a continuous monitoring and exhaustive risk management were necessary.

2 THE NEW BÖZBERG TUNNEL

The new Bözberg Tunnel is 2'693 m long and runs between Schinznach-Dorf and Effingen almost parallel to the existing one (Figure 2). From Schinznach-Dorf to Effingen, the tunnel can be subdivided in following sections: cut-and-cover tunnel (36 m), tunnel section in soil (quaternary deposits) and heavily weathered rock (175 m, excavated conventionally), tunnel section in rock (2'438 m, excavated by means of a Herrenknecht single shielded TBM, Ø 12.36 m) and cut-and-cover-tunnel (44 m).

The construction works started in March 2016. The conventional excavation took 7 months (from September 2016 to April 2017). The TBM drive started in Mai 2017 and ended in November of the same year. Advance rates of up to 40 m/day have been achieved working with two excavation shifts (2 x 8 h/day) and one maintenance shift (1 x 8 h/day). Commercial railway operation in the new tunnel will start in October 2020. The conversion of the existing tunnel to a service and escape tunnel is expected to be completed in April 2022.

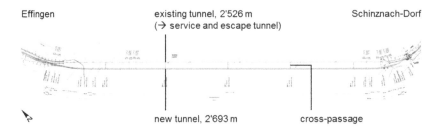

Effingen

existing tunnel, 2'526 m
(→ service and escape tunnel)

Schinznach-Dorf

new tunnel, 2'693 m

cross-passage

Figure 2.   Plan view of the Bözberg Tunnel.

3   TUNNEL SECTION IN ROCK

The 2'438 m long tunnel section in rock runs through the challenging geology of the so-called "Swiss Jura" (Figure 3). Worth to mentioning is the crossing of the "Hauptmuschenkalk" (a shell limestone), which belongs to the catchment area of the thermal spa Bad Schinznach. Likewise important and particularly challenging is the crossing of zones with heavy swelling rock: the Keuper (clay and sulphate swelling), the Anhydrite Group (clay and sulphate swelling) and the Opalinus Clay (clay swelling).

In order to not influence the ground water regime in the catchment area of the thermal spa Bad Schinznach, a so-called "thermal protection zone" with a length of 390 m was defined (Figure 3), where specific ground water protection measures have to be implemented: (a) in order to avoid a "damage" of the rock mass around the tunnel as much as possible, the excavation has to be carried out by means of a single shielded TBM (Figure 4); (b) a double-shell lining with full waterproofing (testable and injectable) has to be installed in order to avoid long-term drainage of the rock mass; (c) longitudinal circulation of ground water along the tunnel has to be prevented (the corresponding countermeasures are filling of the annular gap with mortar and ring-shaped bulkhead grouting between inner lining and segmental lining as well as in the rock mass around the tunnel at selected locations).

The different boundary conditions concerning waterproofing and swelling potential of the rock led to the application of different lining concepts along the tunnel. In the thermal protection zone, where full waterproofing is required, a full double-shell lining (segmental lining + cast-in-situ inner lining in the arch and in the invert, Figure 5) is applied. In this case, the inner lining has to accommodate a water pressure of up to 10 bars. In the rest of the tunnel, where a partial waterproofing is allowed, the lining is double-shell in the arch (segmental lining + cast-in-situ inner lining, Figure 6) and single-shell in the invert (segmental lining, Figure 6). The swelling potential of the rock is taken into account adjusting accordingly the thickness of the lining and the amount of reinforcement. It is

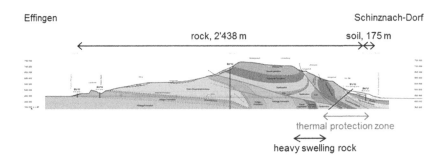

Effingen

Schinznach-Dorf

rock, 2'438 m

soil, 175 m

thermal protection zone

heavy swelling rock

Figure 3.   Geological longitudinal profile (picture: courtesy of Dr. Von Moos AG, modified).

Figure 4. Herrenknecht single shielded TBM, Ø 12.36 m (picture: courtesy of Swiss Federal Railways).

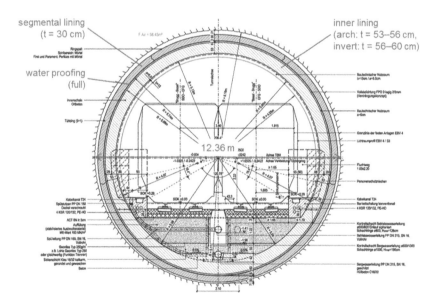

Figure 5. Standard cross-section (tunnel section in rock) with double-shell lining and full water proofing.

worth mentioning that in the case of "clay and sulphate swelling" the lining system has to resist a maximum swelling pressure of 4 MPa (this in combination with the water pressure mentioned above). This results in a heavy reinforcement of the inner lining over a length of 150 m.

Along the entire tunnel the segmental lining consists of five segments and one keystone (placed always in the invert). The segments are 2 m width and are not waterproofed. The segments are part of the load-bearing structural system also in the long-term and, therefore, are made of sulphate-resistant concrete.

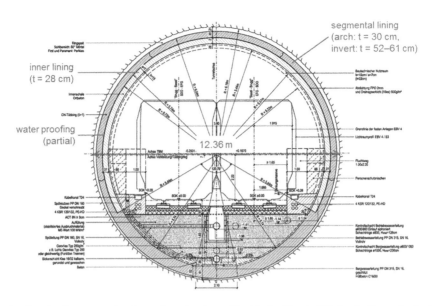

inner lining
(t = 28 cm)

water proofing
(partial)

segmental lining
(arch: t = 30 cm,
invert: t = 52–61 cm)

12.36 m

Figure 6.   Standard cross-section (tunnel section in rock) with double-shell lining in the arch, single-shell lining in the invert and partial waterproofing.

4   TBM BREAK-THROUGH

4.1   *General*

The TBM break-through in Effingen happened on 29 November 2017 and was very challenging. First of all, it occurred very close to the existing tunnel and railway line without stops of the railway operation. Furthermore, along the last 60 m the tunnel runs nearly parallel to a very steep slope with low overburden (Figure 7 and Figure 8). This required the detailed design and the implementation of auxiliary measures, a continuous monitoring and a very comprehensive risk management.

4.2   *Geology/main hazard scenarios*

Figure 9a shows a geological cross-section of the tunnel 4 m ahead the portal wall (TBM break-through-point) in Effingen. In this portal zone the ground consists of (from top to

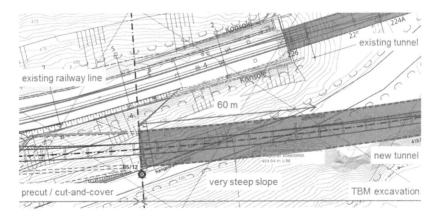

existing tunnel

existing railway line

60 m

new tunnel

precut / cut-and-cover

very steep slope

TBM excavation

Figure 7.   Plan view of the portal zone in Effingen.

Figure 8.   TBM break-through.

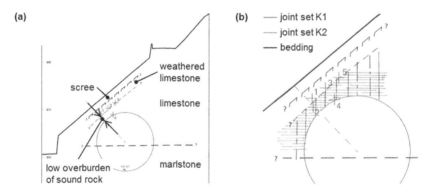

Figure 9.   (a) Geological cross-section 4 m ahead the portal wall; (b) Sketch of the discontinuities (pictures: courtesy of Dr. Von Moos AG, modified).

bottom): scree, weathered limestone, limestone and marlstone. The layer of scree and the layer of weathered limestone runs both parallel to the slope and have both a thickness of 1–2 m. The weathered limestone is intensively jointed (joint distance ≤ 10 cm) and fragmented. In the portal zone, only 1–2 m of sound rock overburdens the tunnel on account of the weathered limestone layer. The overburden of sound rock increases with increasing distance to the portal wall. Beyond the weathered zone, the limestone is less jointed and, in general, of good quality. The same applies for the marlstone. Within the joints, a water circulation is possible.

The stability of the rock mass near to the tunnel profile – in particular, the sector with low overburden of sound rock – depends primarily on the characteristics of the joints in the limestone and on their intensity. As shown schematically in Figure 9b, in the limestone there are two pronounced subvertical joint sets (K1 and K2) with a joint spacing of 0.6–2 m and a persistency of 5–10 m, which together with the bedding planes (with a spacing of 0.2–0.3 m) may lead to the formation of instable rock blocks.

The main hazard scenarios to be considered are related to the potential instability of rock blocks. On the one hand, an instable rock block may lead (depending on its dimensions) to a block of the cutter head or to a jam of the shield (Figure 10a). This is relevant, as – due to the low distance between TBM and slope surface as well as portal wall – the applicable torque and thrust force are limited. On the other hand, such a block may lead to a cave-in up to the surface (Figure 10b), which itself may cause slope instability, thus endangering the existing railway line. Finally, the TBM might push a rock block towards the surface (Figure 10c) also causing instability of the block itself or of the slope.

4.3 Auxiliary measures

The main auxiliary measure which has been implemented is a triple pipe umbrella (Figure 11). Further auxiliary measures were the protection of the existing railway line by means of a temporary rolled-out protection wall (along the last 60 m of the TBM drive) and supplementary anchors in the portal wall.

The triple pipe umbrella is 16 m long and is composed of steel pipes with a diameter of 159 mm, a thickness of 10 mm and a steel quality S355 (Figure 12). The length of the pipe umbrella has been chosen accordingly to the results of block stability analyses and numerical investigations showing that a stabilization of the rock mass is necessary at least along the last 12 m of the TBM drive. The steel pipes have been installed with a spacing of 0.45 m and an inclination of 5°. The accepted boring tolerance was of 2%. After the installation of the steel pipes, the pipes and the boreholes were filled with mortar.

The main function of the steel pipes is to dowel the joints in the rock mass (Figure 13). In this way, the shear strength and the stiffness of the joints and, therefore, of the rock mass

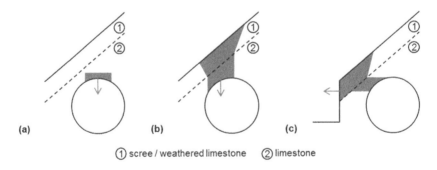

① scree / weathered limestone   ② limestone

Figure 10.   Main hazard scenarios: (a) instable rock block; (b) cave-in up to the surface; (c) "push-out" of a rock block.

Figure 11.   Installation of the triple pipe umbrella (picture: courtesy of Implenia Switzerland).

Figure 12.   Steel tubes of the triple pipe umbrella (picture: courtesy of Implenia Switzerland).

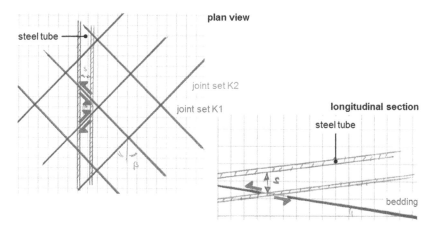

Figure 13.   Sketch of the doweling effect of the steel tubes of the triple pipe umbrella.

around the tunnel profile are increased, thus allowing for the formation of a load-bearing arch (Figure 14) with sufficient bearing capacity to stabilize the opening. The steel pipes avoid not only the instability of rock blocks (Figure 10a) and cave-ins up to the surface (Figure 10b), but also that the TBM might push a rock block towards the surface (Figure 10c).

A load-bearing action of the pipe umbrella in longitudinal direction has not been taken in account. Such a load-bearing action presupposes that the steel pipes are adequately supported in order to limit their span (i.e. the length of the pipes where a load is acting). However, in this case this cannot be assured with sufficient certainty. One the one hand, it cannot be excluded a priori that the rock between pipe umbrella and tunnel profile is removed from the TBM during excavation (in particular considering the presence of weathered rock and joints). On the other hand, the strength of the rock mass is not sufficient to bear the required reaction force.

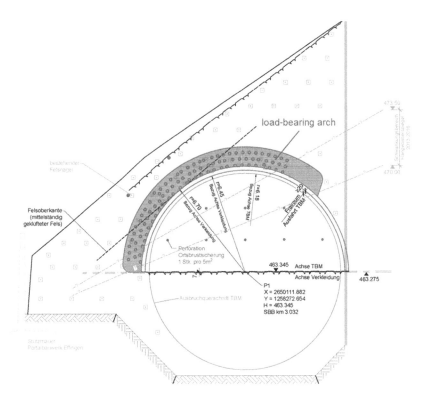

Figure 14. Sketch of the load-bearing arch (idealization).

4.4 *Continuous monitoring*

During the installation of the triple pipe umbrella and the TBM excavation of the last 60 m of the tunnel continuous monitoring was on-going. The slope and the portal wall were surveyed by means of several 3D geodetic points and two 20 m long inclinometers. The triple pipe umbrella was monitored by means of further two 16 m long inclinometers. The measurements have been carried out and evaluated automatically. The results were available real-time on a web-platform. For each measurement a set of limit values (attention value, intervention value and immediate intervention value) was defined. The continuous monitoring has been stopped only after the installation and complete backfilling of the last segment ring.

In addition to the monitoring mentioned above, visual checks of the slope, the temporary rolled-out protection wall, the portal wall and the existing railway line have been carried out. During the installation of the triple pipe umbrella these visual checks was on regular basis; during the last 60 m of TBM excavation security guards inspected the existing railway line continuously.

4.5 *Risk management*

In order to guarantee the maximum possible operational safety of the existing railway line, an exhaustive risk management has been implemented, which comprised the continuous monitoring described in Section 4.4, a specific developed alerting scheme and further auxiliary measures to be executed if required. The risk management involved – with clearly defined tasks and responsibilities – the contractor, the designer, the Owner's engineer and the Swiss Federal Railways (client).

With respect to risk management, it is worth mentioning that an emergency system was also set up in order to be able to close immediately the existing railway line in case of danger or incident. Furthermore, a direct contact with the TBM operator was established in order to

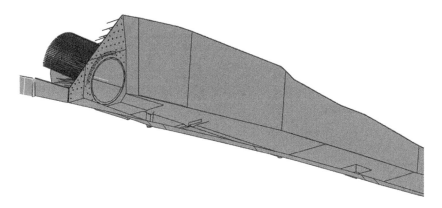

Figure 15. 3D model of the portal zone in Effingen for the detection of conflicts between anchoring of the portal wall and triple pipe umbrella.

stop immediately the TBM drive if necessary. This was because, in spite of the implemented auxiliary measures (Section 4.3), it was not possible to exclude totally the occurrence of undesired events with high damage potential for the existing railway line.

5 CLOSING REMARKS

The new Bözberg Tunnel is a very interesting Swiss tunnelling project – not only because of the challenging geotechnical conditions encountered but also from the project organization point of view, as for this project the overall performance model GU+ is applied. It is particular worth mentioning that – in compliance with one of the goals of the application of the model GU+ – the contractor and his designer implemented numerous project changes increasing quality, robustness and economics of the project.

As expected, the TBM drive was successful. The TBM break-through close to the existing railway line occurred safely without any interference of the railway operation thanks to the careful planned auxiliary measures (Figure 15) and risk management.

REFERENCES

Swiss Federal Railways 2018. The 4-metre corridor on the Gotthard route. https://company.sbb.ch/en/the-company/projects/mittelland-region-and-the-ticino/4-metre-corridor.html, 15.09.2018.
Zieger, T., Bühler, M., Vogelhuber, M., Schmid, W. & Grossauer, K. 2018a. Neubau Bözbergtunnel: Bauliche Massnahmen beim Tunnelbau im quellfähigen Gebirge. *Tunnel* 2018 (2): 14–27.
Zieger, T., Bühler, M., Rick B., Schmid, W. & Grossauer, K. 2018b. Challenges and innovative solutions at the new construction of the Bözberg Tunnel. *Geomechanics and Tunnelling* 11 (1): 62–75.

Tunnels and Underground Cities: Engineering and Innovation meet Archaeology,
Architecture and Art, Volume 8: Public Communication and Awareness/Risk Management,
Contracts and Financial Aspects – Peila, Viggiani & Celestino (Eds)
© 2020 Taylor & Francis Group, London, ISBN 978-0-367-46873-6

Risk management in metro contract procurement procedures. The French example

M. Russo
Icaruss, Paris, France

M.L. Rew & M. Sénéchal
EGIS, Paris, France

ABSTRACT: More than 200 km of new metro lines are planned in the Paris region in the Grand Paris Express scheme. The overall cost of this project is up to 30 billion €. Costs of this level need a rigorous design philosophy and a fair risk sharing procurement procedure. The authors of this article are involved in different roles on different stretches of the project, as design and site engineers, and as contractor's advisors and providers. This global view has allowed the identification of strengths and weaknesses of the procurement procedure adopted for the project. In this article the procedure for risk management and sharing is exposed at the design and procurement stage, and its strengths and weaknesses are outlined. The article reflects the opinions of the Authors.

1 INTRODUCTION

A huge metro project has been going on in Paris since 2009. More than two-hundred kilometres of new metro lines planned are at different phases, from design to construction. This project, named Grand Paris Express, will provide the region with three new lines.

These lines are designed and funded by the Société du Grand Paris (SGP), a public owned company. The Société du Grand Paris, a public agency set up by the French government in 2010, leads this project by supervising the construction of the new lines, the acquisition of rolling stock, the development within and around the stations and the design and building of a pipeline of optic fibres along the new lines.

The Grand Paris Express is currently the largest urban transport project in Europe. Its purpose is to integrate the outer Parisian suburbs to provide the Grand Paris Area with new multimodal transport solutions, thus improving the attractiveness of zones of difficult accessibility.

The final aim is to achieve a new ring around the Paris metropolitan area, to double the link to Charles de Gaulle Airport to the north and to provide a metropolitan link by extending the existing line 14 to Orly Airport (to the south). The latter, funded by SGP, is under the responsibility of a different Client: the RATP (Régie Autonome de Transports Parisiens).

In addition, under SNCF (Société nationale des chemins de fer français, France's national state-owned railway company) ownership, an underground extension of line RER E, an urban express railway, from St Lazare station to La Défense and beyond (aboveground) will provide an east-west link interconnected to the new lines.

The overall estimated cost of the SGP project is 30 billion € and the project will be completed in 2030. From March 2016 to September 2018 more than 6.6 billion € of contracts have been awarded, in the southern and northern portions of the project, see Figure 1.

Due to these numbers, the project required a procurement focusing on risk management at all project stages, from design to construction.

Figure 1. Grand Paris Express construction contract awards and the scheme.

In the following chapters the principles of application of the risk management procedures at different stages of the project, from design to construction are described.

Table 1. Grand Paris Express allotment.

Date	Stretch	Joint Venture	Contract
March 2018	Line 14 Lot 4 4 km tunnel+C&C access+shafts	NGE (leader)+Salini-Impregilo	203 M€
March 2018	Line 14 lot 3 4 km tunnel + 3 stations+3shafts	Razel-Bec (leader)+Eiffage	365 M€
March 2018	Line 14 lot 2 4.6 km tunnel + 1 station+ shafts	Dodin Campenon Bernard (manda-taire)+Vinci Construction France +Vinci Construction Grands Projets+ Botte Fondations+Spie batignolles génie civil+Spie batignolles fondations	400 M€
March 2018	Line 14 lot 1 140 m tunnel + 1 station	Leon Grosse + Subcontractors CIPA and Soletanche-Bachy	155M€
February 2018	Ligne 16 – Lot 1 5 stations + 4 tunnels total lenght 19,3 km +18 shafts + Rail equipment	Eiffage Génie Civil, Razel Bec et TSO (NGE)	1840 M€

(Continued)

Table 1. (*Continued*)

Date	Stretch	Joint Venture	Contract
June 2017	Line 15 South – Lot T3A Île de Monsieur - FIVC (not incl.): 4 km tunnel+2stations, 1 bridge	Bouygues TP (mandataire), Soletanche Bachy France, Soletanche Bachy Tunnels, BESSAC, SADE.	513M€
June 2017	Line 15 South – Lot T2D Noisy – Champs Station: Noisy – Champs station, 370 m C&C	VINCI Construction France (mandataire), Dodin Campenon Bernard, Spie Batignolles TPCI, Botte Fondations, Spie Fondations et VINCI Construction Grands Projets.	156 M€
April 2017	Line 15 South – Lot T2B Créteil l'Échat (not incl.) <-> Bry – Villiers – Champigny (incl.)	Eiffage Génie civil SAS (mandataire) Razel-Bec SAS.	795 M€
February 2017	Line 15 South – Lot T3C Fort d'Issy-Vanves Clamart (not incl.) et Villejuif Louis-Aragon (incl.) 8 km, 5 stations, 8 shafts	Vinci Construction Grands Projets (mandataire), Spie Batignolles TPCI, Dodin Campenon Bernard, Vinci Construction France, Spie Fondations et Botte Fondations.	926 M€
February 2017	Line 15 South – Lot T2A Villejuif Louis-Aragon (not incl.) <-> Créteil l'Échat (incl.)	Bouygues TP (mandataire), Soletanche Bachy France, Soletanche Bachy Tunnels, BESSAC, et SADE	807 M€
Sept. 2016	Line 15 South – Lot T2C Noisy-Champs to Bry-Villiers-Champigny 4.7+2.2 km 8 shafts	ALLIANCE (Demathieu Bard Construction SAS (mandataire), NGE Génie Civil SAS, GTS SAS, Guintoli SAS, Impresa Pizzarotti, Implenia, Franki Foundations Belgium et Atlas Foundations)	363 M€
Sept. 2016	Line 15 South – Lot T2E C&C Noisy-Champs	Léon Grosse TP (mandataire), Dacquin	51 M€
March 2016	Line 15 South – Lot T3B Fort d'Issy Vanves Clamart Station	Bouygues TP (mandataire), Soletanche Bachy France, Soletanche Bachy Pieux et Soletanche Bachy Tunnels	66 M€
	Env. 65 km 27 Stations		6640 M€

2 RISK MANAGEMENT – THE CONTEXT

Underground projects are particularly sensitive to uncertainties. More so than in other engineering disciplines, the input data will contain uncertainties that may have effects on the works and must be identified and managed to successfully achieve project objectives of quality, cost and schedule.

The data input uncertainties must therefore be reduced according to the project phase, so the associated residual risks are better defined. This process should lead every project phase and has the main objective of better defining the costs and work schedule and highlighting uncertainties that may compromise the successful completion of the project.

A risk is thus an event that may affect the development of the project as foreseen in the Contract. If a risk becomes a certain occurrence, it is removed from the risk register and treated in the baseline design.

2.1 *What is risk management and why do we need it?*

Tunnelling risks can have different origins: natural (the geological context), anthropogenic or manmade (quarries, infrastructure, etc.) and contractual (site access, contractual phases, interactions with adjacent contracts, etc.).

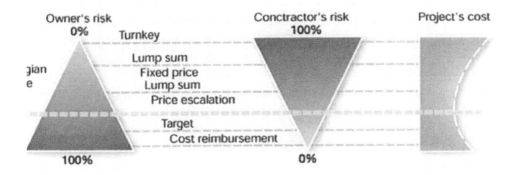

Figure 2. Risk assessment principles [1.].

Risk management practice provides a structured procedure to treat the uncertainties and limit the consequences of residual risks, allowing both the client to identify the extent of the effects of residual uncertainties on his project and the contractor to tune its prices by reducing or withdrawing the provisions for risks.

Indeed, optimal risk sharing will result in a lower cost for owners and a better price calculation for the contractor. The importance of a correct and fair risk management procedure and risk sharing plays thus a major role in securing successful underground projects, see Figure 2.

This risk management procedure must start as soon as possible in the project's life so that enough time is available for the treatments and, for geotechnical risks, the investigations required to reduce the risk by increasing the knowledge of the surrounding environment. The number and criticity of the residual risks will thus be limited by a successful design strategy. The provisional treatment of the residual risks will be defined at the design stage, as well as any monitoring to determine the actual impact of the construction on existing structures.

2.1.1 Risk register

The risk register is the principal element of the SGP risk management procedure. The risk register is created during the feasibility design phase and is regularly updated throughout the design process. It enables the identification and recording of potential risks and uncertainties, as well as their consequences on the project.

The risk register quantifies the likelihood of the risk, as well as its consequences in terms of cost and lead times, the definition and provision of a series of measures implemented to qualify risks more accurately (investigations, etc.) and the definition of the measures to be taken in the event of the risk event occurring. Also, the quantification of provisions for contingencies are estimated.

The risk register includes for each identified risk, the estimation of the uncertainty introducing the risk, the risk probability and its consequences in terms of human loss, delays, costs, image, and any element estimated important by the client.

Also, the measures foreseen to prevent or to treat the risk if it occurs are included.

2.1.2 Evaluation criteria

The risk evaluation criteria are set out by the client, Société du Grand Paris (SGP), to ensure a uniform application over all SGP projects: as Grand Paris Express includes three new lines, each of which is divided into several design contracts and even more construction contracts, a common approach is a fundamental starting point.

The first step is to evaluate the risk ranking based on the probability of occurrence and the gravity of the consequences (schedule impact, cost, safety, public perception, environment, etc.). These elements are rated on a scale from 1 to 4, by applying specific, measurable criteria. A similar approach is proposed by ITA, where the ranking scale is based on 5 points. In

Probabilité	Gravité			
4	4	8	12	16
3	3	6	9	12
2	2	4	6	8
1	1	2	3	4
	1	2	3	4

Critères d'acceptabilité du risque	Action à appliquer
Inacceptable	Doit être éliminé
A traiter	Traitement à prévoir
A statuer	Action de traitement ou non à définir par le chef de projet
Acceptable	Aucune action n'est à mettre en œuvre, simple suivi

Figure 3.   Ranking of risks in terms of probability and consequences.

French ranking, the choice of 4 rankings minimises the trend of the risk manager to set the criterion to "average". One key point is the probability of occurrence affected to the rank. For instance, in the SGP system, the rank 3 of probability of occurrence ranges from 25 to 80% of probability of occurrence. This wide interval leads to many risks with a 3 ranking, requiring detailed investigations before being able, possibly, to reduce the risk to a ranking of 2.

The analysis yields a matrix where the probability of each event is multiplied by the gravity of each of the consequences, and the highest score for each event gives the event's criticity, with a maximum possible score of 16.

The further the project progresses, the lower the highest allowable criticity score. Four rankings are thus identified: Acceptable, To be accepted or not by Client, To be treated by designer with eventual countermeasures, Not acceptable at all. The latter must be solved or countermeasures must be proposed to minimise the consequences and thus reduce the criticity to an acceptable level. At each project phase measures must be implemented to reduce the criticity of all risks, but especially the high scoring ones.

As explained in Chapter 4, at the tender phase identified risks that are deemed to be not acceptable have to be deal with to become "acceptable" or "treated".

3   APPLICATION TO THE DESIGN PHASE

3.1   *Grand Paris Express Line 15 East*

The owner, Société du Grand Paris (SGP), appointed Koruseo as design engineer and architect of the infrastructure of Line 15 East. Koruseo is a consortium led by EGIS (with among others Icaruss and Aecom as subcontractors) and includes TRACTEBEL Engineering, INGEROP and 6 architectural practices: BORDAS-PEIRO, GRIMSHAW, BRENAC & GONZALEZ, SCAPE, VEZZONI and EXPLORATIONS ARCHITECTURE.

The preliminary design, carried out in two phases, AVP-a and AVP-b, was completed at the end of 2017. The project design will start in the autumn of 2018. The authors are involved in the design team, and in this chapter some examples are given of the application of the risk management process described above during the preliminary design phase completed in late 2017.

3.1.1   *Geological campaigns, a way to minimise uncertainties*
In the design phases the JV focused more on geology related risks, due to the important consequences of geological uncertainties on the project costs and schedule.

There is a wealth of information on Parisian geology. The client had already carried out quite extensive borehole campaigns. By the start of the preliminary design we had access to 104 boreholes. A further 286 were carried out during our AVP design and another 310 remain to be bored as well as geophysical testing to minimise uncertainties.

There are also other sources of information. From third parties (other development projects) we obtained 48 cored borehole results, 105 pressure meter tests and 112 non-cored

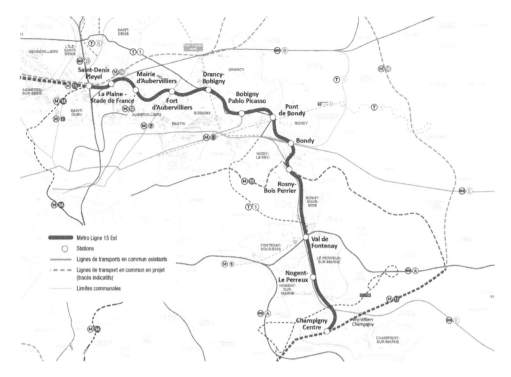

Figure 4. Line 15 East alignment (http://www.metroligne15est.fr/trace/).

boreholes. From Line 15 South, the most advanced of the GPE lines, we obtained 52 cored boreholes, 68 pressure meter tests and 65 non-cored boreholes.

The objective of the coming project geological campaign is to have, in all, 7 cored and 7 pressure meter tests per station, 2 cored and 2 pressure meter tests per shaft, and most importantly 7,3 cored boreholes and 8,2 pressure meter tests per kilometre of tunnel.

This will allow a significant reduction in the current project criticity levels by the time the construction contract goes out to tender.

3.1.2 *A specific example of risk treatment – Voids in some geological units*

We shall look at the example of Gypsum dissolution in soft limestone ("Calcaires de Saint-Ouen") and marly limestone ("Marnes et caillasses") and how this risk was mitigated in the two preliminary design phases AVP-a and AVP-b.

The risk event is surface subsidence caused by a void from Gypsum dissolution in soft or marly limestone resulting in the tunnel boring machine (TBM) losing confinement, or in fluid losses while excavating diaphragm walls.

The context may range from an unforeseen site delay linked to treatments to fill the voids, to costly repairs to buildings, stoppage of tunnelling, evacuation of residents.

The source is Gypsum dissolution prior to tunnelling, or after tunnelling as the water pathways may be altered leading to new dissolution areas.

The risk treatment applied during the AVP-b design stage was:

– Reduce the likelihood by increasing the density of geotechnical testing and by adjusting the alignment where possible to avoid the Gypsum formations
– Limit the risk at the source by grouting from the surface, from the TBM or from the tunnel
– Limit the consequences during construction by installing extensive monitoring 12 months prior to tunnelling at each location, by type of TBM confinement (earth pressure balance, slurry pressure balance or variable density) and by carrying out microgravimetric measuring before invert slab concreting to detect voids

– Limit the consequences during tunnel operation by installing fibre optics in the invert slab concrete to detect deformations and temperature changes (indicative of voids forming), design the invert slab concrete for a 4 m breach by increasing the reinforcement locally, and install 200 mm diameter inserts in the slab concrete under the rails every 3.5 m for grouting in the event of voids being detected

The result of this risk treatment was for the criticity to be reduced from 12 (unacceptable) to 9 (acceptable at this stage of the design). The criticity will be further reduced as additional project-wide information is obtained on geology (stratigraphy, Gypsum, quarries, alluviums, hard bands in marly limestone and sands) and hydrogeology (aquifers, upper ground water levels, superficial ground water, permeability).

3.2 *Anthropogenic risks*

One main application of risk assessment consists in analysing the effects of the works on the existing structures in interface with the project. The geological context and the building data determine the intrinsic sensitivity of each structure, that is, the amount of allowable settlement, rotation and elongation.

The vulnerability study carried out by the designer determines whether the predicted settlement caused by tunnel construction falls within the structure's allowable limits. If it does, the building is not classed as vulnerable. If the predicted settlement exceeds the allowable limits then the building is classed as vulnerable and preventative measures must be taken.

Different ways of implementing preventative measures, or risk mitigation have been foreseen along the alignment. The impact of the project can be reduced by modifying the alignment, by choosing/imposing a suitable tunnelling confinement method or by carrying out ground treatment to improve the geotechnical parameters. If the singular point cannot be avoided it may be possible to mitigate the existing structure's sensitivity by carrying out structural reinforcement, or simply by implementing safety measures such as installing protective netting and repairing any damage after the tunnelling.

3.3 *Current conclusions of the geological and anthropogenic risks on Line 15 East*

The risk analysis has allowed the optimisation of the longitudinal profile in some areas.

The major risk of subsurface gypsum dissolution has been controlled adequately by defining the construction measures to be taken: preferred type of TBM, slab concrete reinforcement, regular inserts in the slab concrete to allow future grouting of cavities detected thanks to fibre optics monitoring. The future geotechnical campaign will additionally be adapted to reflect further requirements.

There are some major risks still to be controlled in the next design stages, due to as yet insufficient geotechnical data for some parts of the project. There is also a remaining risk of alignment redesign in case of undetected/unsuspected deep foundations that will be identified in the next round of building surveys.

4 RISK MANAGEMENT AT THE PROCUREMENT PHASE

Since the early design phase of Line 15 East, the application of the risk management principles at the procurement phase is described based on tender documents published by SGP to date. To date, the procurement strategy of Société du Grand Paris is based on a restricted negotiated procedure according to EU Directive 2014/24.

The main specificities in the legal framework for French construction contracts (CCAG) are:

– The contractor is responsible for the methods and final design
– The contractor is responsible for safety on site

- The Engineer shares at different levels the Contractor's responsibility by accepting final design documents and methodologies
- Unit prices include all possible constraints as described in the tender documents
- The quantities in the bill (BOQ) are indicated as forecast. A variation of more than -20% to +25% in the final quantities permits revision of the unit prices. In the case of lump sum contracts cost revision is allowed for variations outside of -5% to +5%

This legal framework has to cope with the specificity of underground works, where a risk occurrence may lead to a significant variability in quantities and methods or site organisation.

To manage these elements, the procurement procedures adopted include a Risk BOQ, with the cost of countermeasures foreseen at the design phase, whose acceptability is ranked as "to be treated".

This method allows the definition of scenarios that due to low probability of occurrence are not included in the base scenario, but, due to their gravity, will have a ready-made solution in case they occur. The quantities of these solutions are explicitly not included in the base BOQ, thus the limit in variation does not apply. The quantities in the Risk BOQ of the base offer are disclosed to all candidates.

Bidders have to provide a complete base offer with the quantities given in the bid documents and not may not modify the content of any of the BOQ (base and risk). They may propose alternatives, in general limited to some specific points. In this case they can propose modifications to the BOQ, including to the Risk BOQ if appropriate.

One main problem induced by this procedure is that the introduction of alternatives complicates the bid analysis. Often the acceptable alternatives are limited in number and one alternative may include several variations (base alternatives) from the base offer. Nonetheless, in this case the non-acceptability of one of the base alternatives renders the whole alternative unacceptable.

Independent negotiations are held with each candidate and confidentiality is assured by the Client.

4.1 *The risk register and Risk BOQ*

Risk management at procurement phase is based on the risk register refined during the subsequent design phases and linked to a specific Risk BOQ that is proposed to the selected bidders. This includes, for risk event, the type of risk (geological, human, contractual, access, ...), the description, its causes and the residual uncertainties that justify the event to be a risk and not included in the base price, its probability of occurrence, its consequences with countermeasures, and, in the latter case, residual risk and residual treatment. It is also indicated the prices that cover the countermeasures in the Risk BOQ.

Also the risk register is integrated with the measures foreseen to reduce or treat risk's consequences and the bidder's proposal of category of risk (see below)

Risque résiduel									Description du risque RESIDUEL	TYPE Risque	Document de référence du marché en complément des pièces financières	Mesures prévues en cas de survenance du risque RESIDUEL		BPU Risques DE Risques	Description des travaux
Vraisemblance	Objectifs du MOA							Niveau de risque				Préconisées par le Maître d'oeuvre dans les cas où le Maître d'Ouvrage est propriétaire du risque	Proposées par le Titulaire dans les cas où il est propriétaire du risque (cas d'une limite de transfert supplémentaire)		
	Impact Coût	Impact Délai	[illegible]	Environnement	Tiers	Image	Performance								
									- Arrêt du tunnelier	Type 2 Externe	Cahier B	- Ralentissement ou arrêt du tunnelier - Traitement des terrains		famille : RIS 020 RIS 030	- Immobilisation du tunnelier - Traitement de terrain depuis le tunnelier
									- Tassements plus importants que prévus - Ralentissement ou arrêt du tunnelier	Type 2 Externe	Cahier B	- Ralentissement ou arrêt du tunnelier - Traitement des terrains		famille : [illegible]	- Immobilisation du tunnelier - Traitement de terrain depuis la surface et/ou [illegible] - Traitement de terrain depuis le tunnelier
									- Blocs de dimensions supérieures à la roue de coupe	Type 2 Externe	Cahier B	- Intervention hyperbare ou non supplémentaire pour destruction des blocs		famille : RIS 030	- Immobilisation du tunnelier pour intervention dans la chambre d'abattage
									- Risque éliminé par le traitement Quantités réelles de déblais/mortier de bourrage	Type 1 intégré au forfait	Cahier B				

Figure 5. Example of a risk register (risk management by contractor).

In particular, as described in chapter 2.1.2, the three types of risks are ranked, according to risk ownership:

– Type 1: well identified risks (type "acceptable" or accepted by Client in Figure 3): these risks have to be included in the bidder's base prices as an average situation on site (slight differences of soil behaviour not affecting foreseen methods, risks defined in tender documents, etc.). They are owned by the Contractor and are included in the base BOQ.
– Type 2: well identified and for which the countermeasures are known and defined at tender stage (type "to be treated" or not acceptable for the Client in Figure 3). If they occur, these risks are owned by the Client. Nonetheless, their effects and costs are well defined and covered by the Risk BOQ. While completing the Risk BOQ prices, the Bidder has to include in proposed unit prices all the consequences on planning.
– Type 3: unforeseen risks, that are handled by negotiated cost and fee or new prices (contract base).

Risks of type 2 and 3 are to be discussed by Contractor and Engineer in case of occurrence (and eventually by Client). In particular, Risk BOQ prices are supposed to cover all consequences of a type 2 risk (including site organisation, immobilisation, equipment, etc.).

Bidders have the possibility of modifying the risk register in their alternative proposals, either by adding further risks, or by proposing a different threshold for sharing a risk. In this case the Risk BOQ is modified. The aim of this is to optimise at much as possible the base prices by extracting some risks or including explicitly some occurrences. Nonetheless, the transfer of a risk to the Client or the removal of a risk has to be managed as an alternative proposal (in general limited to risks of type 1 and 2).

It is also to be noted that a further main criterion of the contract is that, in case a risk arises following the Contractor's specific construction methods, this won't be owned by the Client.

4.2 Typical award criteria

Typical award criteria account for the risk management and risk sharing proposed by the bidders. The detailed criterion varies for each contract. Typical values are:

• Price:	up to 50% (issued from BOQ + lump sums + Risk BOQ)
• Technical value	up to 30%
• site installations	up to 15%
• Risks management	up to 8%
• Safety/environment	up to 7%
• Planning	up to 20%
• planning delay	up to 9%
• planning robustness	up to 7%
• planning innovations	up to 4%

The procedure allows thus to choose the best bidder taking account of, to a certain extent, the base price, but also the interest of the bid in terms of risk sharing.

5 CONCLUSION

The risk management procedure is adopted throughout the project in Paris new metro line construction. From the early design phases a list of risks is identified, and according to the uncertainties and evaluations a ranking is established based on their probability and gravity.

The risk register is transferred to the contractor at the procurement phase. As bidders may propose a different key in risk sharing, this is accounted for in the award criteria. For instance, a more robust construction technique may result in a higher risk threshold acceptable to the contractor, who is thus rewarded for this effort.

Also a contractor may propose a different risk sharing threshold if this allows more optimised base unit prices.

Although it is too soon to have a return on the efficiency of the procedure and its advantages in contract management, the Risk assessment and management procedure adopted by the Société du Grand Paris, has some undeniable advantages:

– Risk management is a key issue at all project phases, from design to construction
– The robustness of the design is guaranteed by the adoption of this philosophy at every phase of the project
– The overall sum allocated to risks for the whole project may be reduced considering the entire project by managing risks with a global budget
– The bidder's prices may be optimised to take account of minimal costs and lower risk provisions as far as possible
– The procedure complies with the general French legal contractual framework (CCAG).
– The procurement criteria allows the client to reward a more convenient risk sharing proposal, or to obtain lower prices in case he accepts to bear more risks.

On the other hand, some key issues may be further developed such as:

– Definition of the probability and gravity rankings attributed to each risk. Too wide exemptions may lead to a deformed scale.
– A modification in risk allocation by a bidder has to be considered as an alternative, allowing a limited number of alternatives.

REFERENCES

AFTES 2015. Recommandation de l'AFTES N° GT25R3F1. Maîtrise économique & contractualisation. *Tunnels et Espace Souterrain* (249): 170–195.
AFTES 2012. Recommandation de l'AFTES N° GT32R2F1. Caractérisation des incertitudes et risques géologiques hydrogéologiques et géotechniques. *Tunnels et Espace Souterrain* (232):274–314.
Kleivan, E. 1987. NoTCoS – Norwegian Tunnelling Contract System. *Norwegian Tunnelling Today* (5). Oslo: Norwegian Soil and Rock Engineering Association.

Tunnels and Underground Cities: Engineering and Innovation meet Archaeology,
Architecture and Art, Volume 8: Public Communication and Awareness/Risk Management,
Contracts and Financial Aspects – Peila, Viggiani & Celestino (Eds)
© 2020 Taylor & Francis Group, London, ISBN 978-0-367-46873-6

Alternative project delivery based on a risk-based probabilistic approach

P. Sander & M. Spiegl
RiskConsult GmbH, Innsbruck, Tyrol, Austria

J. Reilly
John Reilly International, Framingham, Massachusetts, USA

D. Whyte
Lima Airport Partners, Airport Development Program, Lima, Peru

ABSTRACT: Effective cost and risk management is essential for the success of large infrastructure projects, as demonstrated by a long history of cost overruns. In order to achieve cost transparency, risk-based probabilistic approaches are needed to determine the probability that project delivery can be accomplished within cost and schedule goals. In addition to some owners moving to a more collaborative and incentivized project environment, a significant number of owners and agencies are also considering alternative contracting models to deliver their projects. This paper describes the mechanics of Fixed-Price-Incentive-Fee (FPIF) firm target price contracts, provides a framework for analyzing such contracts, and demonstrates how FPIF pricing arrangements (pain/gain mechanism, target cost, ceiling cost, etc.) can be applied with a risk-based probabilistic approach.

1 INTRODUCTION

Effective cost and risk management is essential for the success of large infrastructure projects, as demonstrated by a long history of significant cost overruns. In order to manage cost to established budgets and to achieve cost transparency, it is necessary to adequately consider cost and schedule uncertainties (risks), which means risk-based probabilistic approaches are needed. This allows us to estimate the probability that project delivery can be accomplished within cost and schedule goals and to define and manage risks that might negatively affect meeting those goals.

In addition, some owners are moving to a more collaborative and incentivized project environment, and a significant number of owners and agencies are also considering alternative contracting models to deliver their projects (Ross 2003, ICE 2018). This paradigm shift is driven by the fact that more traditional delivery methods, e.g., fixed price contracts, often fail to meet objectives due to factors that the authors have described in previous papers.

The purpose of collaborative working agreements and more integrated supply teams is to align the client, design consultants, contractors, sub-contractors, and vendors in a structure, often with incentives, to ensure that everyone works together efficiently to achieve agreed (shared) goals. Such teams are better able to create an environment where outstanding results can be achieved, with incentives leading to improved outcomes for owners and contractors.

This paper describes the mechanics of Fixed-Price-Incentive-Fee (FPIF) firm target price contracts, provides a framework to analyze such contracts, and demonstrates how FPIF pricing arrangements (pain/gain mechanism, target cost, ceiling cost, etc.) can be applied with a risk-based probabilistic approach. Since, in early stages, the project's outturn cost can only be estimated using ranges, estimates of potential profit for the contractor and project price for the owner need to be made using a probability model for total project cost.

An application similar to the example is used for the Lima Airport Extension Program.

2 APPROACH

2.1 FPIF and Delivery Methods

The traditional contracting approach for many project owners is to attempt to transfer as much of the risk as possible to the contractor, e.g., by a Lump-Sum-Turnkey approach (Reilly et al. 2018). This is not necessarily effective for megaprojects. Any attempt to allocate risks of complex project to different parties, no matter how well intentioned, may be little more than an illusion and can give rise to an adversarial culture that may threaten the success of the project (Ross 2003, Reilly et al. 2018).

The FPIF delivery model (using NEC- or FIDIC-type contracts) is a way to implement shared goals, related to an established, negotiated target cost. It is not a full alliancing approach, where the owner, designer, and contractor are jointly bound to meet cost and schedule targets in a pain/gain environment, but it is an option for large, complex projects with a high level of risk and uncertainty and can establish a collaborative working environment using incentives based on a pain/gain mechanism.

Overview Project Delivery Methods

For the following ranges for Complexity, Design and Risk Assumption, the delivery methods below are recommended.

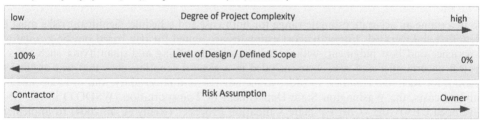

E.g. for High Complexity with Low Design % the Owner assumes Risk and uses Cost-Plus-Percentage-Fee.

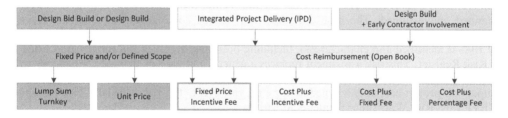

Figure 1. FPIF as hybrid delivery method.

By convention, contracting professionals use share ratios to depict the degree of risk assumed by the owner and contractor. The share ratio for Fixed-Price (FP) contracts is depicted as 0/100. The first number is always the owner's percentage of risk, and the second number is always the contractor's percentage of risk. The 0/100 share ratio means that the contractor assumes 100 percent of risk under an FP contract. Conversely, a Cost-Plus-Fixed-Fee (CPFF) contract share ratio is depicted as 100/0. Since a CPFF contract requires a contractor's "best efforts" and they get paid the fixed fee regardless of their achievement under the contract, the government assumes 100 percent of the risk (Cuskey 2015). The Cost-Plus-Fixed-Fee delivery method allows the owner more control over his or her budget than the Cost-Plus-Percentage-Fee contract. As the total project cost increases, the fee paid to the contractor also increases. Nevertheless, these contracts are more flexible to project changes and reduce the contingency that the contractor has at the time of the bidding (Ibbs et al. 2003).

The differences between the Cost-Plus-Incentive-Fee (CPIF) and FPIF pricing arrangements occur when contract costs are substantially above or below target cost. The CPIF contract pricing arrangement must include a minimum fee and a maximum fee that define the contract range of incentive effectiveness (RIE). When costs are above or below the RIE, the Government assumes full cost risk for each additional dollar spent within the funding or cost limits established in the contract.

While there is no universal optimized FPIF model, contract parameters can be adjusted to best suit both sides. Flexibility does come with greater complexity, but when properly executed, FPIF contracts can be highly effective in motivating contractors to control cost (Hurt et al. 2015).

2.2 Probabilisitc Methods

We believe the reader is familiar with basic concepts of risk, risk management, and risk mitigation and the use of probabilistic cost-risk processes versus deterministic ones (Reilly et al. 2015, Sander et al. 2015). The probabilistic approach, compared to the simpler and more common deterministic approach (unit prices times unit costs plus a contingency), offers more useful information with respect to the range of probable cost as well as cost "drivers" and better quantifies the effects of risks, opportunities, and variability. This improves understanding and leads to a better potential for profit (or loss) for contractors and added value for owners.

2.3 CEVP-RIAAT Process

To determine an accurate estimate range for both cost and schedule, significant risks must be identified and assessed. Formerly, cost estimates accounted for risk based on the estimator's experience and best judgment, without necessarily identifying and quantifying such risks— project uncertainties and risks were included in a general "contingency" that was applied to account for such uncertainties. In order to include risk and uncertainty, and to independently validate costs, the Washington State Department of Transportation (WSDOT) in the USA developed CEVP, the "Cost Estimate Validation Process," (Reilly et. al. 2004) to implement better cost estimating and to include the influence of uncertainty (risk) on project delivery.

In CEVP, estimates consist of two components: the base cost component and the risk component. Base cost is defined as the planned cost of the project if everything materializes as planned and assumed. The base cost does not include contingency but does include the normal variability of prices, quantities, and like units. Once the base cost is established, a list of risks is identified and characterized, including both opportunities and threats, and listed in a Risk Register. This risk assessment replaces a general and vaguely defined contingency with explicitly defined risk events that include the associated probability of occurrence plus the impact on project cost and/or schedule for each risk event. The risk is usually developed in a CEVP Cost Risk Workshop (Sander et al. 2018).

RIAAT (Risk Administration and Analysis Tool – http://riaat.riskcon.at) is an advanced software tool that combines base costs, base variability, risks, opportunities, and schedules to indicate ranges of probable cost and schedule, plus risk management and change tracking and documentation (Sander et al. 2017).

3 FPIF CONTRACTING

As part of the application of CEVP and RIAAT to a major project in South America, the opportunity to include advanced risk management and delivery processes was evaluated. The result was the decision to apply the FPIF process using the CEVP-RIAAT process as input and to help establish an agreed target cost. The FPIF approach is based on US Department of Defense (USDOD) strategies for different types of procurement in different circumstances.

The rationale for selecting this particular contract form, based on the USDOD approach, is that:

1. For projects where there are established historical data regarding outturn costs, the program is stable, and many units are to be delivered with few change requirements, a fixed-price lump sum is appropriate.
2. For projects that are uncertain, with substantial unknowns, such as new weapons systems or components that require significant research and development, a cost-plus negotiated procurement is most appropriate.
3. For projects with some unknowns, but with stable scopes, a process between a fixed-price lump sum and a cost-plus negotiated procurement—a process with characteristics of both approaches—is best. This means a firm upper-cost ceiling, with a defined target cost and a pain/gain mechanism to incentivize reduced cost for the owner, with a defined scope. This is the FPIF form of contract.

FPIF models keep a fixed-price approach but also allow for a certain degree of control over the total price by creating a more collaborative environment with the contractor.

4   DEFINITIONS RELATED TO FPIF

FPIF Contract: Specifies a target cost, a target profit, a ceiling price, and a profit adjustment formula. These elements are all negotiated at the outset. The profit earned by the contractor varies inversely with the project cost by application of a pain/gain mechanism. When the final project cost is negotiated, the contractor's profit is calculated, and the price paid by the owner is the final project cost plus the so-calculated contractor profit. All project transactions and costings are 100% open book and subject to audit.

   Target Cost (TC): Expected total cost of the project (direct plus project-related overheads), excluding contractor profit. It should be reasonably challenging but achievable. It is based on a reasonable best-case scenario of contract performance based on an analysis of available information. It includes the contingency allocated to the risks associated with the delivery of the project, agreed by the parties.

   Target Profit and Target Price (TP): The Target Profit is the profit earned by the contractor for achieving the Target Cost. The Target Price is the sum of the Target Cost plus the Target Profit.

   Share Ratio (S/R): Percentage that each party shares in cost underruns and cost overruns from the negotiated Target Cost. The first number corresponds to the owner, the second to the contractor. For example, an Underrun S/R of 60/40 indicates that the contractor's profit is increased by forty cents for each dollar under the target cost. The same sharing principle applies for an Overrun S/R.

   Pain/Gain Mechanism: Formula applied to calculate the final price paid by the owner, based on the agreed S/R for underruns and overruns. When the final negotiated cost of the project is lower than the target cost (i.e., there has been an underrun), application of the S/R results in a final profit greater than the Target Profit; the price paid by the owner is the final negotiated cost plus the (higher) profit so calculated. Conversely, when the final negotiated cost is higher than the Target Cost (i.e., there has been an overrun), the contractor earns a profit lower than the Target Profit, and the owner pays for the final negotiated cost plus the (lower) profit so calculated.

   Ceiling Price (CP): Maximum price paid by the owner to the contractor, except for any adjustment under other contract clauses.

   Point of Total Assumption (PTA): Overrun cost point at which the Pain/Gain Mechanism results in the owner paying the Ceiling Price, i.e., the negotiated cost of the project plus the profit earned by the contractor as per the Pain/Gain Mechanism equals the Ceiling Price.

$$CP = Cost @PTA + Profit @PTA$$

The S/R becomes 0/100 at the PTA because the owner no longer shares in a cost overrun. Therefore, the contractor is assuming the extra cost at the expense of his profit, dollar per dollar. The formula to calculate the PTA is as follows:

$$PTA = TC + (CP - TP) / (Owner\ Overrun\ Share\ Ratio)$$

Example:
With the data: TC = 100; Target Profit = 10 (therefore TP = 110); CP = 118 and Overrun S/R = 60/40

$$PTA\ results\ in:\ PTA = 100 + (118 - 110) / 0.6 = 113.3$$

Therefore, for this example project where the TC = 100, if the final negotiated cost is 113.3, the owner would pay the CP = 118 and the contractor would make a profit of 118 - 113.3 = 4.7, which is less than the target profit of 10. If the project cost exceeds 113.3, the owner still pays the CP = 118, therefore the contractor reduces his profit one dollar per every additional dollar of project cost.

Figure 2 shows the defined parameters applied to a probability distribution. This basic model is used to define the model for application in a project. The following example is a guide through the steps.

5   APPLICATION EXAMPLE

An application similar to the example is used for the Lima Airport Extension Program.

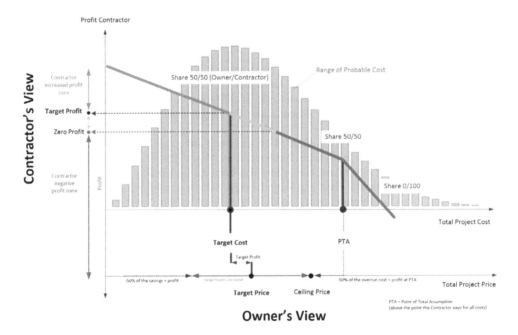

Figure 2.   Visualization of terms.

5.1 *Probable Cost Range*

The Probable Cost Range is the result of the CEVP-RIAAT application. As usual the range depicts the Base Cost + Risk and Escalation. It does not include the contractor's profit. Figure 3 shows a typical result using the probability distribution and the probability function.

5.2 *FPIF Model Set Up*

Table 1 lists all the parameters, formulas, and calculated values that are used to set up the FPIF model for our example.

Figure 4 applies the FPIF on the Probable Cost range (compare to Figure 2).

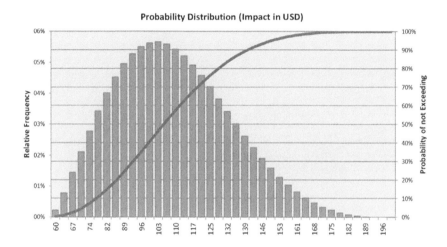

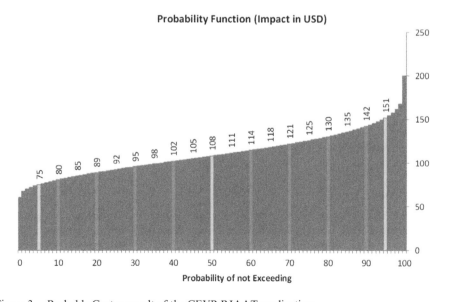

Figure 3. Probable Cost as result of the CEVP-RIAAT application.

Table 1. Calculation of the FPIF parameters.

Parameter	Formula	Value
Probable Cost Range	Is given in a range as result from the CEVP work-shops (see 5.1.).	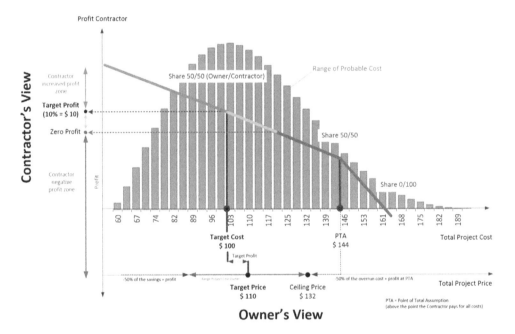
Target Cost (TC)	TC	$100
Profit	10% Profit for Contractor: Profit = TC * 0.1	$10
Target Price (TP)	TC + Profit	$110
Ceiling Price (CP)	Set to $132	$132
Owner Share Ratio (OSR)	Owner/Contractor share ratio: 50/50	50%
Point of Total Assumption (PTA)	PTA = TC + (CP - TP)/(OSR)	$144
	PTA = 1.0 + (1.32 - 1.1)/0.5	

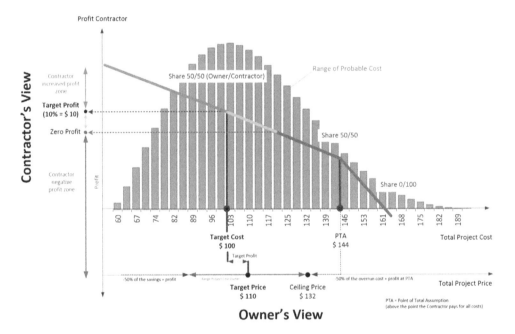

Figure 4. FPIF model applied to the Probable Cost Range.

5.3 Deviation From Target Cost

Since there is a Share Ratio in the case of a cost overrun or underrun, the potential deviation from the Target Cost is essential for calculating the potential pain/gain for the owner and contractor. Figure 5 depicts the probability function that shows the potential deviation. There is a chance of about 38% that the cost will come in below the Target Cost but also a probability of 62% that the final cost will be higher than the Target Cost. For example in 42% (P80 minus Target Cost → P38) of all cases the cost overrun will not exceed $ 30.

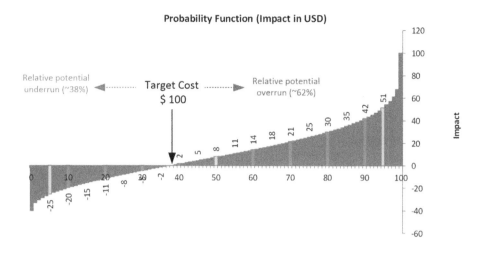

Figure 5.   Potential deviation from the Target Cost.

5.4   *Point of Total Assumption – 0/100 Share Ratio*

If the final cost exceeds the PTA of $144, the contractor takes all the risk. Figure 6 visualizes a cost impact with a 9% probability that the final cost will exceed $144.

5.5   *Contractor's View*

From a contractor's view, there is a probability of 32% that he will drop into the loss zone, but also a probability of 38% that he will have increased profit above $10 (Figure 7 and Figure 8).

    If the PTA is exceeded (9% probability), the contractor takes all the risk, which will rapidly increase his loss. This is depicted by the steep curve in Figure 7 and the flat tail in Figure 8.

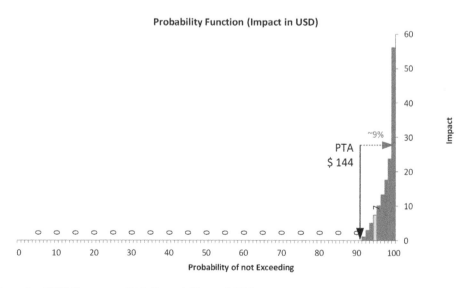

Figure 6.   100% Contractor Risk Potential beyond PTA.

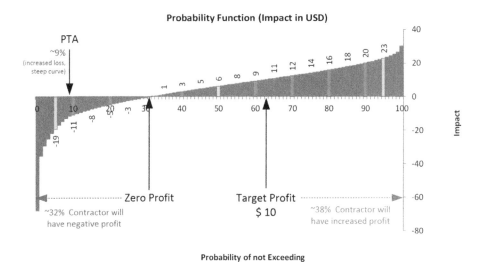

Figure 7.   Contractor's view of Profit/Probability Function.

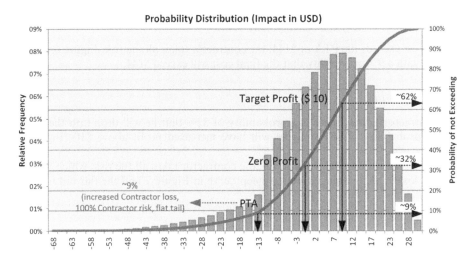

Figure 8.   Contractor's view of Profit/Probability Distribution.

The analysis does not consider the potential increased efficiency by the contractor in order to generate higher profit.

5.6   *Owner's View*

From an owner's perspective, there will be a 38% chance that his cost will be lower than the Target Price (Figure 9 and Figure 10). This chance might be higher if the contractor is incentivized to gain more profit and works with increased efficiency.

   If the PTA is exceeded, the contractor takes all of the risk. This defines the Ceiling Price of $132 for the owner. There is a probability of 9% that the PTA will be exceeded and the Ceiling Price mechanism will be triggered.

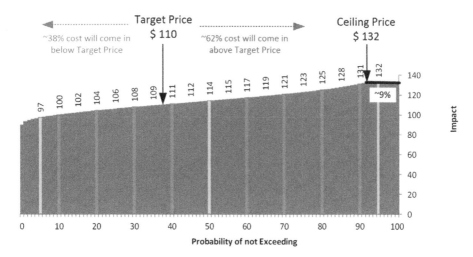

Figure 9.   Owner's view of Cost/Probability Function.

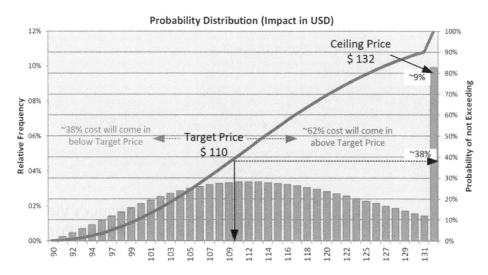

Figure 10.   Owner's view of Cost/Probability Distribution.

6   CONCLUSION

The FPIF contract model is a way to implement shared cost goals and to establish a collaborative working environment, using incentives based on a pain/gain mechanism. One key to the FPIF contract is a consensual agreement on the target cost. The individual risk potential for a chosen target cost for the contractor and owner should be calculated using probabilistic methods. For the contractor, the probability is relevant in estimating the potential to increase his profit or risk to suffer loss. For the owner, the deviation from the target price with the corresponding probability is the basis for the evaluation of the contract. The probabilistic results transparently show the risk potential of both parties (Figure 11), allowing contract negotiations to be conducted from a common basis.

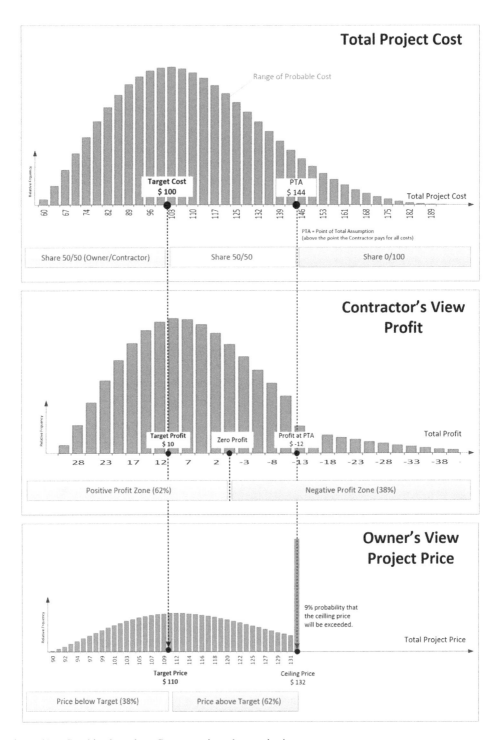

Figure 11. Combined results – Contractor's and owner's view.

REFERENCES

Cuskey, Jeffrey R. (2015), "Understanding the Mechanics of FPIF Contracts," PTACS Montana State University

Fixed Price Incentive Firm Target (FPIF) Contract Type, Acquisition Encyclopedia, https://www.dau.mil/acquipedia/Pages/ArticleDetails.aspx?aid=6794b407-22e0-4d83-aff9-80474fc70014

Hurt, Steven, Elliot, Ryan, (2015), "Modeling Price Outcomes for Complex Government Programs," A.T. Kearney

Ibbs, C. W., Kwak, Y. H., Ng, T., & Odabasi, A. M. (2003), "Project delivery systems and project change: Quantitative analysis." Journal of Construction Engineering and Management, 129(4), 382–387.

ICE UK (2018), Project 13, "Project 13 launch will improve how infrastructure is delivered,"https://www.ice.org.uk/news-and-insight/latest-ice-news/project-13-launches

Reilly, J.J., Essex, R, Hatem, D. (2018), "Alternative Delivery Drives Alternative Risk Allocation Methods," North American Tunnel Conference, Washington DC, June (pending)

Reilly, J.J., McBride, M., Sangrey, D., MacDonald, D, Brown, J. (2004) "The development OF CEVP® - WSDOT's Cost-Risk Estimating Process," Proceedings, Boston Society of Civil Engineers, Fall/Winter

Reilly, J.J., Sander, P., Moergeli, A. (2015), "Construction – Risk Based Cost Estimating," Paper and Presentation, RETC 2015, New Orleans

RIAAT (2018), Risk Administration and Analysis Tool – http://riaat.riskcon.at

Ross, J. (2003), "Introduction to Project Alliancing," April 2003 Update, Alliance Contracting Conference, Sydney, Australia, 30 April

Sander, P., Reilly, J., Entacher, M. (2018), "CEVP-RIAAT Process - Application of an Integrated Cost and Schedule Analysis", North American Tunneling Conference, Washington D.C., June 2018

Sander, P., Reilly, J., Entacher, M., Brady, J. (2017), "Risk-Based Integrated Cost and Schedule Analysis for Infrastructure Projects" Tunnel Business Magazine, August 2017, p. 43–37

Sander, P., Moergeli, A., Reilly, J. (2015), "Quantitative Risk Analysis – Fallacy of the Single Number," Paper and Presentation, WTC 2015, Dubrovnik

USDOD Comparison of major project acquisition types, https://www.acq.osd.mil/dpap/ccap/cc/jcchb/Files/Topical/Contract_Type_Comparison_Table/resources/contract_type_table.docx

Tunnels and Underground Cities: Engineering and Innovation meet Archaeology,
Architecture and Art, Volume 8: Public Communication and Awareness/Risk Management,
Contracts and Financial Aspects – Peila, Viggiani & Celestino (Eds)
© 2020 Taylor & Francis Group, London, ISBN 978-0-367-46873-6

Development of a cost model for a more exact estimation of the construction costs of tunnels

M. Thewes, P. Hoffmann & G. Vollmann
Institute for Tunnelling and Construction Management, Ruhr-University Bochum, Germany

I. Kaundinya & A. Lehan
Federal Highway Research Institute, BASt, Bergisch Gladbach, Germany

ABSTRACT: Cost estimation, as it is carried out at an early stage of a construction project, is based on the classical methods of cost calculation for construction works. In tunnel projects, where every structure is considered a unique item, this often leads to considerable uncertainty in cost estimation due to the project-specific circumstances, accompanying risks of construction and the overall heterogeneous boundary conditions. Funded by the German Federal Highway Research Institute (BASt), the Institute for Tunnelling and Construction Management – together with a group of German consultants – developed new cost models, algorithms and thus new possibilities for a more precise estimation of the construction costs in the course of a research project. This article presents the results and the analyses of recent tunnel projects examined in this context.

1   INTRODUCTION

The calculation of building costs is based on the classic cost calculation of construction services, which is used equally for building construction and underground engineering. Compared to buildings, tunnel construction projects are subject to different boundary conditions. On one hand, this is due to the geotechnical imponderables and challenges of tunnel constructions. On the other hand, tunnel construction projects require different approaches to cost determination, especially regarding the consideration of financial risks.

   In the past, cost estimation approaches for different types of road tunnels have already been developed. This was primarily done in research projects funded by the German Federal Highway Research Institute (BASt), e.g. FE 02.178/1997/FR 1997 (STUVA 1997). Certain limitations in data quality resulted primarily from a low resolution of the cost structure and from the fact that fundamental price-determining components and building materials have been subject to considerable price fluctuations in recent years. Furthermore, the type of operating technology and the level of required safety measures have changed. For this reason, the former studies must be regarded as structurally correct but outdated in terms of content. A new evaluation of current base prices and structures is required.

   In addition, buildings of the transport infrastructure are increasingly being viewed holistically from a cost-related point of view. In addition to the general improvement of cost structures, the possibilities for reducing the total costs of road tunnels by optimising the balance between construction costs (planning and construction costs) and subsequent follow-up costs (operating, maintenance and repair costs) are analysed within the framework of life cycle cost approaches.

   The objective of the study, which is described in the following, was to update and supplement the cost structures for construction costs and operational equipment based on current project data for German road tunnels, also and especially against the background of life cycle-specific considerations. In this context, the uncertainty of construction price development based on statistical analyses had to be considered and to be integrated into a new cost model. In this way, a suitable stochastic simulation model was developed, which can be used in the future to map the ranges of construction price development in cost determination.

The basic principles of this study, the developed cost structures, the forecast models and the results of a consecutive validation are presented below. The focus is on pure cost analyses and forecasts using the example of drill and blast tunnels. Data from 12 tunnel projekts in conventional method were analysed. The projects were chosen because of their construction methods, cross-sections, lengths and diverging boundary conditions to form a representative basic population for the German road tunnels. All corresponding analyses were carried out from here.

2 RESEARCH CONCEPT AND STRUCTURE OF THE COST ANALYSIS

For a targeted development of new cost models for road tunnels, it was initially recommended to look at the emergence of cost increases to adjust further analyses and to be able to track and categorize cost increases within completed projects. This was particularly necessary against the background of the planned cost and risk modelling.

Considering the typical cost increase in construction projects, two specific deviations from the original cost calculations in the planning phase can be easily identified during the planning, awarding and execution phases:

- $\Delta1$ as a cost-specific variance between the original cost estimation from the planning phase to the cost calculation of the awarded contractor in the award phase
- $\Delta2$ as an increase in costs from the cost estimate from the award phase to the as-built sum of the construction project after the final cost determination

Deviation $\Delta1$ usually results from a general increase in raw material prices and the resulting progressive development of the construction price indices, from changes in the market situation, but also from necessary planning changes, e.g. due to new regulations or changes in the boundary conditions. The increase $\Delta2$ is often the result of changes in performance compared to planning and above all of risks occurring during the construction phase (Table 1).

The upper row in Table 1 shows the documents of the individual project phases which are required and used for the analyses. The bottom row assigns the necessary analysis steps to the different phases, which had to be carried out in the course of the investigation.

Table 1. Structure of cost analysis.

	Planning	$\Delta1$ → tender	$\Delta2$ → final invoice
Documents:	- cost estimation and cost calculation - older investigations (STUVA 1997)	- submission results (price comparison list and bill of quantities) - request for bills of quantities	- final invoice or payment on account - supplements (positionally sharp) including basis for claims - Interview with experts
Analyses:	- comparison of costs - analysis of deviations	- comparison of bidder prices/ market prices (position level) - restructuring - unit prices	- analysis of supplements and comparison with the allocation data - determination and evaluation of supplementary risks

2.1 Cost structure developed

For the development of a generally applicable tool for forecasting project costs, it is necessary to develop a uniform structure of positions, which is highly adaptable to a variety of different tunnel projects and can at best serve as a layout for a future general standard tender document. The new structures were developed based on the structures of the available project data, taking into account cost structures and cost groups developed in the past. The results of an earlier research project (STUVA 1997), the instruction for cost determination and estimation of road construction measures of the German Federal Ministry of Transport (BMVI 2015), as well as the corresponding German construction costs standard (DIN 276-4 2009) were used as a basis. The following is an example of the cost structure developed for drill and blast tunnels (Table 2).

Table 2. Developed cost structure of the conventional method (e.g. drill and blast method).

Field	No.	Position	Unit
Preparations	101	Site equipment	rmt tunnel
	102	Time-bound costs	rmt tunnel
	103	Technical processing	rmt tunnel
	104	Other services	rmt tunnel
Main tunnel	105	Excavation of driving class 4 to 4a	m^3
excavation	106	Excavation of driving class 5 to 5a	m^3
incl.	107	Excavation of driving class 6 to 6a	m^3
breakdown	108	Excavation of driving class 7 to 7a	m^3
bay	109	Downtime costs	day
	110	Exploration of probing drilling	m
Excavation	111	Excavation of driving class 4 to 4a C	m^3
crosssection	112	Excavation of driving class 5 to 5a C	m^3
	113	Excavation of driving class 6 to 6a C	m^3
	114	Excavation of driving class 7 to 7a C	m^3
Excavation	115	Excavation calotte foot extension	m^3
others	116	Geological overbreak	m^3
	117	Start procedure and breakthrough of main tunnel	Lump sum (LS)
	118	Start procedure and breakthrough of crosscuts	LS
Securing	119	Heading face sealing, thickness (t) = 3 cm	m^2
main tunnel	120	Heading face sealing, t = 5 cm	m^2
	121	Heading face sealing, t = 10 cm	m^2
	122	Shotcrete lining vault, t = 10–15 cm	m^2
	123	Shotcrete lining vault, t = 20 cm	m^2
	124	Shotcrete lining vault, t = 25 cm	m^2
	125	Shotcrete lining vault, t = 30 cm	m^2
	126	Shotcrete lining breakdown bays, thk = 35 cm	m^2
	127	Shotcrete sole or calotte sole, thk = 20 cm	m^2
	128	Shotcrete sole or calotte sole, thk = 25 cm	m^2
	129	Shotcrete sole or calotte sole, thk = 30 cm	m^2
	130	Grid arches	m
	131	Steel arches	t
	132	Reinforcement steel mashes	t
	133	Steel bars reinforcement	t
	134	Mortar anchor	Pcs.
	135	Self-drilling anchors	Pcs.
	136	System anchor SN-anchor	Pcs.
	137	Injection drilling anchor (pipe anchor)	Pcs.
	138	Self-service injection drill anchor	Pcs.
	139	Mortar spits	Pcs.
	140	Self-service injection spits	Pcs.
Other	141	Concrete geological outbreak	m^3
securing	142	Surcharge for moderate water flow	m^3
	143	Piperoof/reinforcement ahead of the tunnel face	rmt tunnel

(Continued)

Table 2. (*Continued*)

Field	No.	Position	Unit
Drainage	144	Longitudinal drainage	m
system	145	Sole drainage	m
	146	Cheek drainage	m
	147	Inspection shafts	Pcs.
	148	Slotted channel, cleaning pipes, siphon	m
	149	Retention reservoir	LS
Other	150	Other drainage allowance	rmt tunnel
drainage			
Sealing,	151	Sealing carrier	m²
building-	152	Synthetic protection sheet or geotextile	m²
joints	153	Synthetic sealing sheet	m²
	154	Joint tapes	m
	155	Injection hose	m
Other	156	Other sealing work allowance	rmt tunnel
sealing work			
Inner shell	157	Main tunnel vault in-situ concrete	m³
main tunnel	158	Sole vault main tunnel in-situ concrete	m³
	159	Filling concrete	m³
	160	Reinforcement steel bars inner shell	t
	161	Steel mesh reninforcement inner shell	t
Inner shell	162	Vault crosscut in-situ concrete	m³
cross cuts	163	Vault crosscut in-situ concrete	m³
	164	Reinforcement inner shell crosscut	t
Other	165	Formwork carriage vault	LS
concrete,	166	Formwork carriage sole	LS
reinforced	167	Surface protection systems for concrete	m²
concrete	168	Concrete post-treatment	m²
work	169	Partition walls/ceiling	m³
Other shell	170	Extinguishing water pipe	m
work	171	Hydrant, fire-extinguishing, electrical or sink niche	Pcs.
	172	Emergency call niche	Pcs.
	173	Other services	rmt tunnel
Dewatering	174	Dewatering construction phase	rmt tunnel
construction			
Runway and	175	Reinforced concrete carriageway	m²
cable ducts	176	Gravel base layer	m³
	177	Asphalt carriageway	m²
	178	Frost protection layer	m³
	179	Cable conduits	m
	180	Cable ducts	Pcs.
Portal	181	Portal construction	LS
construction			
Operation	182	Operation building	LS
building			

Taking German guidelines ZTV-ING, Part 5 and RABT 2006 or the draft version of RABT 2016 into account, basic variants of possible tunnels were defined for the respective construction methods (BAST 2018, FGSV 2016, FGSV 2006). A differentiation was made in particular regarding the various cross-section and equipment variants.

3 PRELIMINARY ANALYSIS OF PRICE DIFFERENCES OF SELECTED PROJECTS

The preliminary analysis of the price differences between the individual bidder prices on the basis of the list of bidder's estimates served as the basis for the scientific investigation and the cost calculation. Under German law, tenders should be granted to the "most economic bidder", which mostly equals the bidder with the cheapest offer. That said, bidders sometimes tend to go

Table 3. Bandwidths of the submission results of the tunnel projects in the conventional method (12 projects).

Project No.	Δmin	Δmax	distribution of the most economical price
1	-3.84 %	4.02 %	
2	-10.44 %	8.20 %	
3	-8.51 %	7.74 %	
4	-4.84 %	10.98 %	
5	-1.63 %	1.85 %	
6	-7.61 %	8.97 %	
7	-5.62 %	8.40 %	
8	-13.02 %	16.79 %	
9	-11.00 %	19.66 %	
10	-12.57 %	6.91 %	
11	-6.84 %	7.76 %	
12	-13.26 %	21.36 %	
average:	-8.26 %	10.22 %	-8.26 %

with a so called "strategic price", hoping for winning the tender phase and gaining chances for possible claims during the construction phase. Bearing this in mind, the analysis showed, that a goal-oriented calculation of the costs is possible only with large uncertainty by statically well-founded prices, as detected in earlier studies. To this end, the costs for a selected group of dominant positions were initially recorded on a project-by-project basis and compared qualitatively and quantitatively in relation to the bidder. In the process, it was not possible to gain generally valid insights. Often the bids of different bidders within a position and also across the entire project varied so strongly that modelling on the basis of pure averages already at this early stage seemed to be unsuitable.

In addition, the differences in the net bid sums of all bidders were examined across all projects. The differences between the most economical bid and the average of all bidder prices (Δmin) and between the average and the maximum bid (Δmax) were compared (Table 3).

Results of the preliminary analyses:

- Although it is theoretically possible to develop a forecast model based on static prices, the result is blurred and even error-prone.
- The dominant factors influencing price fluctuations in construction costs are on one hand the length of the tunnel and on the other hand the cross-sectional size.
- Pure averaging does not allow a forecast of the "most economical" bidder.
- An approach should be chosen, in which the complete specifications are recorded and aggregated to main items.
- "Strategic" pricing is a real influence on the construction price, which is realized in the form of cost increases through supplements. Such prices must therefore not be taken out of consideration.
- To reflect the fluctuation between the different bidder prices, a forecast model is to be established based on a population of individual prices, which must be as comprehensive as possible.

Based on these findings, a detailed analysis was carried out across all bidders and projects.

4 COST INCREASE ANALYSES OVER THE COURSE OF THE PROJECT

The cost increase analysis is the analysis of the cost differences between the first available engineer's estimates up to the award allocations (Δ1) and between the award allocations and the final costs of the project (Δ2) (see Table 1). This analysis shows how costs change within the individual project phases. This enables a direct comparison of the cost increase in the planning phase with the cost increase in the execution phase. In addition, an average cost increase for Δ1 and Δ2 can be calculated or mapped using a statistical distribution (Table 4).

Table 4. Comparison of costs in the project phases for the conventional method (12 tunnels/projects).

Project No.	Comparison of engineer's estimate with the award allocations (Δ1)	Comparison of award allocations with final costs (Δ2)	Statistical distribution of Δ2 (Costs increase in the execution phase)
1	28.05%	5.33%	
2	50.12%	4.25%	
3	12.49%	16.45%	
4	-27.39%	0.33%	
5	4.59%	8.82%	
6	63.04%	29.48%	
7	-4.69%	7.55%	
8	4.48%	14.10%	
9	35.28%	20.18%	
10	58.36%	29.72%	
11	3.08%	40.55%	
12	26.98%	29.33%	
average:	21.20%	17.17	17.17%

It should be noted here that within the scope of this analysis only the construction costs of the tunnel structure from portal to portal in the individual phases were considered, in order to exclude possible project-specific peculiarities, such as terrain cuts or trough structures. In addition, where necessary, costs were converted to the respective reference year 2016 using the construction price index (BPI) (Destatis 2017). As a result, the mean value of the difference between the first engineer's estimation and the award allocation (Δ1) is 21.2% and between the award allocation and the final costs of the projekt (Δ2) 17.17%. Δ2 is included in the cost calculation tool as a statistical distribution and is therefore shown in Table 4.

In a study from France, a comparison of the award results of a total of 12 tunnel projects with the final invoice was also carried out and a comparable cost increase of an average of approx. 12 % was determined (Humbert & Robert 2017).

A central cause of these cost increases in both the planning and execution phases of a construction project is the realization of risks that have a monetary impact. The risk is traditionally always defined as a product of probability of occurrence and extent of damage, whereby both variables can be estimated either empirically, statistically or by simulations and are therefore subject to a natural dispersion.

In the course of the research project, possible risks were first identified by means of a qualitative analysis and then quantified by means of supplementary analyses. However, the analysis showed that there is currently still a lack of sufficiently large data bases for stochastically reliable modelling of the effects of supplements on cost increases. Nevertheless, risk catalogues and risk registers were developed for the individual construction methods, which are not presented and discussed here.

5 TOOLS FOR COST ESTIMATION AND CALCULATION INCLUDING INITIAL VALIDATION

Based on the project analyses, various tools for forecasting the costs of road tunnel projects were developed. These are explained in the following sections, their functionality is explained and the course of a cost forecast is presented using the specific tool on the basis of a validation project.

In detail, these are the following aids, which can be assigned to the specified service phases of the German Fee Structure for Architects and Engineers (HOAI 2013) and were developed for cost forecasting:

- cost-diagrams, to determine a rough cost framework, in the phase of basic determination (according to Project phase 1 of the HOAI)
- cost-estimation-tool for rough estimation of construction costs on the basis of first site investigations and first preliminary planning (project phases 1 and 2 of the HOAI)

- cost-calculation-tool based on comprehensive analyses of the investigated tunnel projects for the respective construction methods as well as the operational equipment. Prepared for the phases of preliminary and design planning for cost estimation and cost calculation of the shell and equipment costs of road tunnels (project phases 2 and 3 of the HOAI)

The validation project provided, a road tunnel constructed using conventional tunnelling methods, for the road deck has a standard cross section 10.5T (11t) (see Figure 1; 10.5T is the designation according to the valid RABT 2006 and 11t in brackets is the designation according to the draft version RABT 2016), a total length of less than 500 metres and is excavated and secured to 16.5% with excavation class 5 - 5a and 83.5% with excavation class 7 - 7a (DIN 18315 2016). Other special features or difficulties resulting from the geological conditions, hydrogeological conditions or the location of the construction project are not known.

5.1 Cost-diagrams

As an aid for determining an initial rough cost framework, cost diagrams were prepared based on an evaluation of the total costs of the tunnel structure available as a data basis. Only the tunnel structure from portal to portal and the net prices were considered. It should be noted that such diagrams are subject to the aforementioned uncertainties, as they are based on the evaluation of static mean values.

The cost diagrams were prepared based on the existing projects and an average BPI of 2016 for civil engineering structures (Destatis 2017). The cost diagram shown here represents a standard cross-section 10.5T (11t) (Figure 1).

The diagram is divided into three levels of difficulty:

- Simple, stands for geological conditions where long excavation round lengths with large excavation cross-sections are possible and only light securing devices are required (>80 % in excavation classes 4 to 5)
- Medium, geological conditions where only shorter round lengths are possible and smaller excavation cross-sections are required. The need for support measures is also increased accordingly (>80 % in excavation classes 5 to 6)

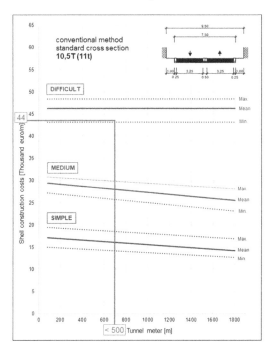

Figure 1.   Cost diagram for standard cross-section 10,5T (11t), conventional method with example project.

- Difficult, demanding geological conditions with corresponding round lengths and extensive support measures (>80 % in excavation classes 6 to 7)

Using values in the direction of the maximum or minimum should always be in accordance with an assessment of other difficulties.

The validation project used in the presented research is classified as "difficult", as the share of excavation class 7 - 7a is greater than 80 %. Further difficulties are not known. Therefore, the lower of the three lines ("min. difficult") should be selected. The forecast of a total price for the validation project deviates by approx. 43 % from the actual award price. This shows the high degree of uncertainty inherent to extremely simplified estimation tools, such as Figure 1. It also underlines the need for more differentiated estimation tool as described hereafter.

5.2 Cost-estimation-tool

Another possibility to estimate the shell costs (net) of the project is the use of the developed cost estimation tools. These tools offer a much higher variability without the need for specific determination of masses and volumes. Compared to the cost diagrams, larger bandwidths can be mapped. This leads to an improvement of the estimation result.

One way to individualize the view is to enter the exact percentage distribution of the degree of difficulty and the number of formwork carriages and operating buildings. The possibility of entering the number of formwork carriages makes it possible to influence the construction price by reducing the construction time. If the percentage distribution of the level of difficulty is changed, all costs associated with this change are influenced accordingly within the tool. In addition, each main cost group and the entire estimation can be adjusted at any time using an adjustment factor, preset at 1.0 (BPI 2016).

For the validation project, the result of the forecast with the cost estimation tool deviates approx 23 % from the actual award price. This deviation is lower than in the diagrams.

5.3 Cost-calculation-tool

The cost-calculation-tool is a standard Excel table including programmed modules based on Visual Basic for Applications (VBA). Palisade's @Risk software is required to statistically evaluate the input data and unit prices for the individual positions of the developed structure (see Chapter 2) and to use the tool. Only by using this statistical risk software the sometimes extreme differences in the bidder prices can be recorded statistically (see chapter 3), approximated by appropriate distribution functions and used for cost forecasting based on Monte-Carlo analyses.

The tool is divided into an area for entering the standard cross section and the tunnel length and an area for entering the quantities of the individual positions. The corresponding data (unit prices) available at 2016 is stored as cost distributions for each individual item. Further data can be added in the future, after appropriate processing and sorting, to any extent (limited only by the capacity of the Excel software).

To evaluate the data and compensate for larger price differences in the bidder prices, the respective value series is approximated by a suitable distribution function. For this purpose, a fitting of the curve and the underlying data is evaluated with the @Risk software based on the Akaike information criterion (AIC). Figure 2 shows the result of the analysis of an exemplary series of values. The x-axis shows the values or prices and the y-axis shows the probability of occurrence. According to AIC, the triangular distribution (1) is the one that allows an optimal reproduction of the value series. Nevertheless, a triangular distribution as a potential value space for possible pricing is unrealistic. For this reason, a Pert distribution (2) is used retrospectively in this example, since it is stochastic and more likely to correspond to the expected prices for future projects.

The result of the tool is the frequency distribution of the average total project costs. Based on a Monte-Carlo analysis, a probable price is calculated for each item, multiplied by the item-related quantities and these individual prices are then added up to total project costs (net). An example of 10,000 runs produces 10,000 potential totals that reflect the probability-

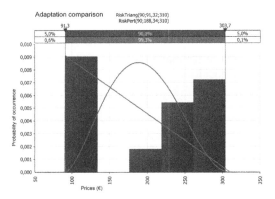

Figure 2. Adaptation comparison of an exemplary frequency distribution with triangle or pert distribution.

distributed unit prices and their dispersion. In addition, it is possible to change each individual item by entering a factor, for example to adjust the construction price index.

The displayed value is the most common value or the mean value of the distribution function. This value roughly corresponds to the expected average bid price from the tender. The most economical bid price is now calculated using the mean difference between the most economical price and the mean value of the bidder prices. The average difference between the average bid price and the most economical price is defined at Δmin and stored as a distribution function (Table 3).

The total project price including possible risks is also calculated automatically. The data stored for this surcharge represents the percentage cost increase between the project phases of the cost estimate and the cost determination (see Table 4). The difference between the cost estimate and the cost determination is defined as Δ2 and also stored as a distribution function. Figure 3 shows the overall result of the cost calculation tool based on the frequency distribution including the delta described. For the validation project, in addition to the cross-section and the tunnel length, the quantities of 63 positions had to be determined. The comparison of the most economical price predicted using the cost calculation tool with the real most economical (most favorable) offer from the tender results in a deviation of approx. + 3 %.

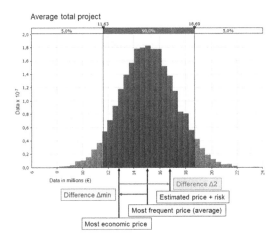

Figure 3. Result of the cost calculation tool as frequency distribution of the mean total sum of the validation project, including most economical price, mean and forecasted final invoice sum.

Table 5. Tabular comparison of the individual calculation result.

Comparison of results	Deviation
Cost-diagram with submission result	+43 %
Cost-estimation-tool with submission result	+23 %
Cost-calculation-tool with submission result	+3 %

6 SUMMARY OF THE VALIDATION RESULTS AND OUTLOOK

A comparison of the individual calculation results described with the respective tool is shown in Table 5 below.

Table 5 shows that the newly developed cost model, in the form of the cost calculation tool based on an initial validation, delivers very accurate results. It also provides an accurate determination of the most economical price and a rough estimate of a price including risk costs. The deviation of the forecast most economical price from the most economical offer from the submission of the validation project amounts to approx. 3 % and is thus far below the 20 % fluctuation range mention in technical literature. The developed cost diagrams and cost estimation tools do not achieve this accuracy due to the lack of the possibility of taking into account the construction price fluctuations, but can be helpful for an initial rough cost estimate. The quality of the cost calculation tool correlates with the quality of the underlying data and therefore further maintenance of the developed database by evaluating future projects is recommended.

REFERENCES

BASt (Bundesanstalt für Straßenwesen) 2018. ZTV-ING Teil 5, Zusätzliche Technische Vertragsbedingungen und Richtlinien für Ingenieurbauten (Additional Technical Contract Conditions and Guidelines for Civil Engineering Works). Bergisch Gladbach: Germany

BMVI (Bundesministerium für Verkehr und digitale Infrastruktur) 2015. AKVS 2014, Anweisung zur Kostenermittlung und zur Veranschlagung von Straßenbaumaßnahmen (Instruction for cost calculation and estimation of road construction measures). Bonn: Germany

Destatis (Statistisches Bundesamt) 2017. Preisindizes für die Bauwirtschaft (Price indices for construction) – Fachserie 17 Reihe 4 2017 (4. Vierteljahresausgabe), Online: https://www.destatis.de/DE/Publi kationen/Thematisch/Preise/Baupreise/BauwirtschaftPreise2170400173244.pdf?__blob=publication File. Zugriff am: 15. 01.18.

DIN (Deutsches Institut für Normung e.V.) 2009. DIN 276-4, Kosten im Bauwesen – Teil 4: Ingenieurbau (Building costs - Part 4: Civil constructions). Berlin: Germany

DIN (Deutsches Institut für Normung e.V.) 2016. DIN 18312, VOB Vergabe- und Vertragsordnung für Bauleistungen – Teil C: Allgemeine Technische Vertragsbedingungen für Bauleistungen (ATV) – Untertagebauarbeiten (German construction contract procedures (VOB) - Part C: General technical specifications in construction contracts (ATV) - Underground construction work)

FGSV (Forschungsgesellschaft für Straßen- und Verkehrswesen) 2006. RABT 2006, Empfehlung für die Ausstattung und den Betrieb von Straßentunneln (Guidelines for Equipment and Operation of Road Tunnels). Cologne: Germany

FGSV (Forschungsgesellschaft für Straßen- und Verkehrswesen) 2016. E-RABT 2016, Empfehlung für die Ausstattung und den Betrieb von Straßentunneln (Guidelines for Equipment and Operation of Road Tunnels) (draft version). Cologne: Germany

HOAI (Honorarordnung für Architekten und Ingenieure) 2013. Verordnung über die Honorare für Architekten- und Ingenieurleistungen (German Fee Structure for Architects and Engineers), Band 2. Berlin: Germany

Humbert, E & Robert, A. 2017. Key Elements for Keeping the Costs of Tunneling Projekts under Control. In Proceedings of the World Tunnel Congress 2017 – Surface challenges – Underground solutions. Bergen: Norway.

STUVA (Studiengesellschaft für unterirdische Verkehrsanlagen e.V.) 1997. FE 02.178/1997/FR 1997, Straßenquerschnitte in Tunneln, Ermittlung der Tunnelbaukosten (cross-sections in tunnels, estimation of tunnel construction costs). Cologne: Germany

Tunnels and Underground Cities: Engineering and Innovation meet Archaeology,
Architecture and Art, Volume 8: Public Communication and Awareness/Risk Management,
Contracts and Financial Aspects – Peila, Viggiani & Celestino (Eds)
© 2020 Taylor & Francis Group, London, ISBN 978-0-367-46873-6

Resolution of a contractual dispute for tunnel excavation

P. Torta, M. Tutinelli, L. Perino & S. Porrello
P.M. & E. Project Management & Engineering – Roma - Torino, Italy

ABSTRACT: The case study is concerning about the crossing of an Alpine geological Fault with the excavation of the pilot tunnel and followed by the two tunnel lines. During the excavation phases, the Contractor has founded for tunnel lines a significant amount of sprayed concrete, due to the filling of the overbreaks arisen behind the steel ribs support, despite the boundary consolidation. This issue has opened the dispute between Contractor and Owner and it has been resolved through the measurements in situ and by analysis of the related consolidation system planned in the Project. From the analysis, it was found out that the consolidation system, originally planned for the pilot tunnel and then adopted at the tunnels project line, did not give back the expected effect of the upgrading of rock mass. Therefore, the arisen overbreaks have obliged the Contractor to fill them supporting extra costs, refunded later by the Owner.

1   CASE STUDY OF A TUNNEL EXCAVATION THROUGH AN ALPINE GEOLOGICAL FAULT WITH THE PILOT TUNNEL AND TWO TUNNEL LINES

The case study, describes the excavation of tunnels crossing through a shear zone.

The project included the construction of the pilot tunnel and the tunnel lines where the former was located under the two upper tunnel lines.

2   UNDERGROUND EXCAVATION

2.1   *Theory and underground excavation techniques*

The project included the use of the traditional method of excavation like drilling and blasting off in rocks with good geotechnical quality, while in the shear zones the project included the method of mechanized excavation with a mechanical hammer.

During the phase of planning, many drilling investigations were performed to provide drill cores in order to get miscellaneous samples to test. Specific tests, allowed to determine geomechanical parameters as uni-axial compressive, tensile and shearing strength.

The above mentioned drilling investigations also allowed to evaluate the local rock types, the rock boundary conditions and the rock quality (Rock Quality Designation - RQD) in terms of interval, length and orientation in comparison to the planned tunnels, opening, roughness, filling and alteration of the discontinuities.

These parameters and information allowed to assess the rock mass in different classes based on the classification of the Beniawsky (Rock Mass Rating - RMR) and for each of them it was possible to determine the elastic modulus.

In order to get a statistic range of the real elastic modulus values, this last parameter was valued also based on others parameters like the Disturbing factor of the rock mass and the GSI (Geological Strength Index).

All above mentioned tests had allowed to develop a model as realistic as possible to the geological condition on site.

During the planning, different excavation sections were defined (so called "light", "medium" and "heavy"). For each of them, the consolidation systems were chosen depending on the quality and strength parameters of the rock mass.

The engineering technique depending on the quality and strength parameters of the rock mass, allow to define the maximum depth of the blast and the relating consolidation system in case of rocks with good geotechnical quality.

In the same way, in the fault zones, where the use of mechanized excavation with a mechanical hammer, was planned the following consolidation systems:

• In the excavation face: sprayed fiber-reinforced concrete, VTR pipe overlaid with cement grout;
• In progress on the boundary consolidation: self-drilling radial steel rods with cement grout (the number of these elements depends on the quality of the class of the rock mass);
• Radial consolidation: self-drilling radial rods with cement grout (the number of these elements depends on the quality of the class of the rock mass);
• First phase lining: sprayed fiber-reinforced concrete, welded steel mesh, double steel ribs supports with defined pitch;
• Structural lining: sprayed concrete without fiber-reinforced, with double layer of welded steel mesh and a concrete inverted arch with double layer of welded steel mesh.

2.2 *Work in progress and critical issues*

The planning involved that the realisation of the pilot tunnel and the tunnel lines should have been on three sides and with three working shifts.

In particular, the pilot tunnel with function of test had a section of approximately 40 m², and has anticipated the excavation of the tunnel lines, with a section of approximately 80 m², for about 500 m in advance.

In this operative configuration, the pilot tunnel would have reached the massive geological fault before the tunnel lines, providing relevant information for tunnel lines excavation, in terms of the behaviour of the rock mass, supports/consolidations application and the containments of the overbreaks behind the steel ribs support.

During the excavation of the pilot tunnel through the massive fault (extension size about 450 m) with a mechanical hammer, the Contractor didn't encounter particular critical issues.

Instead, when the tunnel lines crossed the aforesaid fault, the Contractor noted a large used quantity of sprayed concrete, which was caused by the presence of the massive boundary overbreaks behind the steel ribs support (Fig. 1). This unforeseen issue has happened in spite of the presence of the boundary consolidation executed in radial and in progress by the self-drilling rods.

This unpredictable boundary condition wasn't detected at all during the excavation of the pilot tunnel. In operative way, the pilot tunnel had the following two functions:

• To investigate the real behaviour of the rock mass before the excavation of the tunnel lines;
• To reveal its real functioning and efficiency of the planned supports and to apply it to a unique scale and/or wideness of the excavation section.

In other words, the planning analysis of the supports were correct as long as the wideness of the excavation was the same as the pilot tunnel, but those planning analysis showed their limits, when the excavation section of the tunnel was wider.

The above described issue, set up a contractual dispute between the Contractor and the Owner because the former one continued the excavation temporarily bearing the increased quantities of sprayed concrete, which were necessary to ensure the excavation stability of the tunnels. Vice versa, the Owner didn't want to acknowledge the need to modify the whole excavation planning of the tunnel lines. Many letters were exchanged between the Contractor and the Owner concerning the above critical issue, in which, moreover the Contractor has highlighted to the Owner the necessity to assess the effective overbreaks behind the steel rib support.

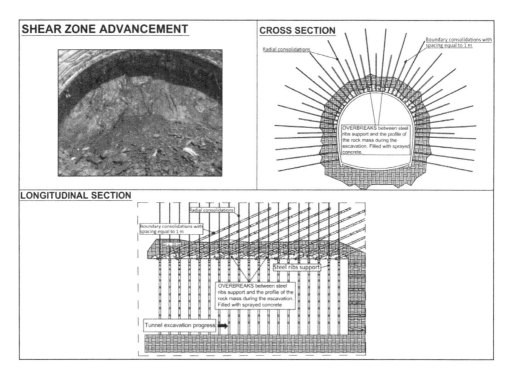

Figure 1. Shear zone advancement – Overbreaks between steel ribs support and the profile of the rock mass.

In order to examine the thickness of the real sprayed concrete placed on the back of steel ribs support by the Contractor, many drill cores were carried out with the direction of the Owner-Supervision, the results have shown an over-thickness of about 25 cm within a 450 m line.

Moreover, the lack of particular problems during the construction of the pilot tunnel through the fault, has involved the analysis of the possible causes which determined the problems which occurred during the excavation of the tunnel lines, concerning both the over-thickness and the massive quantity of sprayed concrete placed.

The result of the analysis highlights that the planning in the same geological context provided:

• A spacing between the boundary consolidation elements is equal to 1,20 m in the pilot tunnel;
• A spacing between the boundary consolidation elements is equal to 1 m in the tunnels line.

The planned spacing which is equal to 1 m situated between the consolidation elements at the boundary excavation of the tunnels line, actually wasn't adequate enough to their wider section (the section of the tunnel lines was about 80 m^2, while the section of the pilot tunnel test was about 40 m^2).

As a consequence, the quick deterioration of the boundary excavation was caused by the underestimation of two factors: the spacing of the consolidation boundary supports and the incidence of the largest excavation area in the same geological context.

Indeed, the reduced spacing between the boundary supports in the tunnel lines, compared to the pilot tunnel ones, leads to the two following aspects:

• The number of self-drilling radial steel rods were adequate to the overall stability of the rock mass;

- On the other hand, the number of self-drilling radial steel rods were instead inadequate to support the boundary surface of the excavation with a consequent overbreaks between the steel ribs support and the profile of the rock mass.

As a consequence the Contractor was forced to fill the overbreaks with sprayed concrete between the ribs support and the profile of the rock mass for a medium thickness of 25 cm within a 450 m line.

2.3 Conclusions

The result of the analysis performed by Contractor on the consolidation system foreseen in the project for the tunnels line (number of self-drilling radial steel rods adequate to the overall stability of the rock mass but inadequate to support the boundary surface of the excavation), had been confirmed from the Amicable Agreement Committee which has recognised the lack of liability of the Contractor related to the unforeseen mentioned overbreaks.

Based on Amicable Agreement Committee proposal, the Owner has recognised the higher operating costs bore by the Contractor.

The issue related to overbreaks, occurred during the works because of the underestimation of the scale-effect between the section of a pilot tunnel and the section of the tunnel lines, could have been contained and/or at least reduced if the "Observational Method" (OM) would have been applied.

The adoption of O.M., would have allowed to monitor the behaviour of the rock mass of the excavation section: 80 m^2 of tunnel lines, and 40 m^2 for pilot tunnel test. So, it is clear that the application of a minor number of the self-drilling radial steel rods in a large section of excavation such us in the line of the tunnel project was inadequate to support the boundary surface of the excavation.

Therefore, the scale-effect has to keep into consideration when evaluating the number of boundary supports and the spacing between them, but not only. In fact, the previous criteria must be monitored with the O.M. during the excavation in order to contain the overbreaks and consequently to minimize the contractual disputes between the Owner and Contractor.

REFERENCES

Bieniawski, Z.T.1989. *Engineering Rock Mass classifications:a complete manual for engineers and geologists in mining, civil, and petroleum engineering.* New Jersey:WILEY-INTERSCIENCE.

P.M.& E. Ltd., 2007–2018. *Professional Training Course "Management of Industrial Disputes" at Polytechnic of Turin.* Turin: Project Management & Engineering, Ltd.

Tunnels and Underground Cities: Engineering and Innovation meet Archaeology,
Architecture and Art, Volume 8: Public Communication and Awareness/Risk Management,
Contracts and Financial Aspects – Peila, Viggiani & Celestino (Eds)
© 2020 Taylor & Francis Group, London, ISBN 978-0-367-46873-6

Analysis of contractual approaches in Italian railway tunnels

L. Zaccaria & R. Ferro
Italferr S.p.A. (Italian State Railways Group), *Rome, Italy*

ABSTRACT: In recent years, various new tunnels have been built along the Italian railway system that, although operating in a common regulatory framework, have adopted different contractual approaches, that in some cases can be even alternative.

The purpose of the article is to: analyze the different solutions adopted in defining contractual provisions; identify the relevant consequences in terms of contract total amount, duration of the works and amount of claim issued by the Contractors; define possible solutions that can reduce the effects for the Client and outline some criteria to be adopted in future contracts.

In order to carry out the above-mentioned evaluation, four tunnel projects, excavated with conventional methods have, been identified and, for each one of them, have been considered some indicators of the contractual approach and have been detected the outcomes during the development of the works.

1   INTRODUCTION

In the implementation of big and complex public works the definition of an adequate and specific contract strategy is certainly one of the factor that can affect the outcome of the project

The aim of this paper is to analyse the different alternatives adopted prior to starting tendering procedures, detect the effects in terms of variation of time and costs and identify possible helpful solutions or criteria that could be adopted in future projects.

In order to make the afore mentioned assessments, four projects for the construction of new tunnels excavated along the Italian railway system, using the conventional method and completed in recent years were identified.

Italferr S.p.A, the Engineering Company of the Ferrovie dello Stato Italiane group (FS), that has been operating since 1984 in Italy and abroad in all fields of civil and railway engineering, including underground works, has performed, in all of the projects taken into consideration, Project Management, Work Supervision and Site Safety Coordination services on behalf of RFI Rete Ferroviaria Italiana that is the company in charge of managing the railway system and act as client in new investments in the railway infrastructure.

Despite the amount of uncertainty regarding the geological and geomechanical context intrinsic to underground work, we believe that the analysis of several specific contract models should allow to control, or anyhow lessen, the critical aspects connected to the management of the construction phase of a tunnel project.

2   THE REGULATORY AND CONTRACTUAL CONTEXT AND DESIGN LEVELS

Contracts for the execution of works carried out for Ferrovie dello Stato Italiane (FS) fall within the following Italian regulatory regime:

- Codice dei Contratti Pubblici (Public Procurement Code): as amended made over the years, also in function of the implementation of European Directives (Leg. Decree No. 163/2006 or Leg. Decree No. 50/2016). To this regard, it should be noted that, due to its being a 'subject operating in the transport sector', Gruppo Ferrovie dello Stato Italiane, is subject to the regulation pertaining to 'special sectors' and, therefore, because of these 'specialty' elements, some aspects of the afore mentioned framework law are subjected to a specific regulatory regime.

- Condizioni Generali di Contratto (FS General Terms and Conditions): these are the contractual tools of Gruppo Ferrovie dello Stato Italiane uses to regulate the specific aspects of the framework law and that any contractor entering a contract with the companies of the Group must comply with.

Moreover, without prejudice to the provisions of the paragraph above, for each contract performed on behalf of Ferrovie dello Stato Italiane, a specific Agreement is entered with the contractor that wins the public call for bids. The Ferrovie dello Stato Italiane group has typological templates that are periodically reviewed by the legal departments of the various companies of the group, getting feedback from other projects, so as to change clauses and/or add further prescriptions.

In particular, within the context of contracts managed by Italferr S.p.A., the Project Manager, prior to the issuance of the invitation to bid, examines the content of the particular condition of contract and its annexes with regard to the context of the works to be constructed, and has the faculty to propose to the client any changes in function of the specific context in which the project is located. Some examples of what may cause changes to the contract are given here below:

- peculiar work contexts (e.g.: urban environment) or interference with other work projects;
- specific environmental problems or problems linked to disposal of excavated material;
- special aspects regarding work site safety;
- significant interference with operating railway lines.

The Public Procurement Code mentioned earlier envisages, in both versions, various levels of design with progressively advanced technical and economic details, developing therefore from the 'preliminary' level (or 'technical and economic feasibility study') through the 'detailed' design level on to the 'execution' design level. The latter is the most detailed and must be such as to allow for the construction of the work in all of its details and must give precise quantities of work costs.

Within this framework, the contracting authority can place an 'executive' design as base for tender procedures and therefore a contract the subject of which is solely the construction of the structure or, as an alternative, create a 'detailed' design and put in the call for bids and as subject-matter of the contract with the contractor, in addition to the construction works, also the drafting of the execution design that will anyhow be subject to the client's verification and approval.

To this regard, it should be noted that the provision given in Art. 59 of Leg. Decree No. 50/ 2016 that forbids the use, in certain cases, of so-called 'integrated' contracts (for both the execution design and the construction), is not applicable to the awards made by subjects operating in the 'special sectors', one of which is Gruppo Ferrovie dello Stato Italiane.

3 THE PROJECTS CONSIDERED

In order to perform an analysis as representative as possible of the experience acquired and therefore an evaluation of the effects in terms of development and management, four construction projects were singled out for new tunnels, all excavated with conventional method and completed in recent years.

It should be noted that, because these are projects intended to expand the Italian railway network, the tunnels do not represent the entire subject matter of the contract but are part of a more extensive line section that sometimes includes other civil infrastructures, electrification equipment and signalling & telecommunication systems.

The following table (Table 1) shows the main technical characteristics of the tunnels examined in the projects selected, in terms of geological context, type of section and tunnelling method.

Table 1. Main technical characteristics of the projects selected.

PROJECT			TECHNICAL CHARACTERISTICS		
Project No.	Geographical location	Overall length of tunnel sections (m)	General characteristics of rock	Section type (Single track/ Double track)	Tunnelling method
I	Emilian Apennines	4,185	Argillites and flysch consisting of sequences of calcarenites and argillaceous marls	Double	Conventional
II	Abruzzi Apennines	395	Thickened sands, with presence of silts and clays	Single	Conventional
III	Monti Sicani (Sicily)	810	Clays and clayey marls with presence of sand or sandstone or variably cemented sandstone	Single	Conventional
IV	Lombard Alpine foothills	920	Alternation of silty sands and sandy gravels alternated with cemented gravels and sands and sandy clays	Double	Conventional

The examination of the design macro-data given above reveals how the projects selected cover a sufficiently wide range of geological formations crossed through as well as of excavation section sizes. The only common feature is the conventional tunnelling method because mechanised tunnelling excavation (TBM) methods would have made the basic data excessively heterogeneous and difficult to compare, thereby excessively complicating any kind of analysis.

4 IDENTIFICATION OF SIGNIFICANT ELEMENTS

After finding the four representative projects to be considered, the basic data of the contract, such as the overall price and the expected duration of works, together with the incidence of tunnelling works, are identified. The tunnel works, in particular, although as mentioned before are the largest structures of the contract, are still only a part of more extensive rail infrastructure projects.

Table 2 here below shows the key data of the contract of each project selected and the status of work progress up to the end of July 2018.

A quick look at the elements shown in the table reveals that the works selected represent a varied range of contract schemes also in terms of contract price and duration, while the incidence of tunnel works is on average about 40% of the total amount of the contract.

At the same time, for each one of the contract schemes adopted in the four projects, several elements deemed significant in qualifying the client's approach to the completion of the specific work have been focused on.

The elements are:

I. The subject placed in charge of Execution design. This subject can be the client, if necessary through an engineering company assigned by the client, consequently entering into a contract for construction only (e.g.: FIDIC Red Book) or it can be the contractor who is awarded a contract for both the execution design and construction (e.g.: Design & Build – FIDIC Yellow Book), in a, so-called, 'integrated contract'.

II. The methods of payment of the contract prices for the underground works and their adjustment when necessary in function of changes in section type during excavation due to results of the investigations of the tunnel face.

In the four case studies selected, three main methods were identified:

Table 2. Basic contract data for the four projects selected.

	PROJECT		CONTRACT KEY DATA			WORK PROGRESS STATUS
No.	Geographical location	Contract terms (calendar days)	Initial contract price(Mn Euros)	Incidence (%) of bored tunnel works w.r.t. the total contract	Current work progress status (July 2018)	
I	Emilian Apennines	4,185	165.5	37%	100%	
II	Abruzzi Apennines	395	12.2	48%	95%	
III	Monti Sicani (Sicily)	810	23.3	56%	100%	
IV	Lombard Alpine foothills	920	113.9	20%	100%	

- Payment of prices entirely on a 'unit-price' basis, based on the contract's price list and on the quantities actually constructed;
- Payment on a 'unit-price' basis but via the application of prices 'consolidated' per metre of tunnel and/or by predefined constructed portions (e.g.: bore, lining, finishing, etc.) defined for each one of the section types and therefore settled based on actual percentage of completion of the work;
- Payment on a 'lump sum' basis, excluding only compensation for the recovery or disposal of any material qualifying as waste pursuant to the laws in force.

III. Management of underground work execution time and any adjustments relating to variations in section type during excavation. In this case, unlike in the previous cases, no different contract approaches were identified because none of the contracts envisaged any adjustment of tunnel completion time and of useful terms or of other automatisms that allow to vary the number of days initially established in function, for example, of the variation of excavation sections adopted with respect to the design provisions. The contractor can however apply for an extension of the contractual useful terms, but the request must have adequate motive and anyhow is subordinate to inspection by the Work Supervisor and, if deemed necessary, to a shortening of the requested terms and finally to the Client's final authorization.

Table 3 summarises the elements indicated above for each project selected.

As indicated above, due to the fact that the works are conducted underground, every work phase brings with it some uncertainty linked to the geological and geomechanical behaviour of the rock formations and the degree of agreement with design forecasts.

5 ANALYSIS OF RESULTS

After identifying the elements deemed useful for giving an overview of the contractual approach, for each project the development stages as work progressed were examined.

Although it should be said that, just like when identifying the basic elements, the factors that allow to qualify the more or less positive outcome of the development of any project can be many and, most importantly, can vary according to the framework in which one operates, it was deemed useful to define a limited number of elements that along with being undoubtedly significant in assessing the 'performance' of the contract can be defined in a reasonably simple way.

Table 3. Specific contract elements of the four projects selected.

PROJECT			CONTRACT FEATURES	
No.	Geographical location	Subject in charge of Execution designing	Method of payment and management of section changes	Management of tunnel work time extension
I	Emilian Apennines	Contractor	Unit-price basis, with prices consolidated by section type	No time adjustment mechanism envisaged
II	Abruzzi Apennines	Client	Lump sum basis, except for recovery or waste disposal	No time adjustment mechanism envisaged
III	Monti Sicani (Sicily)	Client	Unit-price basis	No time adjustment mechanism envisaged
IV	Lombard Alpine foothills	Client	Lump sum basis, except for recovery or waste disposal	No time adjustment mechanism envisaged

The elements identified are the following:

a) Increase in contract costs with respect to the contract price initially underwritten between the parties with respect to the contractual forecasts shared at the end of the execution design phase, considering the amounts linked either directly or indirectly to tunnelling.
b) Variation in work execution deadlines with respect to those initially forecast at the end of the execution design phase, considering those linked to issues linked either directly or indirectly to tunnelling.
c) Amount of increase of contract costs requested by the contractor during the contract by claims (in Italy called 'riserva') with special reference to that due to tunnelling issues. In this respect, it is important to know that the 'riserva' is a faculty envisaged by the Italian Public Procurement Code based on which a contractor, via a correct and timely application, can put forth claims for any increased compensation he thinks he is entitled to. Each claim is then analysed and assessed by the client. The data given in the table therefore refer to contractor claims and not to compensation already paid because as yet not known, because the definition of the disputes has not been accomplished yet.

In order to carry out a macro-analysis of the results, first of all it should be noted that, in view of the differences in size and type of the projects selected, it was deemed best not to take into consideration absolute values but rather percentages assessed with reference to contract forecasts regarding completion time and amounts shared by the parties at the end of the execution design phase (regardless of who the designing subject was). Similarly for the parameter c) was considered the percentage respect to the contract's economic value, defined at project kick-off.

To this regard, 3 range of deviation from the reference value were defined:

• Deviation < 20%: this value indicates a not excellent but certainly acceptable performance level with respect to a design and contract system that is complex and having elements of uncertainty such as those described in the foregoing.
• Deviation between 20% and 50%: this interval is deemed an indicator of medium-high critical issues that significantly impact project development and can cause extensive delays in management and resolution, thereby causing remarkable costs for the contractor, due to the extended presence of personnel and means and for the client too due to higher supervision costs and delays in taking over of the infrastructure.
• Deviation > 50%: any value above this threshold indicates a highly negative performance of the contract, such as to significantly jeopardize initial forecasts and to result in high direct and indirect costs and financial charges both for the contractor – in terms of construction management – and for the client, as well as being disruptive for the entire region's infrastructure plans.

Table 4. Results of contract performance assessment of the four projects selected

PROJECT		ELEMENTS USED TO ASSESS CONTRACT PERFORMANCE		
No.	Geographical location	a) Extended completion time (%) with respect to Execution design forecast	b) Increase in costs (%) with respect to Execution design forecast	c) Litigation costs related to tunnelling works (%) with respect to Execution design contract amounts
I	Emilian Apennines	31%	18%	11%
II	Abruzzi Apennines	190%	15%	37%
III	Monti Sicani (Sicily)	81%	52%	45%
IV	Lombard Alpine foothills	25%	3%	17%
	Average value	82%	22%	28%

Table 4 shows, for each project, the deviations in % of each parameter from contract forecasts.

The table clearly shows that the most critical parameter seems to be the work completion time, which has always high percentage values that are anyhow higher than those of the other parameters.

This indicator should be adequately analysed because, as mentioned earlier, the extension of completion time of a public works project first of all translates into economic and financial damage for the construction company, that is often extended to the administration in terms of litigation costs, and also produces greater effort for the client in terms of contract management activities and of Work Supervision and finally, when the delay is very extensive, it brings further damage to society that will be able to take advantage from the infrastructure only much later than the planned completion deadline.

The other two parameters show lower percentage values, although they are on average only slightly above the limits of acceptability. Despite the fact that the first of the two remaining values regards higher costs defined in the contract and therefore already consolidated within the project's economic framework, while the second relates to amounts still being disputed, it should be noted that, because these come from public funds, these economic increases also impact society. Consequently, any consideration made to establish technical/contractual measures aimed at reducing their impact should be verified and applied.

6 CONSIDERATIONS

The following table (Table 5) shows, for each project, an overview of the percentage changes regarding each factor, compared to the reference value, and the calculation of the average value compared to each identified element, for the purpose of qualifying the approach adopted by the client for the execution of the project.

The above data inspire some considerations.

The decision to make contractors responsible for the execution design process as well has improved performance, compared to when the execution design was the responsibility of the client. Of course, this should not be interpreted as a comparison between the technical capabilities of the respective design teams, but should be viewed within the peculiar context of underground works, in which the tunnelling method and progress and the site organisation are closely related to design decisions. Since these operational aspects must necessarily be decided

Table 5.   Assessment of contract performance for the selected projects.

PROJECT	CONTRACT FEATURES		PERFORMANCE ASSESSMENT ELEMENTS		
			a) Increased time %	b) Increased costs %	c) Litigation % per contract
I Emilian Apennines	Preparation of Execution Design	Contractor	31%	18%	11%
	Payment method	Unit-price basis (with prices consolidated by section type)			
	Management of time extensions	No adjustment mechanism			
II Abruzzi Apennines	Preparation of Execution Design	Client	190%	15%	37%
	Payment method	Lump sum basis (except for waste management)			
	Management of time extensions	No adjustment mechanism			
III Monti Sicani (Sicily)	Preparation of Execution Design	Client	81%	52%	45%
	Payment method	Unit-price basis			
	Management of time extensions	No adjustment mechanism			
IV Lombard Alpine foothills	Preparation of Execution Design	Client	25%	3%	17%
	Payment method	Lump sum basis (except for waste management)			
	Management of time extensions	No adjustment mechanism			
Mean value			82%	22%	28%
Preparation of Execution Design Payment method	Average value in the case of design by contractor ('integrated contract')		31%	18%	11%
	Mean value in the case of design by Client		99%	23%	34%
	Average value for contracts predominantly on a unit-price basis		56%	35%	28%
	Average value for contracts predominantly on a lump sum basis		108%	9%	27%

by the contractor, it ensures that by also giving the contractor the opportunity to develop the final detailed design level it is possible to improve the planning of the works and the organisation of the site, ensuring a more regular progress in construction.

Furthermore, in the case of the so-called 'integrated contracts', the time gap between the execution design and actual construction stages is narrowed, because the bidding process has already been completed and, therefore, reduces the risk of changes to the conditions of the locations and the related circumstances, although this aspect has a lower impact on tunnelling projects in non-urban areas, compared to other works or urban tunnelling projects.

Last but not least, for completeness of information, it should be noted that the option of design & build contracts goes against the guidelines set out in the current version of the Italian Public Procurement Code, which contains a preference for construction-only contracts, probably as a result of the different experience in the field of public works, compared to the contracts considered in this analysis.

In the case of contracts in which the tunnelling works are paid for on a 'lump sum' basis, on average, the prices deviate less from initial project prices, in terms of contract price increases and of ensuing litigation, as may be inferred from the mean values of elements b) and c).

This type of contract, based on the guidelines of the Italian Public Procurement Code, should be the preferred option, because it requires the parties to agree to the price at the start of the contract works, which price cannot then be changed based on the amount of works effectively carried out. Lump sum contracts, therefore - theoretically at least - provide a higher guarantee that the agreed price will be respected, with less risk of the price spiralling out of control during construction.

In the case of unit-price contracts, instead, the cost of the works is presumptive and the price may change depending on the quantities actually required for the construction work. Unlike lump sum contracts, the Italian Public Procurement Code considers unit-price contracts an exception, to be entered into only in the presence of specific circumstances expressly provided in the Code.

The time required to carry out the works, compared to the contract schedule, is a different matter. Based on the collected data, besides the frequency with which the initial forecasts are exceeded, it is not possible to make considerations on the different types of contract because none of the examined cases feature procedures for adjusting tunnel completion time based on specific factors encountered during the works.

Despite this, time is a significant factor in public contracts and the constant exceeding of the initial forecasts requires certain considerations.

Although the contractual regulation of construction time, in relation to the technical elements of the project, does not mean that this can be reduced, it is nevertheless this author's opinion that, especially with regard to conventional tunnelling, managing the extension of the deadlines solely through the process of a motivated application for extension, which, moreover, may be lodged up to the day before the expiry of the deadline, and subject to the client's examination and authorisation, should be reconsidered. This is because the procedure necessarily entails a reasonable response time and, therefore, on the one hand determines the deferral between the time in which the cause for time extension emerged and that in which the extra time is granted, and, on the other, the possibility for the contractor to interpret and extend the reasons on which his application is grounded, claiming a time extension greater than that actually necessary.

Therefore, solely with regard to tunnelling works, there could be a different approach based on the greater flexibility in contract schedule management, which is often applied in the contracts concluded in other European countries, such as Norway and Switzerland, for example.

In the case of Norway, the construction schedule is regulated on the basis of predefined 'standard capacities' ('time equivalents') for the different types of work.

These 'standard capacities' are established as a result of a technical negotiation between contractor and client, and may then be updated, in connection with any technological developments, although they generally continue to be valid for several years.

These assumptions, which may be assimilated to the contract references established at the conclusion of the execution design phase, are automatically adapted during the progress of the works, in the event that the balance of the construction activities in a certain section of the tunnel entails a greater or a lesser amount of work than that agreed by the parties as 'standard', and the ensuing adjustment of the relevant works schedule.

These factors, to the extent that they are reasonably realistic, provide a fair instrument for regulating the construction schedule and the date of completion of the works.

A similar approach, but based on a different method, is adopted in certain cases in Switzerland. Time is viewed as a factor to be quantified on both a 'unit-price' basis and, therefore, is adapted to the amount of work identified by the client, as well as on the basis of the performance elements assessed by the contractor.

In this respect, the following specific time factors are established:

- the theoretical duration of the works: this is contract-based, determined on the basis of the type and amount of works envisaged at the conclusion of the contract, and the contract performance offered by the contractor;
- the determinative duration of the works: calculated on the basis of the final bill of quantities of a section of the work and the contract performance offered by the contractor, considering any approved alterations and the contractually acknowledged suspensions;
- the effective duration of the works: the time actually employed to complete the works.

During the execution of the works, the Work Supervision team and the contractor compile quarterly Tables showing the (positive or negative) updates, in respect of compliance with the works schedule (tunnelling, lining, etc.), based on the actual work carried out. Based on these tables, the 'determinative' deadlines are adjusted and, if necessary, the critical contract process - and related penalties - are redefined.

In conclusion, however, it should be noted that, in order to ensure actual effectiveness, both of the above-mentioned proposals should be associated with a clear and predefined allocation of risks and responsibilities between client and contractor within the contract. For example, under the Swiss standard SIA 118/98, the client is responsible for the risks arising from unforeseen soil characteristics, including the presence of gas or the discovery of contaminated sites, while the contractor is made responsible for all operating risks associated with the rock formations, within the limits set out in the contract.

7 CONCLUSIONS

The analysis of the development of some projects for the construction of new tunnels for the Italian state railways (FS), falling under the provisions of the Italian Public Procurement Code and the FS General Term and Conditions, can be summarised as follows:

- The opportunity of assigning, where allowed by the applicable regulations, the preparation of the Execution Design to the contractor, so as to ensure work scheduling and site organization that are more suited to the specifics of tunnelling work and therefore more flexible in the event of the onset of conditions deviating from the initial design forecasts.
- Confirmation of the opportunity of foreseeing the payment of the work on a 'lump sum' basis according to which the price is agreed upon at the start of contract works in order to guarantee compliance with the price initially agreed upon. This approach seems even more reasonable in combination with the previous point, i.e. when execution design is assigned to the contractor.
- The systematic criticality in defining the duration of the works and the ensuing proposal for adjustment of current contract terms and conditions which envisage the approval of time extensions only when the contractor applies for them, via management mechanisms that are more flexible than contract terms, based on the examples of other mechanisms being applied by other European countries.

In closing, although the tunnelling works context is extremely complex, I believe that the use of an adequate contractual approach together with a set of risk sharing principles, can result in the optimal management of construction progress, such as to allow to effectively tackle any unforeseen events and diminish the repercussions on production activities and the economic and temporal impact on the contractor as well as on the public client paying for the works.

REFERENCES

E. Grov. (2016) 'Risk Sharing Principles in Tunnel Contracts' Paper and Presentation, WTC San Francisco.

Cary Hirner et al. (2016). Black & Veatch 'Decision Analysis of Alternative Contracting Strategies for Tunnel Projects' WTC San Francisco.

M. Neuenschwander - Neuenschwander Consulting Engineers, Bellinzona, Switzerland & C. Nairac - White & Case, Paris, France. (2017) 'New Contract Form for Underground Works, a joint endeavor of ITA-AITES and FIDIC' WTC Paris.

P. Sander (2014). Continuous Cost and Risk Management for Major Projects in the Infrastructure Sector,' Brenner Congress.

Swiss Standard SIA 118/198: 2007 Genio civile. General Conditions for Underground Construction – General contract conditions applicable to code SIA 198 Underground Construction - Execution.

Contracts in Norwegian tunnelling. Publication No. 21 Norwegian Tunnelling Society 2012 ISBN 978-82-92641-24-8.

Tunnels and Underground Cities: Engineering and Innovation meet Archaeology,
Architecture and Art, Volume 8: Public Communication and Awareness/Risk Management,
Contracts and Financial Aspects – Peila, Viggiani & Celestino (Eds)
© 2020 Taylor & Francis Group, London, ISBN 978-0-367-46873-6

Author Index

Printed and bound by CPI Group (UK) Ltd, Croydon, CR0 4YY

18/10/2024

01776250-0003